Astrophysical Techniques

Fifth Edition

PUBLICATIONS BY THE SAME AUTHOR
Books

Early Emission Line Stars, Adam Hilger Ltd., 1982

Astrophysical Techniques, Adam Hilger Ltd., 1984

Translation from the French of *Astronomie; Methodes et Calculs* (by A. Acker and
 C. Jaschek), John Wiley & Sons, 1986

Stars, Nebulae and the Interstellar Medium; Observational Physics and Astrophysics,
 Adam Hilger Ltd., 1987

Journeys to the Ends of the Universe, IOP Publications, 1990

Astrophysical Techniques (second edition), IOP Publications, 1991

Telescopes and Techniques, Springer-Verlag, 1995

Optical Astronomical Spectroscopy, IOP Publications, 1995

Seeing Stars (coauthor, R. Forrest), Springer-Verlag, 1997

*A Photo Guide to the Constellations: A Self-Teaching Guide to Finding Your Way
 around the Heavens*, Springer-Verlag, 1997

Astrophysical Techniques (third edition), IOP Publications, 1998

Solar Observing Techniques, Springer-Verlag, 2001

Dictionary of Practical Astronomy, Springer-Verlag, 2002

Telescopes and Techniques (second edition), Springer-Verlag, 2002

Astrophysical Techniques (fourth edition), IOP Publishing, 2003

*Galaxies in Turmoil—Active and Starburst Galaxies and the Black Holes that
 Drive Them*, Springer, 2007

Astrophysical Techniques

Fifth Edition

C.R. Kitchin

University of Hertfordshire Observatory
UK

CRC Press
Taylor & Francis Group
Boca Raton London New York

CRC Press is an imprint of the
Taylor & Francis Group, an **informa** business

A TAYLOR & FRANCIS BOOK

CRC Press
Taylor & Francis Group
6000 Broken Sound Parkway NW, Suite 300
Boca Raton, FL 33487-2742

© 2009 by Taylor & Francis Group, LLC
CRC Press is an imprint of Taylor & Francis Group, an Informa business

No claim to original U.S. Government works
Printed in the United States of America on acid-free paper
10 9 8 7 6 5 4 3 2

International Standard Book Number-13: 978-1-4200-8243-2 (Hardcover)

Library of Congress Cataloging-in-Publication Data

Kitchin, C. R. (Christopher R.)
  Astrophysical techniques / C.R. Kitchin. -- 5th ed.
    p. cm.
  Includes bibliographical references and index.
  ISBN 978-1-4200-8243-2 (hardback : alk. paper)
    1. Astrophysics--Technique. 2. Astronomy--Technique. 3. Astronomical instruments. 4. Imaging systems in astronomy. I. Title.

QB461.K57 2008
522--dc22   2008019647

Visit the Taylor & Francis Web site at
http://www.taylorandfrancis.com

and the CRC Press Web site at
http://www.crcpress.com

Dedication

For Christine

Contents

Preface

The aim of this book is to provide a coherent state-of-the-art account of the instruments and techniques used in astronomy and astrophysics today. While every effort has been made to make it as complete and up-to-date as possible, the author is only too well aware of the many omissions and skimpily treated subjects throughout the book. For some types of instrumentation, it is possible to give full details of the instrument in its finally developed form. However, for the "new astronomies" and even some aspects of established fields, development is occurring at a rapid pace and the details will change between the writing and publishing of this edition. For those areas of astronomy therefore, a fairly general guide to the principles behind the techniques is given and this should enable the reader to follow the detailed designs in the scientific literature.

In this fifth edition, many new instruments and techniques are included for the first time, and some topics have been eliminated on the grounds that they have not been used by either professional or amateur astronomers for many years. Other topics, while no longer employed by professional observers for current work, are included because archive material that is still in use was obtained using them or because amateur astronomers use the techniques. A few references to Web sites have been included but not many because the sites change so frequently and search engines are now so good. However, this resource is usually the point of first call for most scientists when they have a query, and much material, such as large sky surveys, is only available over the Internet. Furthermore, it is used for the operation of some remote telescopes and it forms the basis of "virtual" telescopes discussed in the last chapter. Since this edition will come out around the time of the fourth centennial of the (probable) invention of the telescope, a brief discussion of how the telescope came to be invented and how it developed subsequently has been added in Section 1.1. A new section has been added that describes the attempts to detect dark matter and dark energy and the discussions of computer, Internet, and space-craft-based observations and research brought up-to-date and extended to reflect their increasing importance.

As in earlier editions, another aim has always been to try and reduce the trend toward fragmentation of astronomical studies and this is retained in this edition. The new techniques that are required for observing in exotic regions of the spectrum bring their own concepts and terminology with them. This can lead to the impression that the underlying processes are quite different, when in fact they are identical but are merely being discussed from differing points of view. Thus, for example, the Airy disk and its rings and the polar diagram of a radio dish do not at first sight look related, but they are simply two different ways of presenting the same data. As far as possible, therefore, broad regions of the spectrum are dealt with as a single area, rather than as many smaller disciplines. The underlying unity of all astronomical observations is also emphasized in the layout of the book; the pattern of detection—imaging—ancillary instruments has been adopted so that one stage of an observation is encountered together with the similar stages required for all other information carriers. This is not an absolutely hard and fast rule, however, and in some places it seemed appropriate to deal with topics out of sequence, either to prevent a multiplicity of very small sections, or to keep the continuity of the argument going.

The treatment of the topics is at a level appropriate to science-based undergraduate students. As far as possible, the mathematics or physics background that may be needed for a topic is developed or given within that section. In some places, it was felt that some astronomy background would be needed as well, so that the significance of the technique under discussion could be properly realized. Although aimed at an undergraduate level, most of the mathematics should be understandable by anyone who has attended a competently taught mathematics course in their final years at school, and some sections are nonmathematical. Thus, many amateur astronomers will find aspects of the book useful and of interest. The fragmentation of astronomy, which has already been mentioned, means that there is a third group of people who may find the book useful, and that is professional astronomers themselves. The treatment of the topics in general is at a sufficiently high level, yet in a convenient and accessible form, for those professionals seeking information on techniques in areas of astronomy with which they might not be totally familiar.

Last, I must pay a tribute to the many astronomers and other scientists whose work is summarized here. It is not possible to list them by name, and studding the text with detailed references would have ruined the readability of the book. I would, however, like to take this opportunity to express my deepest gratitude to all of them.

Clear skies and good observing to you all!

C.R. Kitchin

Standard Symbols

Most of the symbols used in this book are defined when they are first encountered and the definition then applies throughout the rest of the section concerned. In a few cases, the symbol may have a different meaning in another section—W, for example, is used as the symbol for the Wiener filter in Section 2.1 and for the linear spectral resolution in Section 4.1. Such duplication though is avoided as far as possible and the meaning in any remaining cases should be clear enough. Some symbols, however, have acquired such a commonality of use that they have become standard symbols among astronomers. Some of these are listed below, and the symbol will not then be separately defined when it is encountered in the text.

amu	atomic mass unit $= 1.6605 \times 10^{-27}$ kg
c	velocity of light in a vacuum $= 2.9979 \times 10^8$ m s^{-1}
e	charge on the electron $= 1.6022 \times 10^{-19}$ C
e^-	symbol for an electron
e^+	symbol for a positron
eV	electron volt $= 1.6022 \times 10^{-19}$ J
G	gravitational constant $= 6.670 \times 10^{-11}$ N m^2 kg^{-2}
h	Planck constant $= 6.6262 \times 10^{-34}$ J s
k	Boltzmann constant $= 1.3806 \times 10^{-23}$ J K^{-1}
m_e	mass of the electron $= 9.1096 \times 10^{-31}$ kg
n	symbol for a neutron
p^+	symbol for a proton
U, B, V	magnitudes through the standard UBV photometric system
$^{12}_6\text{C}$	symbol for nuclei (normal carbon isotope given as the example), the superscript is the atomic mass (in amu) to the nearest whole number and the subscript is the atomic number
γ	symbol for a gamma ray

ε_o permittivity of free space $= 8.85 \times 10^{-12}$ F m^{-1}

λ symbol for wavelength

μ symbol for refractive index (also, n, is widely used)

μ_o permeability of a vacuum $= 4\pi \times 10^{-7}$ H m^{-1}

ν symbol for frequency (also, f, is widely used)

CHAPTER 1

Detectors

1.1 OPTICAL AND INFRARED DETECTION

1.1.1 Introduction

In this and the immediately succeeding sections, the emphasis is upon the detection of the radiation or other information carriers, and upon the instruments and techniques used to facilitate and optimize that detection. There is inevitably some overlap with other sections and chapters, and some material might arguably be more logically discussed in a different order from the one chosen. Particularly in this section, telescopes are included as a necessary adjunct to the detectors themselves. The theory of the formation of an image of a point source by a telescope, which is all that is required for simple detection, takes us most of the way through the theory of imaging extended sources. Both theories are therefore discussed together even though the latter should perhaps be in Chapter 2. There are many other examples such as x-ray spectroscopy and polarimetry that appear in Section 1.3 instead of Sections 4.2 and 5.2. In general, the author has tried to follow the pattern of detection–imaging–ancillary techniques throughout the book, but has dealt with items out of this order when it seemed more natural to do so.

The optical region is taken to include the near infrared (NIR) and ultraviolet, and thus roughly covers the region from 10 nm to 100 μm. The techniques and physical processes employed for investigations over this region bear at least a generic resemblance to each other and so may be conveniently discussed together.

1.1.2 Detector Types

In the optical region, detectors fall into two main groups: thermal and quantum (or photon) detectors. Both these types are incoherent, that is to say, only the amplitude of the electromagnetic wave is detected, the phase information is lost. Coherent detectors are common at long wavelengths (Section 1.2), where the signal is mixed with that from a local oscillator (heterodyne principle). Only recently have optical heterodyne techniques been developed in the laboratory, and these have yet to be applied to astronomy. We may therefore safely regard all optical detectors as incoherent in practice. With optical aperture synthesis (Section 2.5.5), some phase information may be obtained provided three or more telescopes are used.

In quantum detectors, the individual photons of the optical signal interact directly with the electrons of the detector. Sometimes, individual detections are monitored (photon counting), and at other times, the detections are integrated to give an analogue output like that of the thermal detectors. Examples of quantum detectors include the eye, photographic emulsion, photomultiplier, photodiode, and charge-coupled devices (CCD).

Thermal detectors, by contrast, detect radiation through the increase in temperature that its absorption causes in the sensing element. They are generally less sensitive and slower in response than quantum detectors, but have a very much broader spectral response. Examples include thermocouples, pyroelectric detectors, and bolometers.

1.1.3 Eye

This is undoubtedly the most fundamental of the detectors to a human astronomer although it is a very long way from being the simplest. It is now rarely used for primary detection, though there are still a few applications in which it performs comparably with or possibly even better than other detection systems. Examples of this could be very close double-star work and planetary observation, in which useful work can be undertaken even though the eye may have been superseded for a few specific observations by, say, interferometry, speckle interferometry, and planetary space probes. Much more commonly, it is necessary to find and/or guide objects visually while they are being studied with other detectors. Plus there are, of course, millions of people who gaze into the skies for pleasure, which includes most professional astronomers. Thus, there is some importance in understanding how the eye and especially its defects can influence these processes.

It is vital to realize that a simple consideration of the eye on basic optical principles will be misleading. The eye and brain act together in the visual process, and this may give better or worse results than, say, an optically equivalent camera system depending upon the circumstances. Thus, the image on the retina is inverted and suffers from chromatic aberration (see below) but the brain compensates for these effects. Conversely, high-contrast objects or structures near the limit of resolution, such as planetary transits of the Sun and the Martian "canals," may be interpreted falsely.

The optical principles of the eye shown in Figure 1.1 are too well-known to be worth discussing in detail here. The receptors in the retina (Figure 1.2) are of two types: cones for color detection and rods for black and white reception at higher sensitivity. The light passes through the outer layers of nerve connections before arriving at these receptors. In the rods, a pigment known as rhodopsin or, from its color, visual purple, absorbs the radiation. It is a complex protein with a molecular weight of

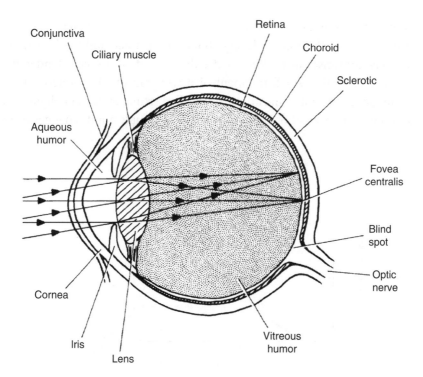

FIGURE 1.1 Optical paths in a horizontal cross section of the eye.

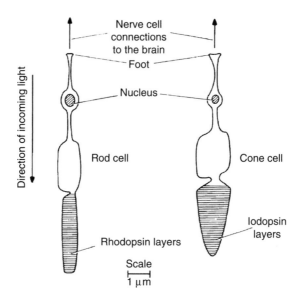

FIGURE 1.2 Retinal receptor cells.

about 40,000 amu, and its absorption curve is shown in Figure 1.3. It is arranged within the rods in layers about 20 nm thick and 500 nm wide, and may comprise up to 35% of the dry weight of the cell. Under the influence of light, a small fragment of it will split off. This fragment, or chromophore, is a vitamin A derivative called retinal or retinaldehyde and has a molecular weight of 286 amu. One of its double bonds changes from

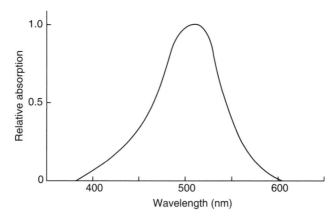

FIGURE 1.3 Rhodopsin absorption curve.

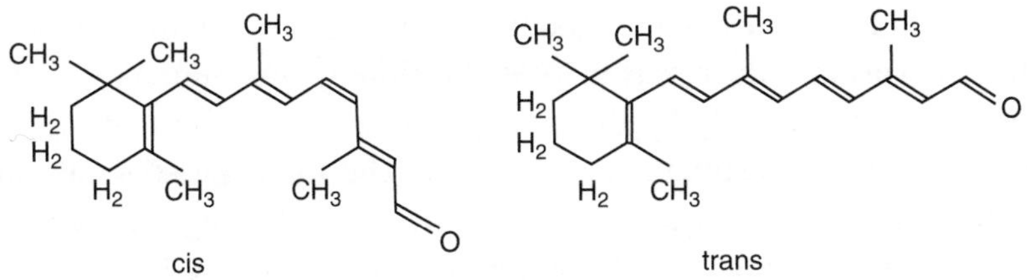

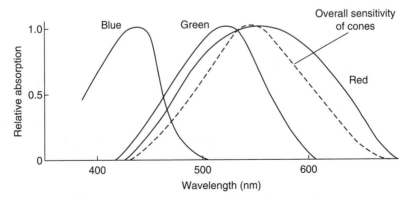

FIGURE 1.4 Transformation of retinaldehyde from cis to trans configuration by absorption of a photon near 500 nm.

a cis to a trans configuration (Figure 1.4) within a picosecond of the absorption of the photon. The portion left behind is a colorless protein called opsin. The instant of visual excitation occurs at some stage during the splitting of the rhodopsin molecule, but its precise mechanism is not yet understood. The reaction causes a change in the permeability of the receptor cell's membrane to sodium ions and thus a consequent change in the electrical potential of the cell. The change in potential then propagates through the nerve cells to the brain. The rhodopsin molecule is slowly regenerated. The response of the cones is probably due to a similar mechanism. A pigment known as iodopsin is found in cones and this also contains the retinaldehyde group. Cone cells, however, are of three varieties with differing spectral sensitivities, and this appears to arise from the presence of much smaller quantities of other pigments. The absorption curves of these other pigments are shown in Figure 1.5, together with the overall sensitivity curve of the cones for comparison.

FIGURE 1.5 Absorption curves of pigments involved in cone vision.

In bright light, much of the rhodopsin in the rods is split into opsin and retinaldehyde, and their sensitivity is therefore much reduced. Vision is then provided primarily by the cones, although their sensitivity is only about 1% of the maximum for the rods. The three varieties of cones combine their effects to produce color vision. At low light levels, only the rods are triggered by the radiation and vision is then in black and white. The overall spectral sensitivities of the rods and cones differ (Figures 1.3 and 1.5) with that of the rods peaking at about 510 nm and that of the cones at 550 nm. This shift in sensitivity is called the Purkinje effect. It can be a problem for double-star observers since it may cause a hot and bright star to have its magnitude underestimated in comparison with a cooler and fainter star, and vice versa. Upon entering a dark observatory from a brightly illuminated room, the rhodopsin in the rods slowly re-forms over about half an hour; their sensitivity therefore improves concurrently. Thus, we have the well-known phenomenon of dark adaptation, whereby far more can be distinguished after a few minutes in the dark than can be detected initially. If sensitive vision is important for an observation, then, for optimum results, bright lights should be avoided for half an hour before the observation is due to be made. Most observatories are only faintly illuminated and usually with red light to try and minimize any loss of dark adaptation.

Astronomical observation is generally due to rod vision. Usually with a dark-adapted eye, between 1 and 10 photons are required to trigger an individual rod. However, several rods must be triggered to result in a pulse being sent to the brain. This arises because many rods are connected to a single nerve fiber. Most cones are also multiply connected, although a few, particularly those in the *fovea centralis*, have a one-to-one relationship with nerve fibers. The total number of rods is about 10^8 with about 6×10^6 cones and about 10^6 nerve fibers, so that upwards of a hundred receptors can be connected to a single nerve. In addition there are many cross-connections between groups of receptors. The cones are concentrated toward the fovea centralis, and this is the region of most acute vision. Hence, the rods are more plentiful toward the periphery of the field of view, and so the phenomenon called averted vision, whereby a faint object only becomes visible when not looked at directly, arises. Its image is falling onto a region of the retina richer in rods when the eye is averted from its direct line of sight. The combination of differing receptor sensitivities, change in pigment concentration, and aperture adjustment by the iris mean that the eye is usable over a range of illuminations differing by a factor of 10^9 or 10^{10}.

The Rayleigh limit (see later) of resolution of the eye is about $20''$ when the iris has its maximum diameter of 5–7 mm. But for two separate images to be distinguished, they must be separated on the retina by at least one unexcited receptor cell. So that even for images on the fovea centralis, the actual resolution is between $1'$ and $2'$. This is much better than elsewhere on the retina since the fovea centralis is populated almost exclusively by small tightly packed cones, many of which are singly connected to the nerve fibers. Its slightly depressed shape may also cause it to act as a diverging lens producing a slightly magnified image in that region. Away from the fovea centralis, the multiple connection of the rods, which may be as high as a 1000 receptors to a single nerve fiber, degrades the resolution far beyond this figure. Other aberrations and variations between individuals in their retinal structure means that the average resolution of the human eye lies between $5'$ and $10'$ for point sources. Linear sources such as an illuminated grating can be resolved down to 1 minute of arc fairly commonly. The effect of the granularity of the retina is minimized in images by rapid oscillations of the eye through a few tens of seconds of arc with a frequency of a few hertz, so that several receptors are involved in the detection when averaged over a short time.

With areas of high contrast, the brighter area is generally seen as too large, a phenomenon that is known as irradiation. We may understand this as arising from stimulated responses of unexcited receptors due to their cross-connections with excited receptors. We may also understand the eye fatigue that occurs when staring fixedly at a source (e.g., when guiding on a star) as due to depletion of the sensitive pigment in those few cells covered by the image. Averting the eye very slightly will then focus the image onto different cells, reducing the problem. Alternatively, the eye can be rested momentarily by looking away from the eyepiece to allow the cells to recover.

The response of vision to changes in illumination is logarithmic (the Weber–Fechner law). That is to say, if two sources A and B are observed to differ in brightness by a certain amount, and a third source C appears midway in brightness between them, then the energy from C will differ from that from A by the same factor as that of B differs from C. In other words, if we use L to denote the perceived luminosity and E to denote the actual energy of the sources, we have if

$$\tfrac{1}{2}(L_A + L_B) = L_C \tag{1.1}$$

$'$ Refers to minutes of arc.
$''$ Refers to seconds of arc.

then

$$\frac{E_A}{E_C} = \frac{E_C}{E_B} \tag{1.2}$$

This phenomenon is the reason for the nature of the magnitude scale used by astronomers to measure stellar brightness (Section 3.1), since that scale had its origins in the eye estimates of stellar luminosities by the ancient astronomers. The faintest stars visible to the dark-adapted naked eye from a good observing site on a good clear moonless night are of about magnitude six. This corresponds to the detection of about 3×10^{-15} W. Special circumstances or especially sensitive vision may enable this limit to be improved upon by about one to one-and-a-half stellar magnitudes (two to four times). Conversely, the normal aging processes in the eye mean that the retina of a 60 year old person receives only about 30% of the amount of light seen by a person half that age. Eye diseases and problems, such as cataracts, may reduce this much further. Thus, observers should expect a reduction in their ability to perceive very faint objects as time goes by.

1.1.4 Semiconductors

The photomultiplier, CCDs, and several of the other detectors considered later derive their properties from the behavior of semiconductors. Thus, some discussion of the relevant aspects of these materials is a necessary prerequisite to a full understanding of detectors of this type.

Conduction in a solid may be understood by considering the electron energy levels. For a single isolated atom, they are unperturbed and are the normal atomic energy levels. As two such isolated atoms approach each other, their interaction causes the levels to split (Figure 1.6). If further atoms approach, then the levels develop further splits. So that for N atoms in close proximity to each other, each original level is split into N sublevels (Figure 1.7). Within a solid therefore each level becomes a pseudo-continuous band of permitted energies since the individual sublevels overlap each other. The energy level diagram for a solid thus has the appearance as shown in Figure 1.8. The innermost electrons remain bound to their nuclei, while the outermost electrons interact to bind the atoms together. They occupy an energy band called the valence band.

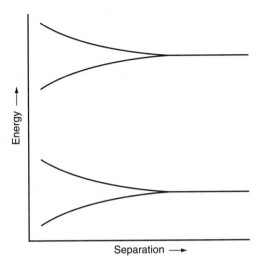

FIGURE 1.6 Schematic diagram of the splitting of two of the energy levels of an atom due to its proximity to another similar atom.

To conduct electricity through such a solid, the electrons must be able to move. We see from Figure 1.8 that free movement could occur within the valence and higher bands. However, if the original atomic level that

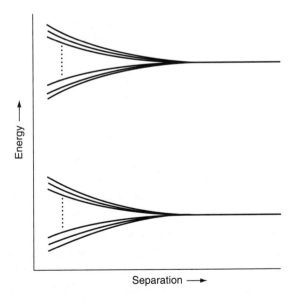

FIGURE 1.7 Schematic diagram of the splitting of two of the energy levels of an atom due to its proximity to many similar atoms.

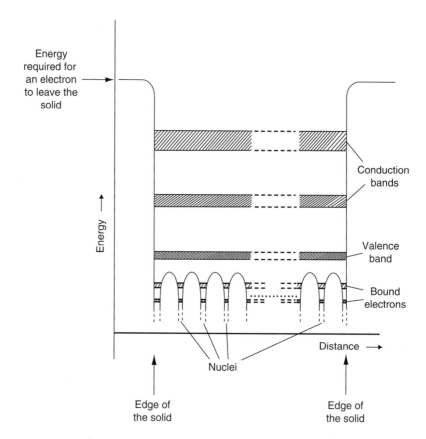

FIGURE 1.8 Schematic energy level diagram of a solid.

became the valence band upon the formation of the solid was fully occupied, then all the sublevels within the valence band will still be fully occupied. If any given electron is to move under the influence of an electric potential, then its energy must increase. This it cannot do, since all the sublevels are occupied, and so there is no vacant level available for it at this higher energy. Thus, the electron cannot move after all. Under these conditions, we have an electrical insulator. If the material is to be a conductor, we can now see that there must be vacant sublevels that the conduction electron can enter. There are two ways in which such empty sublevels may become available. Either the valence band is unfilled, for example, when the original atomic level had only a single electron in an s subshell, or one of the higher energy bands becomes sufficiently broadened to overlap the valence band. In the latter case, at a temperature of absolute zero, all the sublevels of both bands will be filled up to some energy that is called the Fermi level. Higher sublevels will be unoccupied. As the temperature rises, some electrons will be excited to some of these higher sublevels, but will still leave room for conduction electrons.

A third type of behavior occurs when the valence and higher bands do not actually overlap, but have only a small energy separation. Thermal excitation may then be sufficient to push a few electrons into some of the higher bands. An electric potential can now cause the electrons in either the valence or the higher band to move. The material is known as a semiconductor since its conductance is generally better than that of an insulator but considerably worse than that of a true conductor. The higher energy bands are usually known as the conduction bands. A pure substance will have equal numbers of electrons in its conduction bands and of spaces in its valence band. However, an imbalance can be induced by the presence of differing atomic species. If the valence band is full, and one atom is replaced by another that has a larger number of valence electrons (a donor atom), then the excess electrons usually occupy new levels in the gap between the valence and conduction bands. From there they may more easily be excited into the conduction band. The semiconductor is then an n-type since any current within it will largely be carried by the (negative) electrons in the conduction band. In the other case, when an atom is replaced with the one that has fewer valence electrons (an acceptor atom), spaces will become available within the valence band and the movement of electrons within this band will carry an electric current. Since the movement of an electron within the valence band is exactly equivalent to the movement of a positively charged hole in the opposite direction, this type of semiconductor is called p-type, and its currents are thought of as being transported by the positive holes in the valence band.

1.1.4.1 Photoelectric Effect

The principle of the photoelectric effect is well known; the material absorbs a photon with a wavelength less than the limit for the material and an electron is emitted. The energy of the electron is a function of the energy of the photon, while the number of electrons depends upon the intensity of the illumination. In practice, the situation is somewhat more complex, particularly when we are interested in specifying the properties of a good photoemitter.

The main requirements are that the material should absorb the required radiation efficiently, and that the mean free paths of the released electrons within the material should be greater than that of the photons. The relevance of these two requirements may be most simply understood from looking at the behavior of metals in which neither condition is fulfilled. Since metals are conductors, there are many vacant sublevels near their Fermi levels (see the earlier discussion). After absorption of a photon, an electron is moving rapidly and energetically within the metal. Collisions will occur with other electrons and since these electrons have other nearby sublevels that they may occupy, they can absorb some of the energy from our photoelectron. Thus, the photoelectron is slowed by collisions until it may no longer have sufficient energy to escape from the metal even if it does reach the surface. For most metals the mean free path of a photon is about 10 nm and that of the released electron less than 1 nm, thus the eventually emitted number of electrons is very considerably reduced by collisional energy losses. Furthermore, the high reflectivity of metals results in only a small fraction of the suitable photons being absorbed and so the actual number of electrons emitted is only a very small proportion of those potentially available.

A good photoemitter must thus have low-energy loss mechanisms for its released electrons while they are within its confines. The loss mechanism in metals (collision) can be eliminated by the use of semiconductors or insulators. Then the photoelectron cannot lose significant amounts of energy to the valence electrons because there are no vacant levels for the latter to occupy. Neither can it lose much energy to the conduction electrons since there are very few of these around. In insulators and semiconductors, the two important energy loss mechanisms are pair production and sound production. If the photoelectron is energetic enough, then it may collisionally excite other valence electrons into the conduction band thus producing pairs of electrons and holes. This process may be eliminated by requiring that E_1, the minimum energy to excite a valence electron into the conduction band of the material (Figure 1.9), is larger than E_2, the excess energy available to the photo-electron. Sound waves or phonons are the vibrations of the atoms in the material and can be produced by collisions between the photoelectrons and the atoms especially at discontinuities in the crystal lattice, etc. Only 1% or so of the electron's energy will be lost at each collision because the atom is so much more massive than the electron. However, the mean free path between such collisions is only 1 or 2 nm so that this energy loss mechanism becomes very significant when the photoelectron originates deep in the material. The losses may be reduced by

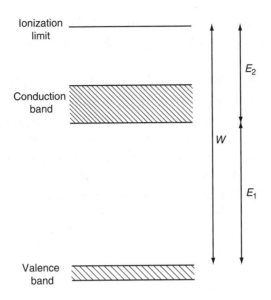

FIGURE 1.9 Schematic partial Grotrian diagram of a good photoemitter.

cooling the material since this then reduces the available number of quantized vibration levels in the crystal and also increases the electron's mean free path. —time b|w collisions on avg

The minimum energy of a photon if it is to be able to produce photoemission is known as the work function and is the difference between the ionization level and the top of the valence band (Figure 1.9). Its value may, however, be increased by some or all of the energy loss mechanisms mentioned above. In practical photoemitters, pair production is particularly important at photon energies above the minimum and may reduce the expected flux considerably as the wavelength decreases. The work function is also strongly dependent upon the surface properties of the material; surface defects, oxidation products, impurities, etc. can cause it to vary widely even among samples of the same substance. The work function may be reduced if the material is strongly p-type and is at an elevated temperature. Vacant levels in the valence band may then be populated by thermally excited electrons, bringing them nearer to the ionization level and so reducing the energy required to let them escape. Since most practical photoemitters are strongly p-type, this is an important process and confers sensitivity at longer wavelengths than the nominal cutoff point. The long-wave sensitivity, however, is variable since the degree to which the substance is p-type is strongly dependent upon the presence of impurities and so is very sensitive to small changes in the composition of the material.

1.1.5 Detector Index

After the natural detector formed by the eye, there are numerous types of artificial optical detectors. Before looking at some of them in more detail, however, it is necessary to place them in some sort of logical framework or the reader is likely to become confused rather than enlightened by this section. We may idealize any detector as simply a device wherein some measurable property changes in response to the effects of electromagnetic radiation. We may then classify the types of detector according to the property that is changing, and this is shown in Table 1.1.

Other properties may be sensitive to radiation but fail to form the basis of a useful detector. For example, forests can catch fire, and some human skins turn brown under the influence of solar radiation, but these are extravagant or slow ways of detecting the Sun. Yet, other properties may become the basis of detectors in the future. Such possibilities might include the initiation of stimulated radiation from excited atomic states as in the laser and the changeover from superconducting to normally conducting regimes in a material (see Section 1.1.13.2).

TABLE 1.1   Classification Scheme for Types of Detector

Sensitive Parameter	Detector Names	Class
Voltage	Photovoltaic cells	Quantum
	Thermocouples	Thermal
	Pyroelectric detectors	Thermal
Resistance	Blocked impurity band device	Quantum
	Bolometer	Thermal
	Photoconductive cell	Quantum
	Phototransistor	Quantum
	Transition edge sensor	Thermal
Charge	Charge-coupled device	Quantum
	Charge injection device	Quantum
Current	Superconducting tunnel junction	Quantum
Electron excitation	Photographic emulsion	Quantum
Electron emission	Photomultiplier	Quantum
	Television	Quantum
	Image intensifier	Quantum
Chemical composition	Eye	Quantum

1.1.6 Detector Parameters

Before resuming discussion of the detector types listed earlier, we need to establish the definitions of some of the criteria used to assess and compare detectors. The most important of these are listed in Table 1.2.

TABLE 1.2   Some Criteria for Assessment and Comparison of Detectors

QE (quantum efficiency)	Ratio of the actual number of photons that are detected to the number of incident photons
DQE (detective quantum efficiency)	Square of the ratio of the output signal/noise ratio to the input signal/noise ratio
τ (time constant)	This has various precise definitions. Probably, the most widespread is that τ is the time required for the detector output to approach to within $(1 - e^{-1})$ of its final value after a change in the illumination; that is, the time required for about 63% of the final change to have occurred
Dark noise	Output from the detector when it is unilluminated. It is usually measured as a root-mean-square (RMS) voltage or current
NEP (noise equivalent power or minimum detectable power)	Radiative flux as an input, which gives an output signal-to-noise ratio of unity. It can be defined for monochromatic or black body radiation, and is usually measured in watts
D (detectivity)	Reciprocal of NEP. The signal-to-noise ratio for incident radiation of unit intensity
D^* (normalized detectivity)	Detectivity normalized by multiplying by the square root of the detector area, and by the electrical bandwidth. It is usually pronounced "dee star" $$D^* = \frac{(a\,\Delta f)^{1/2}}{NEP} \qquad (1.3)$$ The units, cm Hz$^{1/2}$ W^{-1}, are commonly used and it then represents the signal-to-noise ratio when 1 W of radiation is incident on a detector with an area of 1 cm^2, and the electrical bandwidth is 1 Hz
R (responsivity)	Detector output for unit intensity input. Units are usually volts per watt or amps per watt
Dynamic range	Ratio of the saturation output to the dark signal. Sometimes only defined over the region of linear response
Spectral response	Change in output signal as a function of changes in the wavelength of the input signal. Usually given as the range of wavelengths over which the detector is useful
λ_m (peak wavelength)	The wavelength for which the detectivity is a maximum
λ_c (cutoff wavelength)	There are various definitions. Among the commonest are wavelengths at which the detectivity falls to zero, wavelengths at which the detectivity falls to 1% of its peak value, and wavelengths at which D^* has fallen to half its peak value

For purposes of comparison D^* is generally the most useful parameter. For photomultipliers in the visible region it is around 10^{15} to 10^{16}. Figures for the eye and for photographic emulsion are not directly obtainable, but values of 10^{12} to 10^{14} can perhaps be used for both to give an idea of their relative performances.

1.1.7 Cryostats

The noise level in many detectors can be reduced by cooling them to below ambient temperature. Indeed for some detectors, such as superconducting tunnel junctions (STJs) and TESs (see Section 1.1.13.2), cooling to very low temperatures is essential for their operation. Small CCDs produced for the amateur market are usually chilled by Peltier-effect coolers (see below), but almost all other approaches require the use of liquid nitrogen or liquid helium as the coolant. Since these materials are both liquids, they must be contained in such a way that the liquid does not spill out as the telescope moves. The container is called a cryostat, and while there are many different detailed designs, the basic requirements are much the same for all detectors. In addition to acting as a container for the coolant, the cryostat must ensure that the detector, and sometimes preamplifiers, filters, optical components, etc. are cooled whatever the position of the telescope, that the coolant is retained for as long as possible, that the evaporated coolant can escape, and that the detector and other cooled items do not ice up. These requirements invariably mean that the container is a Dewar (vacuum flask) and that the detector is behind a window within a vacuum or dry atmosphere. Sometimes the window is heated to avoid ice forming on its surface. Most cryostats are of the "bath" design, and are essentially simply tanks containing the coolant with the detector attached to the outside of the tank, or linked to it by a thermally conducting rod. Such devices can only be half filled with the coolant if none is to overflow as they are tilted by the telescope's motion. Although if the movement of the telescope is restricted, higher levels of filling are possible. Hold times between refilling with coolant of a few days can currently be achieved. When operating at Nasmyth or Coudé foci or if the instrument is in a separate laboratory fed by fiber optics from the telescope, so the detector is not tilted, continuous flow cryostats can be used, where the coolant is supplied from a large external reservoir, and hold times of weeks, months, or longer are then possible.

Bolometers, STJs, and TESs require cooling to temperatures well below 1 K. Temperatures down to about 250 mK can be reached using liquid $^3_2$He. The $^3_2$He itself has to be cooled to below 2 K before it liquefies, and this is achieved by using $^4_2$He under reduced pressure. Temperatures down to a few mK require a dilution refrigerator. This uses a mix of $^3_2$He and $^4_2$He at a temperature lower than 900 mK. The two isotopes partially separate out under gravity, but the lower $^4_2$He layer still contains some $^3_2$He. The $^3_2$He is removed from the lower layer distilling it off at 600 mK in a separate chamber. This forces some $^3_2$He from the upper layer to cross the boundary between the two layers to maintain the equilibrium concentration. However, crossing the boundary requires energy, and this is drawn from the surroundings, thus cooling them. SCUBA (submillimeter common user bolometer array), for example (see Section 1.1.13.2), used a dilution refrigeration system to operate at 100 mK. An alternative route to mK temperatures is the adiabatic demagnetization refrigerator. The ions in a paramagnetic salt are first aligned by a strong magnetic field. The heat generated in this process is transferred to liquid helium via a thermally conducting rod. The rod is then moved from contact with the salt, the magnetic field reduced, and the salt cools adiabatically.

1.1.8 Charge-Coupled Devices

Willard Boyle and George Smith invented CCDs in 1969 at the Bell Telephone Labs for use as a computer memory. The first application of CCDs within astronomy as optical detectors occurred in the late 1970s. Since then, they have come to dominate completely the detection of optical radiation at professional observatories, and are very widely used among amateur astronomers. Their popularity arises from their ability to integrate the detection over long intervals, their dynamic range ($>10^5$), high quantum efficiency, and the ease with which arrays can be formed to give two-dimensional imaging. In fact, CCDs can only be formed as an array; a single unit is of little use by itself.

The basic detection mechanism is related to the photoelectric effect. Light incident on a semiconductor (usually silicon) produces electron–hole pairs as we have already seen. These electrons are then trapped in potential wells produced by numerous small electrodes. There they accumulate until their total number is read out by charge coupling the detecting electrodes to a single read-out electrode.

An individual unit of a CCD is shown in Figure 1.10. The electrode is insulated from the semiconductor by a thin oxide layer. In other words,

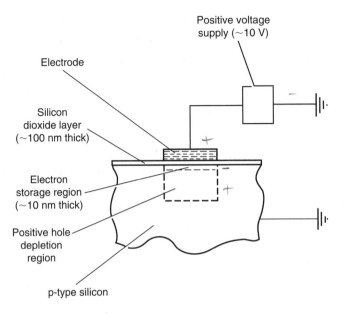

FIGURE 1.10 Basic unit of a CCD.

the device is related to the metal oxide–silicon (MOS) transistors. It is held at a small positive voltage that is sufficient to drive the positive holes in the p-type silicon away from its vicinity and to attract the electrons into a thin layer immediately beneath it. The electron–hole pairs produced in this depletion region by the incident radiation are thereby separated and the electrons accumulate in the storage region. Thus, an electron charge is formed whose magnitude is a function of the intensity of the illuminating radiation. In effect, the unit is a radiation-driven capacitor.

Now if several such electrodes are formed on a single silicon chip and zones of very high p-type doping insulate the depletion regions from each other, then each will develop a charge that is proportional to its illuminating intensity. Thus, we have a spatially and electrically digitized reproduction of the original optical image (Figure 1.11). All that remains is to retrieve this electron image in some usable form. This is accomplished by charge coupling. Imagine an array of electrodes such as those we have already seen in Figure 1.10, but without their insulating separating layers. Then if one such electrode acquired a charge, it would diffuse across to the nearby electrodes. However, if the voltage of the electrodes on either side of the one containing the charge were reduced, then their hole depletion regions would disappear and the charge would once again be contained within two p-type insulating regions (Figure 1.12). This time, however, the

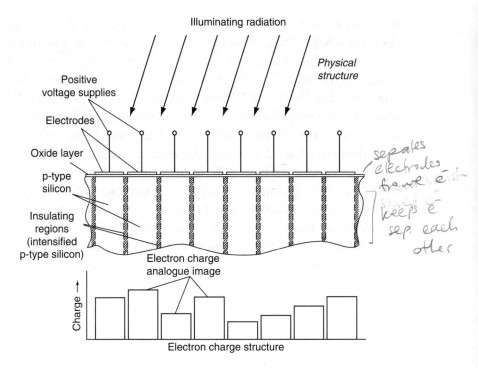

FIGURE 1.11 Array of CCD basic units.

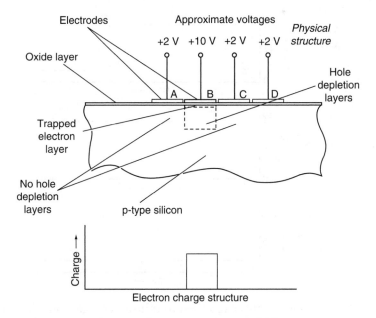

FIGURE 1.12 Active electron charge trapping in CCD.

insulating regions are not permanent but may be changed by varying the electrode voltage. Thus, the stored electric charge may be moved physically through the structure of the device by sequentially changing the voltages of the electrodes. Hence, in Figure 1.12, if the voltage on electrode C is changed to about +10 V, then a second hole depletion zone will form adjacent to the first. The stored charge will diffuse across between the two regions until it is shared equally. Now if the voltage on electrode B is gradually reduced to +2 V, its hole depletion zone will gradually disappear and the remaining electrons will transfer across to be stored under electrode C. Thus, by cycling the voltages of the electrodes as shown in Figure 1.13, the electron charge is moved from electrode B to electrode C. With careful design, the efficiency of this charge transfer (or coupling) may be made as high as 99.9999%. Furthermore, we may obviously continue to move the charge through the structure to electrodes D, E, F, etc. by continuing to cycle the voltages in an appropriate fashion. Eventually,

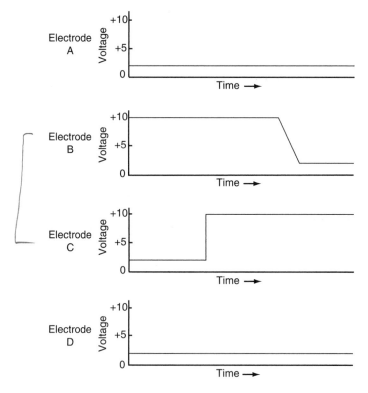

FIGURE 1.13 Voltage changes required to transfer charge from electrode B to electrode C in the array shown in Figure 1.12.

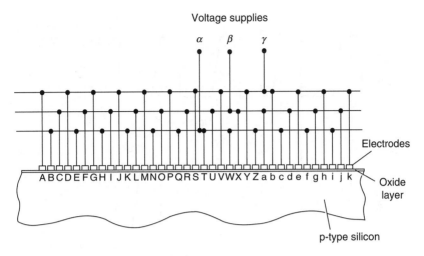

FIGURE 1.14 Connection diagram for a three-phase CCD.

the charge may be brought to an output electrode from whence its value may be determined by discharging it through an integrating current meter or some similar device. In the scheme outlined here, the system requires three separate voltage cycles to the electrodes to move the charge. Hence, it is known as a three-phase CCD. Most CCDs used in astronomy are three-phase devices. Two-phase and virtual-phase CCDs are discussed below. Three separate circuits are formed, with each electrode connected to those three before and three after it (Figure 1.14). The voltage supplies, α, β, and γ (Figure 1.14), follow the cycles shown in Figure 1.15 order to move charge packets from the left toward the right. Since only every third electrode holds a charge in this scheme the output follows the pattern shown schematically at the bottom of Figure 1.15. The order of appearance of an output pulse at the output electrode is directly related to the position of its originating electrode in the array. Thus, the original spatial charge pattern and hence the optical image may easily be inferred from the time-varying output.

The complete three-phase CCD is a combination of the detecting and charge transfer systems. Each pixel has three electrodes and is isolated from pixels in adjacent columns by insulating barriers (Figure 1.16). During an exposure, electrodes B are at their full voltage and the electrons from the whole area of the pixel accumulate beneath them. Electrodes A and C meanwhile are at a reduced voltage and so act to isolate each pixel from its neighbors along the column. When the exposure is completed, the

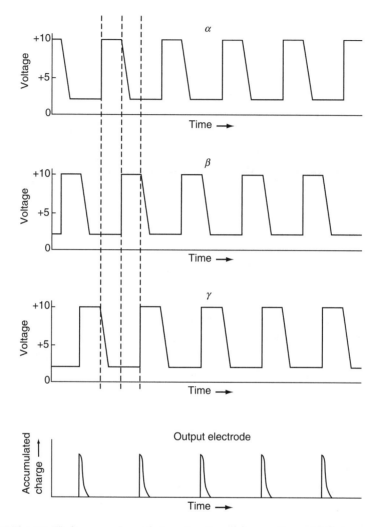

FIGURE 1.15 Voltage and output cycles for a three-phase CCD.

voltages in the three electrode groups are cycled as shown in Figure 1.15 until the first set of charges reaches the end of the column. At the end of the column, a second set of electrodes running orthogonally to the columns (Figure 1.16) receives the charges into the middle electrode for each column. That electrode is at the full voltage, and its neighbors are at reduced voltages, so that each charge package retains its identity. The voltages in the read-out row of electrodes are then cycled to move the charges to the output electrode where they appear as a series of pulses. When the first row of charges has been read out, the voltages on the

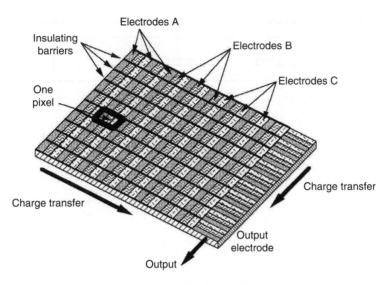

Insulating barriers
Electrodes A
Electrodes B
Electrodes C
One pixel
Charge transfer
Charge transfer
Output electrode
Output

FIGURE 1.16 Schematic structure of a three-phase CCD.

column electrodes are cycled to bring the second row of charges to the read-out electrodes, and so on until the whole image has been retrieved. With the larger format CCDs, the read-out process can take some time (typically several hundred milliseconds). To improve observing efficiency some devices therefore have a storage area between the imaging area and the read-out electrodes. This is simply a second CCD that is not exposed to radiation or half of the CCD is covered by a mask (frame transfer CCD). The image is transferred to the storage area in less than 0.1 ms, and while it is being read out from there, the next exposure can commence on the detecting part of the CCD. Even without a separate storage area, reading the top half of the pixels in a column upward, and the other half downward can halve the read-out times. Recently column parallel CCDs (CPCCDs) have been constructed that have independent outputs for each column of pixels thus allowing read-out times as short as 50 μs.

A two-phase CCD requires only a single clock, but needs double electrodes to provide directionality to the charge transfer (Figure 1.17). The second electrode, buried in the oxide layer, provides a deeper well than that under the surface electrode and so the charge accumulates under the former. When voltages cycle between 2 and 10 V (Figure 1.18), the stored charge is attracted over to the nearer of the two neighboring surface electrodes and then accumulates again under the buried electrode. Thus, cycling the electrode voltages, which may be done from a single clock,

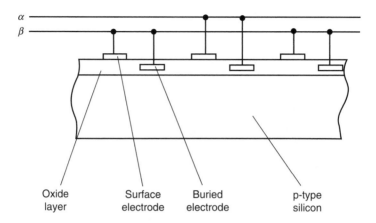

FIGURE 1.17 Physical structure of a two-phase CCD.

causes the charge packets to move through the structure of the CCD, just as for the three-phase device.

A virtual-phase CCD requires just one set of electrodes. Additional wells with a fixed potential are produced by p and n implants directly into the silicon substrate. The active electrode can then be at a higher or lower potential as required to move the charge through the device. The active electrodes in a virtual-phase CCD are physically separate from each other leaving parts of the substrate directly exposed to the incoming radiation. This enhances their sensitivity, especially at short wavelengths.

Nontracking instruments such as the Carlsberg Meridian (Section 5.1.3), liquid mirror, and Hobby–Eberly telescopes (see below) can follow the motion of objects in the sky by transferring the charges in their CCD detectors at the same speed as the image drifts across their focal planes

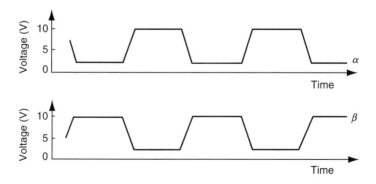

FIGURE 1.18 Voltage cycles for a two-phase CCD.

(time delayed integration or TDI). To facilitate this, orthogonal transfer CCDs (OTCCDs) have recently been made. These can transfer the charge in up to eight directions (up/down, left/right, and at 45° between these directions). OTCCDs can also be used for active image motion compensation arising from scintillation, wind shake, etc. Other telescopes use OTCCDs for the nod and shuffle technique. This permits the almost simultaneous observation of the object and the sky background through the same light path by a combination of moving the telescope slightly (the nod) and the charge packets within the CCD (the shuffle).

A combination of photomultiplier/image intensifier (Sections 2.1 and 2.3) and CCD, known as an electron bombarded CCD (EBCCD), electron multiplying CCD (EMCCD), or an intensified CCD (ICCD) has recently been developed. This places a negatively charged photocathode before the CCD. The photoelectron is accelerated by the voltage difference between the photocathode and the CCD, and hits the CCD at high energy, producing many electron–hole pairs in the CCD for each incident photon. This might appear to give the device a quantum efficiency of over 100% but it is in effect merely another type of amplifier; the signal-to-noise ratio remains that of the basic device (or worse), and so no additional information is obtained.

Interline transfer CCDs have an opaque column adjacent to each detecting column. The charge can be rapidly transferred into the opaque columns and read out from there more slowly while the next exposure is obtained using the detecting columns. This enables rapid exposures to be made, but half the detector is dead space. They are mainly used for digital video cameras, and rarely find astronomical applications.

The electrodes on or near the surface of a CCD can reflect some of the incident light, thereby reducing the quantum efficiency and changing the spectral response. To overcome this problem, several approaches have been adopted. Firstly, transparent polysilicon electrodes replace the metallic electrodes used in early devices. Secondly, the CCD may be illuminated from the back so that the radiation does not have to pass through the electrode structure at all. This, however, requires that the silicon forming the CCD be very thin so that the electrons produced by the incident radiation are collected efficiently. Nonetheless, even with thicknesses of only 10–20 μm, some additional cross talk (see below) may occur. More importantly, the process of thinning the CCD chips is risky and many devices may be damaged during the operation. Successfully thinned CCDs are therefore very expensive in order to cover the cost of the failures. They are also very fragile and can become warped, but they do have the advantage of being less affected by cosmic rays than thicker CCDs (see below). With a suitable antireflection coating, a back-illuminated CCD can now reach a quantum efficiency of 90% in the red and NIR regions. Other methods of reducing reflection losses include using extremely thin electrodes, or an open electrode structure (as in a virtual-phase CCD) that leaves some of the silicon exposed directly to the radiation. At longer wavelengths, the thinned CCD may become semitransparent. Not only does this reduce the efficiency as fewer photons are absorbed but also interference fringes may occur between the two faces of the chip. These can become very obtrusive and have to be removed as a part of the data reduction process.

The CCD, as so far described, suffers from loss of charge during transfer because of imperfections at the interface between the substrate and the insulating oxide layer. This affects the faintest images worst since the few electrons that have been accumulated are physically close to the interface. Artificially illuminating the CCD with a low-level uniform background during an exposure will ensure that even the faintest parts of the image have the surface states filled. This offset or "fat zero" can then be removed later in signal processing. Alternatively, a positively charged layer of n-type silicon between the substrate and the insulating layer forces the charge packets away from the interface, producing a buried-channel CCD. Charge transfer efficiencies for CCDs now approach 99.9999%, leading to typical read-out noise levels of one to two electrons per pixel. Using a nondestructive read-out mechanism and repeating and averaging many read-out sequences can reduce the read-out noise further. To save time, only the lowest intensity parts of the image are repeatedly read out, with the high intensity parts being skipped. The devices are therefore known as skipper CCDs.

The spectral sensitivity of CCDs ranges from 400 to 1100 nm, with a peak where the quantum efficiency can approach 90% near 750 nm. The quantum efficiency drops off further into the infrared as the silicon becomes more transparent. Short-wave sensitivity can be conferred by coating the device with an appropriate phosphor (see below) to convert the radiation to the sensitive region. The CCD regains sensitivity at very short wavelengths because x-rays are able to penetrate the device's surface electrode structure.

For integration times longer than a fraction of a second, it is usually necessary to cool the device to reduce its dark signal. Using liquid nitrogen for this purpose and operating at around $-100°C$, integration times of minutes to hours are easily attained. Small commercial CCDs produced for the amateur market that nonetheless often find applications at professional observatories for guiding, etc., usually use Peltier coolers to get down to about 50° below the ambient temperature. Subtracting a dark frame from the image can further reduce the dark signal. The dark frame is in all respects (exposure length, temperature, etc.) identical to the main image, excepting that the camera shutter remains closed while it is obtained. It therefore just comprises the noise elements present within the detector and its electronics. Because the dark noise itself is noisy, it may be better to use the average of several dark frames obtained under identical conditions.

Two-dimensional arrays are made by stacking linear arrays. The operating principle is unchanged, and the correlation of the output with position within the image is only slightly more complex than before. The largest single CCD arrays currently produced are $2k \times 4k$ (2048×4096) pixels in size. $10k \times 10k$ CCDs are being envisaged, but this is probably approaching the practical limit because of the time taken to read out such large images. For an adaptive optics telescope (see later in this section) operating at $0.2''$ resolution, a $2k \times 4k$ CCD covers only $200'' \times 400''$ of the sky if the resolution is to be preserved. Many applications require larger areas of the sky than this to be imaged, so several such CCDs must then be formed into a mosaic. However, the normal construction of a CCD with electrical connections on all four edges means that there will then be a gap of up to 10 mm between each device, resulting in large dead-space gaps in the eventual image. To minimize the dead space, three-edge-buttable CCDs are now made. These have all the connections brought to one of the short edges, reducing the gaps between adjacent devices to about 0.2 mm on the other three edges. It is thus possible to form mosaics with two rows and as many columns as desired with a minimum of dead space. Currently, the largest such mosaics use twelve $2k \times 4k$ CCDs to give an $8k \times 12k$ total area. Larger mosaics but with dead spaces operate on the 2.5 m Apache point telescope for the Sloan Digital Sky Survey using thirty $2k \times 2k$ CCDs in six columns and the 1.2 m Oschin–Schmidt camera on Mount Palomar started working in 2007 with the QUEST (Quasar Equatorial Survey Team) large area camera that has one hundred and twelve $0.6k \times 2.4k$ CCDs in a 4×28 array. A 4×8 mosaic of $2k \times 4k$ CCDs (OmegaCAM) is expected to receive first light during 2008 on European Southern Observatory's (ESO's) 2.6 m VLT (very large telescope) survey telescope. This will cover a $1°$ square field of view, but the two outer rows are separated from the central rows by wide dead spaces. European Space Agency's (ESA's) *Gaia** spacecraft, due for launch in 2010, will have a 1.5 gigapixel detector for its camera made up from one hundred and seventy 9 megapixel individual CCDs.

For typical noise levels and pixel capacities, astronomical CCDs have dynamic ranges of 100,000 to 500,000 (usable magnitude range in accurate brightness determination of up to 14.5^m). This therefore compares very favorably with the dynamic range of less than 1000 (brightness range of less than 7.5^m) available from a typical photographic image.

A major problem with CCDs used as astronomical detectors is the noise introduced by cosmic rays. A single cosmic ray particle passing through one of the pixels of the detector can cause a large number of ionizations. The resulting electrons accumulate in the storage region along with those produced by the photons. It is usually possible for the observer to recognize such events in the final image because of the intense "spike" that is produced. Replacing the signal from the affected pixel by the average of the eight surrounding pixels improves the appearance of the image, but does not retrieve the original information. This correction often has to be done by hand and is a time-consuming process. When two or more images of the same area are available, automatic removal of the cosmic ray spikes is possible with reasonable success rates.

Another serious defect of CCDs is the variation in background noise between pixels. This takes two forms. There may be a large-scale variation of 10%–20% over the whole sensitive area, and there may be individual pixels with permanent high background levels (hot spots). The first problem has been much reduced by improved production techniques and may largely be eliminated during subsequent signal processing (flat fielding), if its effect can be determined by observing a uniform source. Commonly used sources for the flat field include the twilit sky and a white screen inside the telescope dome illuminated by a single distant light source. The flat field image is divided into the main image after dark frame subtraction from both images to reduce the large-scale sensitivity variations. The effect of a single hot spot may also be reduced in signal processing by replacing its value with the mean of the four or eight surrounding pixels. However, the hot spots are additionally often poor

---

* The name originated as an acronym for global astrometric interferometer for astronomy. It is not now planned to use interferometry for the project but the name is retained as "just" a name.

transferors of charge and there may be other bad pixels. All preceding pixels are then affected as their charge packets pass through the hot spot or bad pixel and a spurious line is introduced into the image. There is little that can be done to correct this last problem, other than buying a new CCD. Even the good pixels do not have 100% charge transfer efficiency, so that images containing very bright stars show a "tail" to the star caused by the electrons lost to other pixels as the star's image is read out.

Yet another problem is that of cross talk or blooming. This occurs when an electron strays from its intended pixel to the one nearby. It affects rear-illuminated CCDs because the electrons are produced at some distance from the electrodes, and is the reason why such CCDs have to be thinned. It can also occur for any CCD when the accumulating charge approaches the maximum capacity of the pixel. Then the mutual repulsion of the negatively charged electrons may force some over into adjacent pixels. The capacity of each pixel depends upon its physical size and is around half-a-million electrons for 25 μm sized pixels. Some CCDs have extra electrodes to enable excess charges to be bled away before they spread into nearby pixels. Such electrodes are known as drains, and their effect is often adjustable by means of an anti-blooming circuit. Anti-blooming should not be used if you intend making photometric measurements on the final image, but otherwise it can be very effective in improving the appearance of images containing both bright and faint objects.

The size of the pixels in a CCD can be too large to allow a telescope to operate at its limiting resolution. In such a situation, the images can be "dithered" to provide improved resolution. Dithering consists simply of obtaining multiple images of the same field of view, with a shift of the position of the image on the CCD between each exposure by a fraction of the size of a pixel. The images are then combined to give a single image with subpixel resolution. A further improvement is given by drizzling (also known as variable pixel linear reconstruction) in which the pixel size is first shrunk leaving gaps between the pixels. The images are then rotated before mapping onto a finer scale output grid.

One of the major advantages of a CCD over a photographic emulsion is the improvement in the sensitivity. However, widely different estimates of the degree of that improvement may be found in the literature, based upon different ways of assessing performance. At one extreme, there is the ability of a CCD to provide a recognizable image at very short exposures, because of its low noise. On the basis of this measure, the CCD is perhaps 1000 times faster than photographic emulsion. At the other extreme, one may use the time taken to reach the midpoint of the dynamic range. Because of the much greater dynamic range of the CCD, this measure suggests CCDs that are about 5–10 times faster than photography. The most sensible measure of the increase in speed, however, is based upon the time required to reach the same signal-to-noise ratio in the images (i.e., to obtain the same information). This latter measure suggests that in their most sensitive wavelength ranges (around 750 nm for CCDs and around 450 nm for photographic emulsion), CCDs are 20–50 times faster than photographic emulsions.

1.1.9 Photography

This is dealt with as a part of imaging in Section 2.2. Here it is sufficient to point out that the basic mechanism involved in the formation of the latent image is pair production. The electrons excited into the conduction band are trapped at impurities, while the holes are removed chemically. Unsensitized emulsion is blue-sensitive as a consequence of the minimum energy required to excite the valence electrons in silver bromide.

1.1.10 Photomultipliers

Electron multiplier phototubes (or photomultipliers as they are more commonly but less accurately known) were at one time the workhorses of optical photometry (Chapter 3). They continue sometimes to be used when individual photons need to be detected as in the neutrino and cosmic ray Čerenkov detectors (Sections 1.3.2.9 and 1.4.3.3) or when very rapid responses are required as in the observation of occultations (Section 2.7). They may also be used on board spacecraft for ultraviolet measurements in the 10–300 nm region where CCDs are insensitive. They are also used within neutrino detectors like Super Kamiokande (Section 1.5) and cosmic ray detectors (Section 1.4).

Photomultipliers detect photons through the photoelectric effect. The basic components and construction of the device are shown in Figure 1.19. The photoemitter is coated onto the cathode and this is at a negative potential of some 1000 V. Once a photoelectron has escaped from the photoemitter, it is accelerated by an electric potential until it strikes a second electron emitter. The primary electron's energy then goes into pair production and secondary electrons are emitted from the substance in a manner analogous to photoelectron emission. Since typically 1 eV (1.6×10^{-19} J) of energy is required for pair production and the primary's energy can reach

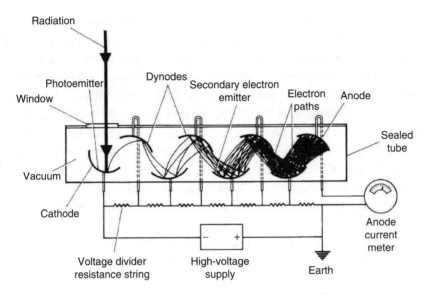

FIGURE 1.19 Schematic arrangement for a photomultiplier.

100 eV or more by the time of impact, several secondary electron emissions result from a single primary electron. This multiplication of electrons cannot, however, be increased without bound, since if the primary's energy is too high it will penetrate too far into the material for the secondary electrons to escape. The upper limit, however, is a factor of 100 or more and is not usually a restriction in practical devices in which multiplications by factors of 10 or less are used. The properties of a good secondary electron emitter are the same as those of a good photoemitter except that its optical absorption and reflection are no longer important.

The secondary emitter is coated onto dynodes that are successively more positive than the cathode by 100 V or so for each stage. The various electrodes are shaped and positioned so that the electrons are channeled toward the correct next electrode in the sequence after each interaction. The final signal pulse may contain 10^6 electrons for each incoming photon, and after arriving at the anode it may be further amplified and detected in any of the usual ways. This large intrinsic amplification of the photomultiplier is one of the major advantages of the device.

The electrode may be opaque as illustrated in Figure 1.19 so that the electrons emerge from the illuminated face, or it may be semitransparent, when they emerge from the opposite face. In the latter case, some change to the spectral sensitivity occurs due to the absorption in the material, but there are the compensating advantages of higher quantum efficiency and convenience since the cathodes may be coated on the inside of the window. Semitransparent cathodes are usually only about 30 nm thick so that it is very easy for small thickness changes to occur and to alter their sensitivities. Their usual application is for long-wave radiation detection.

The micro-channel plate (MCP) (Figure 1.71) can also be used at optical wavelengths. With an array of anodes to collect the clouds of electrons emerging from the plate, it provides an imaging detector with a high degree of intrinsic amplification. Such devices are often referred to as MAMA (multi-anode micro-channel array) detectors.

The commonest photoelectric and secondary emitter materials currently in use are listed in Table 1.3.

Noise in the signal from a photomultiplier arises from many sources. The amplification of each pulse can vary by as much as a factor of 10

TABLE 1.3   Photoelectron Emitting Substances

Substance	Long Wavelength Cutoff Point (nm)
Sodium chloride (NaCl)	150
Potassium bromide (KBr)	155
Rubidium iodide (RbI)	185
Cuprous chloride (CuCl)	190
Cesium iodide (CsI)	200
Copper/beryllium (Cu/Be)	200
Copper iodide (CuI)	210
Rubidium telluride ($RbTe_2$)	300
Cesium telluride (Cs_2Te)	350
Cesium antimonide ($Cs_{2.9}Sb$)	600–700
Bi-alkali ((K_2Cs)Sb)	670
Tri-alkali ((Cs)Na_2KSb)	850
Gallium arsenide (GaAs(Cs))	1000
Silver/oxygen/cesium (Ag/Cs_2O)	1000–1100
Secondary electron emitting substances	
Beryllium oxide (BeO(Cs))	
Cesium antimonide (Cs_3Sb)	
Gallium phosphide (GaP(Cs))	
Magnesium oxide (MgO(Cs))	
Potassium chloride (KCl)	
Silver/magnesium (Ag/Mg)	

through the sensitivity variations and changes in the number of secondary electrons lost between each stage. The final registration of the signal can be by analogue means and then the pulse strength variation is an important noise source. Alternatively, individual pulses may be counted and then it is less significant. Indeed when using pulse counting even the effects of other noise sources can be reduced by using a discriminator to eliminate the very large and very small pulses, which do not originate in primary photon interactions. Unfortunately, however, pulse counting is limited in its usefulness to faint sources otherwise the individual pulses start to overlap. Electrons can be emitted from either the primary or secondary electron emitters through processes other than the required ones, and these electrons then contribute to the noise of the system. The most important of such processes are thermionic emission and radioactivity. Cosmic rays and other sources of energetic particles and γ rays contribute to the noise in several ways. Their direct interactions with the cathode or dynodes early in the chain can expel electrons that are then amplified in the normal manner. Alternatively, electrons or ions may be produced from the residual gas or from other components of the structure of the photomultiplier. The most important interaction, however, is generally Čerenkov radiation produced in the window or directly in the cathode. Such Čerenkov pulses can be up to a 100 times more intense than a normal pulse. They can thus be easily discounted when using the photomultiplier in a pulse counting mode, but make a significant contribution to the signal when simply its overall strength is measured.

1.1.11 Superconducting Tunnel Junction Detectors

A possible replacement for the CCD in a few years is the STJ detector. The STJ can operate from the ultraviolet to long-wave infrared, and also in the x-ray region, can detect individual photons, has a very rapid response, and provides intrinsic spectral resolution of perhaps 500 or 1000 in the visible region. Its operating principle is based upon a Josephson junction. This has two superconducting layers separated by a very thin insulating layer. Electrons are able to tunnel across the junction because they have a wave-like behavior as well as a particle-like behavior, and so a current may flow across the junction despite the presence of the insulating layer. Within the superconductor, the lowest energy state for the electrons occurs when they link together to form Cooper pairs. The current flowing across the junction due to paired electrons can be suppressed by a magnetic field.

The STJ detector therefore comprises a Josephson junction based upon tantalum, hafnium, niobium, etc. placed within a magnetic field to suppress the current and having an electric field applied across it. It is cooled to about a tenth of the critical temperature of the superconductor— normally less than 1 K. A photon absorbed in the superconductor may split one of the Cooper pairs. This requires energy of a milli-electron-volt or so compared with about an electron volt for pair production in a CCD. Potentially therefore the STJ can detect photons with wavelengths up to a millimeter. Shorter wavelength photons will split many Cooper pairs, with the number split being dependent upon the energy of the photon, hence the device's intrinsic spectral resolution. The free electrons are able to tunnel across the junction under the influence of the electric field, and produce a detectable burst of current. STJ detectors and arrays made from them are still very much under development at the time of writing, but have recently been tried on the William Herschel telescope. The S Cam used a 6×6 array of 25 μm square STJs and successfully observed the light curve of the Crab Nebula pulsar and the color changes in the rapidly variable binary, UZ For.

1.1.12 Other Types of Detectors

Detectors other than the ones considered above may still find use in particular applications or have historical interest. Some of these devices are briefly surveyed below.

1.1.12.1 Photovoltaic Cells

These are also known as photodiodes, photoconductors, and barrier junction detectors. They rely upon the properties of a p–n junction in a semiconductor. The energy level diagram of such a junction is shown in Figure 1.20. Electrons in the conduction band of the n-type material are at a higher energy than the holes in the valence band of the p-type material. Electrons therefore diffuse across the junction and produce a potential difference across it. Equilibrium occurs when the potential difference is sufficient to halt the electron flow. The two Fermi levels are then coincident, and the potential across the junction is equal to their original difference. The n-type material is positive and the p-type negative, and we have a simple p–n diode. Now if light of sufficiently short wavelength falls onto such a junction then it can generate electron–hole pairs in both the p- and the n-type materials. The electrons in the conduction band in

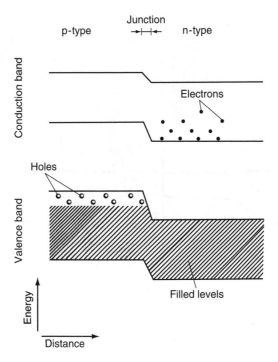

FIGURE 1.20 Schematic energy level diagram of a p–n junction at the instant of its formation.

the p region will be attracted toward the n region by the intrinsic potential difference across the junction, and they will be quite free to flow in that direction. The holes in the valence band of the p-type material will be opposed by the potential across the junction and so will not move. In the n-type region, the electrons will be similarly trapped while the holes will be pushed across the junction. Thus, a current is generated by the illuminating radiation and this may simply be monitored and used as a measure of the light intensity. For use as a radiation detector, the p–n junction often has a region of undoped (or intrinsic) material between the p and n regions in order to increase the size of the detecting area. These devices are then known as p–i–n junctions. Their operating principle does not differ from that of the simple p–n junction.

The response of a p–n junction to radiation is shown in Figure 1.21. It can be operated under three different regimes. The simplest, labeled B in Figure 1.21, has the junction short-circuited through a low-impedance meter. The current through the circuit is the measure of the illumination.

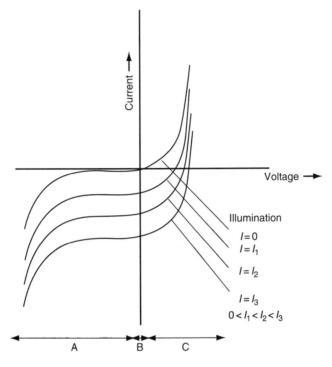

FIGURE 1.21 Schematic V/I curves for a p–n junction under different levels of illumination.

In regime C the junction is connected to a high impedance so that the current is very close to zero, and it is the change in voltage that is the measure of the illumination. Finally in regime A, the junction is back biased, that is, an opposing voltage is applied to the junction. The voltage across a load resistor in series with the junction then measures the radiation intensity. In this mode the device is known as a photoconductor (see Section 1.1.13).

The construction of a typical photovoltaic cell is shown in Figure 1.22. The material in most widespread use for the p and n semiconductors is silicon that has been doped with appropriate impurities. The solar power cells found on most artificial satellites are of this type. The silicon-based cells have a peak sensitivity near 900 nm and cutoff wavelengths near 400 and 1100 nm. Their quantum efficiency can be up to 50% near their peak sensitivity and D^* can be up to 10^{12}.

Indium arsenide, indium selenide, indium antimonide, gallium arsenide, and indium gallium arsenide can all be formed into photodiodes.

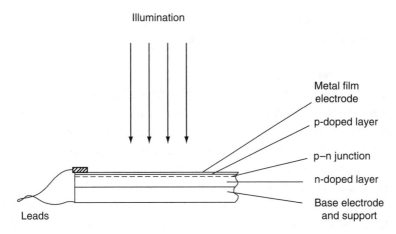

FIGURE 1.22 Cross section through a p–n junction photovoltaic detector.

They are particularly useful in the infrared where germanium doped with gallium outperforms bolometers (see Section 1.1.13.2) for wavelengths up to 100 μm.

If a p–n junction is reverse-biased to more than half its breakdown voltage, an avalanche photodiode (APD) results. The original electron–hole pair produced by the absorption of a photon will be accelerated by the applied field sufficiently to cause further pair production through inelastic collisions. These secondary electrons and holes can in their turn produce further ionizations and so on (cf. Geiger and proportional counters; Sections 1.3.2.1 and 1.3.2.2, respectively). Eventually an avalanche of carriers is created, leading to an intrinsic gain in the signal by a factor of 100 or more. The typical voltage used is 200 V, and the avalanche is quenched (because otherwise the current would continue to flow once started) by connecting a resistor of several hundred kilo-ohms in series with the diode. Then as the current starts to flow, the voltage across the diode is reduced and the breakdown is quenched. A major drawback for APDs is that the multiplication factor depends very sensitively upon the bias. The power supply typically has to be stable to a factor of 10 better than the desired accuracy of the output. The devices also have to be cooled to reduce noise. Using, for example, gallium arsenide, gains of over a factor of 1000 are possible, enabling individual photons to be detected. APDs are used in the University of Hertfordshire's extra-solar planet polarimeter (Section 5.2.3) and in the wavefront sensor for the ESO's multi-application curvature adaptive optics (MACAO) systems for the

VLT telescopes, but have found few other applications in astronomy since CCDs and now STJs are generally to be preferred.

1.1.12.2 Thermocouples

As is well known from school physics, two dissimilar metals in contact can develop a potential difference across their junction. This is called the Seebeck effect. It arises from the differing Fermi levels of the two metals. Electrons flow from the one with the higher level until the potential across the junction is sufficient to halt the transfer. Now the Fermi level at temperatures above absolute zero is the energy level that has a 50% chance of being occupied. As the temperature is increased, more and more of the electrons become thermally excited. However, since the holes will be densely packed near the top of the unexcited electrons while the excited electrons can be much more widely dispersed among the upper levels, the position of the Fermi level will change with temperature. The change in the Fermi level will not necessarily be identical for two different metals, and so at a junction the difference between the two Fermi levels will vary with temperature. Thus, the Seebeck potential at the junction will change to continue producing a balance. In a thermocouple therefore two dissimilar metals are joined into a circuit that incorporates a galvanometer. When one junction is at a different temperature from the other, their Seebeck potentials differ and a current flows around the circuit.

A practical thermocouple for radiation detection is made from two metals in the form of wires that are twisted together and blackened to improve their absorption. The other junction is kept in contact with something with a large thermal inertia so that it is kept at a constant temperature. The sensitive junction absorbs the radiation and so heats up slightly. The junctions are then at different temperatures so that a current flows as described above. A sensitive galvanometer can detect this current and it can be used to measure the intensity of the radiation. Several thermocouples are usually connected serially so that their potentials combine. Such an array is called a thermopile.

Practical thermocouples and thermopiles are usually made from antimony and bismuth or from nickel with various admixtures of copper, silicon, chromium, aluminum, etc. They are useful wideband detectors, especially for infrared work. Their simplicity of operation and their robustness have led to many applications being found for them in satellite-borne instrumentation, in spite of their relatively low sensitivity (they have values of $D^\star$ only up to 10^9 to 10^{10}).

1.1.12.3 Phototransistors

These are of little direct use in astronomy because of their low sensitivity. They find wide application, however, within the apparatus used by astronomers, for example, in conjunction with a light-emitting diode (LED) to generate a switching signal. They consist simply of a pnp or npn transistor with the minority current carriers produced by the illumination instead of the normal emitter. Thus, the current rises with increasing radiation and provides a measure of its intensity. The photovoltaic cell discussed earlier when operated in mode A is acting as a phototransistor. It is then sometimes called a photodiode.

1.1.13 Infrared Detectors

Many of the detectors just considered have some infrared sensitivity, especially out to 1 μm. However at longer wavelengths, other types of detectors are needed, though if the STJ fulfills its promise, it may replace many of these devices in the future. The infrared region is conventionally divided into three: the NIR, 0.7–5 μm; the mid-infrared (MIR), 5–30 μm; and the far infrared (FIR), 30–1000 μm. All infrared detectors need to be cooled, with the longer the operating wavelength the colder the required temperature. Thus, in the NIR, liquid nitrogen (77 K) generally suffices, in the MIR, liquid helium (4 K) is needed, while in the FIR, temperatures down to 100 mK are used. Currently, there are two main types of infrared detectors: the photoconductor for the near- and mid-infrared and somewhat into the FIR and the bolometer for the FIR. The recent development of a superconducting transistor (Quasiparticle Trapping Transistor or Quatratran) has the potential to act as an infrared and an x-ray detector, but it does not yet seem to have been so used.

The earth's atmosphere is opaque over much of the infrared region, although there are narrow wavelength ranges (windows) where it becomes transparent to a greater or lesser degree. The windows can be enhanced by observing from high altitude, dry sites, or by flying telescopes on balloons or aircraft. Nonetheless, the sky background can still be sufficiently high that images have to be read out several hundred times a second so that they do not saturate. Much of the observing, however, has to be done from spacecraft. Conventional reflecting optics can be used for the telescope and the instruments, though the longer wavelength means that lower surface accuracies are adequate. Refractive optics, including achromatic lenses, can be used in the NIR, using materials such as barium, lithium, and strontium fluoride, zinc sulfate or selenide, and infrared-transmitting glasses.

1.1.13.1 Photoconductive Cells

Photoconductive cells exhibit a change in conductivity with the intensity of their illumination. The mechanism for that change is the absorption of the radiation by the electrons in the valence band of a semiconductor and their consequent elevation to the conduction band. The conductivity therefore increases with increasing illumination, and is monitored by a small bias current. There is a cutoff point determined by the minimum energy required to excite a valence electron over the band gap. A very wide variety of materials may be used, with widely differing sensitivities, cutoff wavelengths, operating temperatures, etc. The semiconductor may be intrinsic, such as silicon, germanium, mercury cadmium telluride, lead sulfide, or indium antimonide. However, the band gaps in intrinsic semiconductors tend to be large, restricting their use to the NIR. Doping of an intrinsic semiconductor produces an extrinsic semiconductor with the electrons or gaps from the doping atom occupying isolated levels within the band gap. These levels can be just above the top of the valence band, or close to the bottom of the conduction band so that much less energy is needed to excite electrons to or from them. Extrinsic semiconductors can therefore be made that are sensitive across most of the infrared. Doping is normally carried out during the melt stage of the formation of the material; however, this can lead to variable concentrations of the dopant, and so to variable responses for the detectors. For germanium doped with gallium (Ge(Ga)), the most widely used detector material at wavelengths longer than 50 μm, extremely uniform doping has recently been achieved by exposing pure germanium to a flux of thermal neutrons in a nuclear reactor. Some of the germanium nuclei absorb a neutron and become radioactive. The $^{70}_{32}$Ge nucleus transmutes to $^{71}_{32}$Ge that in turn decays to $^{71}_{31}$Ga via β decay. Arsenic, an n-type dopant, is also produced from $^{74}_{32}$Ge during the process, however, only at 20% of the rate of production of the gallium. The process is known as neutron transmutation doping (NTD). The response of Ge(Ga) detectors may be changed by applying pressure or stressing the material along one of its crystal axes. The pressure is applied by a spring, and can change the detectivity range of the material, which is normally from ~40 to ~115 μm to ~80 to ~240 μm.

The *InfraRed Astronomy Satellite* (*IRAS*), launched in 1983, carried a cooled 0.57 m telescope and used small arrays of Ge(Ga) detectors to survey 98% of the sky at 12, 25, 60, and 100 μm. For the recently completed 2MASS (2 Micron All Sky Survey) project, three 256×256 mercury cadmium telluride arrays were used to observe at 1.25, 1.65, and 2.17 μm. The *Spitzer* spacecraft (previously *Space InfraRed Telescope Facility, SIRTF*), launched in August 2003, carries an NIR/MIR camera using 256×256 indium antimonide arrays to observe at 3.6 and 4.5 μm and 256×256 Si(As) blocked impurity band (BIB) (see below) detectors for 5.8 and 8.0 μm. It also carries a 32×32 array of unstressed Ge(Ga) detectors to observe at 70 μm, and a 2×20 array of stressed Ge(Ga) detectors to observe at 160 μm.

Ge(Ga) along with silicon doped with arsenic (Si(As)) or antimony (Si(Sb)) is also one of the materials used in the relatively recently developed BIB* detectors. These use a thin layer of very heavily doped semiconductor to absorb the radiation. Such heavy doping would normally lead to high dark currents but a layer of undoped semiconductor blocks these. Ge(Ga) BIB detectors are sensitive out to about 180 μm, and are twice as sensitive as normal stressed Ge(Ga) photoconductors around the 140 μm region. The VLT imager and spectrometer for mid-infrared (VISIR), which is used on ESO's VLT for imaging and spectroscopy between 8 and 13 μm and 16.5 and 24.5 μm has two 256×256 BIB detectors. While SOFIA (Stratospheric Observatory for Infrared Astronomy) planned to start operating in 2009 will use a Boeing 747 aircraft to fly a 2.5 m telescope at high altitudes. Its MIR detector will use 256×256 Si(As) and Si(Sb) BIB arrays to observe over the 5–8, 17–25, and 25–40 μm ranges.

Photoconductive detectors that do not require the electrons to be excited all the way to the conduction band have been made using alternating layers of gallium arsenide (GaAs) and indium gallium arsenide phosphide (InGaAsP) or aluminum gallium arsenide (AlGaAs), each layer being only 10 or so atoms thick. The detectors are known as QWIPs (quantum well infrared photodetectors). The lower energy required to excite the electron gives the devices a wavelength sensitivity ranging from 1 to 12 μm. The sensitivity region is quite narrow and can be tuned by changing the proportions of the elements. Recently, National Aeronautics and Space Administration (NASA) has produced a broadband 1k × 1k QWIP with sensitivity from 8 to 12 μm by combining over a hundred different layers

---

* Also known as impurity band conduction (IBC) detectors.

ranging from 10 to 700 atoms thick. QWIPs have been used for terrestrial and atmospheric remote sensing and for commercial and military purposes. A 256×256 QWIP array has been used on the 5 m Hale telescope, and others in search for extraterrestrial intelligence (SETI) but they have found few other astronomical applications so far.

Large (2048×2048) arrays can now be produced for some of the NIR detectors and 1024×1024 arrays for some MIR detectors, though at the long-wave end of the MIR, arrays are 256×256 at maximum. The United Kingdom infrared telescope's (UKIRT's) wide-field camera (WFCAM), which started operating early in 2005 and is currently undertaking the UKIRT Infrared Deep Sky Survey (UKIDSS), uses four $2k \times 2k$ mercury–cadmium–telluride detector arrays operating between 1 and 2.5 μm wavelengths. ESO's high-acuity, wide-field K-band imaging (HAWK-I) camera used on the VLT saw first light in mid-2007 and similarly uses four $2k \times 2k$ mercury–cadmium–telluride detector arrays to cover a 7.5′ square field of view over the 0.9–2.5 μm region, as does wide-field infrared camera (WIRCam) on the Canada–France–Hawaii telescope (CFHT). While the United Kingdom's 4 m VISTA (visible and infrared survey telescope for astronomy), planned to start operating from mid-2008, will use sixteen $2k \times 2k$ mercury–cadmium–telluride detector arrays to cover 0.6 square-degrees of the sky in a single exposure over the 0.84–2.5 μm region. VISTA's arrays are separated by gaps almost as large as their own sizes so six images (called pawprints) are needed with slight shifts of the telescope between each exposure to ensure complete coverage of the field whose total diameter is 1.65 degrees. In August 2008, a servicing mission is due to replace the *Hubble Space Telescope*'s (*HST*'s) wide-field planetary camera 2 (WFPC2) with the wide-field camera (WFC3). WFC3 will use a $1k \times 1k$ mercury–cadmium–telluride array for imaging over the 800–1700 nm region and will also have a ultraviolet channel (see below).

In the FIR, a region that overlaps the short wavelength end of the radio region (Section 1.2), array sizes are still only up to 80×80 (see SCUBA2 below). Unlike CCDs, infrared arrays are read out pixel by pixel. Although this complicates the connecting circuits, there are advantages; the pixels are read out nondestructively, and so can be read out several times and the results averaged to reduce the noise, there is no cross talk (blooming) between the pixels, and one bad pixel does not affect any of the others. The sensitive material is usually bonded to a silicon substrate that contains the read-out electronics.

In the NIR, large format arrays have led to the abandonment of the once common practice of alternately observing the source and the background (chopping), but at longer wavelengths the sky is sufficiently bright that chopping is still needed. Chopping may be via a rotating windmill whose blades reflect the background (say) onto the detector, while the gaps between the blades allow the radiation from the source to fall onto the detector directly. Alternatively, some telescopes designed specifically for infrared work have secondary mirrors that can be oscillated to achieve this switching.

Details of some of the materials used for infrared photoconductors are listed in Table 1.4. Those in current widespread use are given in italics.

1.1.13.2 Bolometers

A bolometer is simply a device that changes its electrical resistivity in response to heating by illuminating radiation. At its simplest, two strips of the material are used as arms of a Wheatstone bridge. When one is heated by the radiation, its resistance changes and so the balance of the bridge alters. Two strips of the material are used to counteract the effect of slower

TABLE 1.4 Some Materials Used for Infrared Photoconductors

Material	Cutoff Wavelength or Wavelength Range (μm)
Silicon (Si)	1.11
Germanium (Ge)	1.8
Gold-doped germanium (Ge(Au))	1–9
Mercury–cadmium–telluride (HgCdTe)	1–12
Gallium arsenide QWIPS (GaAs + InGaAsP or AlGaAs)	1–12
Lead sulfide (PbS)	3.5
Mercury-doped germanium (Ge(Hg))	4
Indium antimonide (InSb)	6.5
Copper-doped germanium (Ge(Cu))	6–30
Gallium-doped silicon (Si(Ga))	17
Arsenic-doped silicon BIB (Si(As))	23
Gallium-doped germanium (Ge(Ga))	~40 to ~115
Gallium-doped germanium stressed (Ge(Ga))	~80 to ~240
Boron-doped germanium (Ge(B))	120
Gallium-doped germanium BIB (Ge(Ga))	~180
Antimony-doped germanium (Ge(Sb))	130

environmental temperature changes, since they will both vary in the same manner under that influence.

Bolometers used in astronomy are of two main varieties: room temperature thermistor bolometers and cooled semiconductor bolometers. Because of their simplicity and sturdiness, the former have found an extensive application in rockets and spacecraft in the past. They consist of mixtures of manganese, nickel, and cobalt oxides sintered together and coated with an appropriate absorber for the wavelength that is to be observed. They have a high negative coefficient of resistance. Values of D^* of up to 10^9 over a very wide range of wavelengths can be obtained.

Cooled semiconductor bolometers were once used as detectors over most of the infrared region. Photoconductive cells have now replaced them for near- and mid-infrared work, but they are still used for the FIR (~100 μm to a few mm). Germanium doped with gallium (a p-type dopant) is widely used for the bolometer material with a metal-coated dielectric as the absorber. The bolometer is cooled to around 100 mK when in operation to reduce the thermal noise. Germanium doped with beryllium, silicon, and silicon nitride are other possible bolometer materials. Bolometers may make a comeback at shorter wavelengths in the future as detectors for signals from spacecraft. Many spacecraft, particularly planetary probes, generate large quantities of data, which the current radio receivers and transmitters can send back to Earth only very slowly. Faster communications requires shorter wavelengths to be used and transmission speeds thousands of times that possible in the radio region could be attained if the 1.55 μm infrared radiation commonly used to transmit broadband signals along optical fibers on the earth could be utilized. A recent development has a superconductor bolometer using a coil of extremely thin wire placed within a mirrored cavity (photon trap) that bounces unabsorbed photons back to the coil of wire to increase their chances of being absorbed. Just such a nanowire detector, produced by the Massachusetts Institute of Technology (MIT), has now achieved a 57% detection efficiency, perhaps enabling lasers with power requirements low enough for them to be used on spacecraft to be employed for communications in the future.

SCUBA on the 15 m James Clerk Maxwell Telescope (JCMT) was decommissioned in 2004. It operated over the 350–450 μm and 750–850 μm bands using NTD germanium bolometer arrays with 91 and 37 pixels, respectively. The individual detectors operated at a temperature of 100 mK and were fed by polished horns whose size matched the telescope's resolution. The current state of the art uses an absorber that is a mesh of metallized silicon nitride like a spider's web, with a much smaller bolometer bonded to its center. This minimizes the noise produced by cosmic rays, but since the mesh size is much smaller than the operating wavelength, it still absorbs all the radiation. Arrays of bolometers up to 30 × 30 pixels in size can now be produced. The *Herschel Space Observatory* scheduled for launch in 2008 with a 3.5 m telescope, will carry an FIR detector known as the spectral and photometric imaging receiver (SPIRE). This will use spider-web NTD-germanium bolometer arrays with feed horns to observe at 250, 350, and 500 μm. Each array will cover a 4′ × 8′ of the sky and contain 149, 88, and 43 detectors, respectively. The JCMT has had the astronomical thermal emission camera (AzTEC) which uses 144 silicon nitride spider-web bolometers cooled to 250 mK and supplied by conical feed horns to observe at 1.1 and 2.1 mm as a stand-in for SCUBA. AzTEC has recently started to be replaced by SCUBA2 (see below). The large APEX bolometer camera (LABOCA) on the 12 m Atacama pathfinder experiment (APEX) telescope situated at the Atacama large millimeter array (ALMA) (Section 2.5.5) site on the Chajnantor plateau in Chile comprises 295 NTD germanium bolometers mounted on silicon-nitride membranes to observe around 870 μm.

A recent development that promises to result in much larger arrays is the TES. These detectors are thin films of a superconductor, such as tungsten, held at their transition temperature from the superconducting to the normally conducting state. There is a very strong dependence of the resistivity upon temperature in this region. The absorption of a photon increases the temperature slightly, and so increases the resistance. The resistance is monitored through a bias voltage. The replacement for SCUBA (SCUBA2), due to start operating in early 2008, will use two 80 × 80 TES arrays. These should have twice the sensitivity of the SCUBA detectors and be able to produce images up to 2000 times faster.

1.1.13.3 Other Types of Detectors

For detection from the visual out to 30 μm or so, a solid-state photomultiplier can be used. This is closely related to the APD. It uses a layer of arsenic-doped silicon on a layer of undoped silicon. A potential difference is applied across the device. An incident photon produces an electron–hole pair in the doped silicon layer. The electron drifts under the potential difference toward the undoped layer. As it approaches the latter, the increasing potential difference accelerates it until it can ionize further atoms. An electrode collects the resulting avalanche of electrons on the far side of the undoped silicon layer. A gain of 10^4 or so is possible with this device. It has found little application in astronomy to date.

Platinum silicide acting as a Schottky diode can operate out to 5.6 μm. It is easy to fabricate into large arrays, but is of low sensitivity compared with photoconductors. Its main application is for terrestrial surveillance cameras.

1.1.14   Ultraviolet Detectors

Some of the detectors that we have reviewed are intrinsically sensitive to short-wave radiation, although modification from their standard optical forms may be required. For example, photomultipliers will require suitable photoemitters (Table 1.3) and windows that are transparent to the required wavelengths. Lithium fluoride and sapphire are common materials for such windows. Thinned, rear illuminated CCDs have a moderate intrinsic sensitivity into the long-wave ultraviolet region. EBCCDs (see Section 1.1.8) with an appropriate ultraviolet photocathode can also be used. At shorter wavelengths MCPs (Section 1.3.2.11) take over. Detectors sensitive to the visible region as well as to the ultraviolet need filters to exclude the usually much more intense longer wavelengths. Unfortunately, the filters also absorb some of the ultraviolet radiation and can bring the overall quantum efficiency down to a few percent. The term solar blind is used for detectors or detector/filter combinations that are only ultraviolet sensitive.

Another common method of short-wave detection is to use a standard detector for the visual region, and to add a fluorescent or short glow phosphorescent material to convert the radiation to longer wavelengths. Sodium salicylate and tetraphenyl butadiene are the most popular of such substances since their emission spectra are well matched to standard photocathodes. Sensitivity down to 60 nm can be achieved in this manner. Additional conversions to longer wavelengths can be added for matching to CCD and other solid-state detectors whose sensitivities peak in the red and infrared. Ruby (Al_2O_3) is suitable for this, and its emission peaks near 700 nm. In this way, the sensitivity of CCDs can be extended down to about 200 nm.

The EBCCDs were used for some of the first-generation instruments on board the *HST*. It now uses phosphor-coated CCDs for the WFPC2 and MCPs for the space telescope imaging spectrograph (STIS). In August 2008, a servicing mission is due to replace WFPC2 with WFC3 and to add a new ultraviolet spectroscope, the cosmic origins spectrograph (COS). COS will use two windowless 1k × 16k MCPs (see Section 1.3) as the detectors for the far UV region from 115 to 205 nm and a single 1k × 1k cesium telluride MAMA (see Section 1.32.11) for the near UV (170–320 nm). WFC3 will use two 2k × 4k CCD detectors with antireflection coatings for imaging within the 200–1000 nm region and will also have an infrared channel (see above).

1.1.15 Future Possibilities

Many of the devices discussed above are in the early stages of their development and considerable improvement in their characteristics and performance may be expected over the next decade or so. Other possibilities such as laser action initiation and switching of superconducting states have already been mentioned. Another technique currently in the early stages of application to astronomy is optical or infrared heterodyning. A suitable laser beam is mixed with the light from the object. The slow beat frequency is at radio frequencies and may be detected using relatively normal radio techniques (Section 1.2). Both the amplitude and phase information of the original signal can be preserved with this technique. Very high dispersion spectroscopy seems likely to be its main area of application.

1.1.16 Noise

In the absence of noise, any detector would be capable of detecting any source, however faint. Noise, however, is never absent and generally provides the major limitation on detector performance. A minimum signal-to-noise ratio of unity is required for reliable detection. However, most research work requires signal-to-noise ratios of at least 10, and preferably 100 or 1000. Noise sources in photomultipliers and CCDs have already been mentioned, and the noise for an unilluminated detector (dark signal) is a part of the definitions of DQE, NEP, D^*, and dynamic range (see earlier discussion). Now we must look at the nature of detector noise in more detail.

We may usefully divide noise sources into four main types: intrinsic noise, that is, noise originating in the detector; signal noise, that is, noise arising from the character of the incoming signal, particularly its quantum nature; external noise, such as spurious signals from cosmic rays, etc.; and processing noise, arising from amplifiers, etc. used to convert the signal from the detector into a usable form. We may generally assume processing noise to be negligible in any good detection system. Likewise, external noise sources should be reduced as far as possible by external measures. Thus, an infrared detector should be in a cooled enclosure and be allied to a cooled telescope (e.g., *IRAS* and *Spitzer*) to reduce thermal emission from its surroundings. In the NIR, recent developments in fiber optics hold out the hope of almost completely suppressing the sky-background noise. Atmospheric emission from the OH molecule forms about 98% of background noise in the NIR and between the lines, the sky is very clear. The lines, though, are very numerous and closely spaced so that existing filters cannot separate the clear regions from the emission regions. However, by manufacturing a fiber whose refractive index varies rapidly along its length in a sinusoidal manner, a Bragg grating (Section 1.3) can be formed. Such a grating produces a narrow absorption band at a wavelength dependent upon the spacing of the refractive index variations. The absorption band can thus be tuned to center on one of the OH emission lines' wavelengths. By making the refractive index variations aperiodic, many absorption bands can be produced, each aligned with an OH emission line. In this way, the Anglo-Australian Observatory (AAO) has recently manufactured OH-suppressing infrared fibers covering 36 OH emission lines over the 1.5–1.57 μm region. If the potential of this technology can be extended, then NIR observing could become as easy as that in the visible. Similarly, a photomultiplier used in a photon-counting mode should employ a discriminator to eliminate the large pulses arising from Čerenkov radiation from cosmic rays, and so on. Thus, we are left with intrinsic and signal noise to consider further.

1.1.16.1 Intrinsic Noise

Intrinsic noise in photomultipliers has already been discussed and arises from sources such as variation in photoemission efficiency over the photocathode, thermal electron emission from the photocathode and dynode chain, etc. Noise in photographic emulsion, such as chemical fogging, is discussed later (Section 2.2.2), while irradiation, etc. in the eye has also been covered earlier.

In solid-state devices, intrinsic noise arises from four sources.

Thermal noise, also known as Johnson or Nyquist noise (see Section 1.2.2), arises in any resistive material. It is due to the thermal motion of the charge carriers. These motions give rise to a current, whose mean value is zero, but which may have nonzero instantaneous values. The resulting fluctuating voltage is given by Equation 1.77.

Shot noise occurs in junction devices and is due to variation in the diffusion rates in the neutral zone of the junction because of random thermal motions. The general form of the shot noise current is

$$i = (2eI\,\Delta f + 4eI_0\,\Delta f)^{0.5} \tag{1.4}$$

where
e is the charge on the electron
Δf is the measurement frequency bandwidth
I is the diode current
I_0 is the reverse bias or leakage current

When the detector is reverse-biased, this equation simplifies to

$$i = (2eI_0\,\Delta f)^{0.5} \tag{1.5}$$

g–r Noise (generation–recombination) is caused by fluctuations in the rate of generation and recombination of thermal charge carriers, which in turn leads to fluctuations in the device's resistivity. *g–r* noise has a flat spectrum up to the inverse of the mean carrier lifetime and then decreases roughly with the square of the frequency.

Flicker noise, or $1/f$ noise, occurs when the signal is modulated in time, either because of its intrinsic variations or because it is being chopped (i.e., the source and background or comparison standard are alternately observed). The mechanism of flicker noise is unclear but its amplitude follows an f^{-n} power spectrum, where f is the chopping frequency and n lies typically between 0.75 and 2.0. This noise source may obviously be minimized by increasing f. Furthermore, operating techniques such as phase-sensitive detection (Section 5.2) are commonplace especially for infrared work, when the external noise may be many orders of magnitude larger than the desired signal. Modulation of the signal is therefore desirable or necessary on many occasions. However, the improvement of the detection by faster chopping is usually limited by the response time of the detector. Thus, an optimum chopping frequency normally needs to be found for each type of detector. For example, this is about 100 Hz for bolometers and 1000 Hz or more for most photoconductive and photovoltaic cells.

The relative contributions of these noise sources are shown in Figure 1.23.

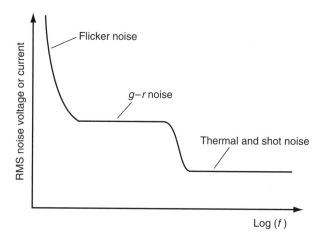

FIGURE 1.23 Relative contributions of various noise sources (schematic).

1.1.16.2 Signal Noise

Noise can be present in the signal for a variety of reasons. One obvious example is background noise. The source under observation will generally be superimposed upon a signal from the sky due to scattered terrestrial light sources, scattered starlight, diffuse galactic emission, zodiacal light, microwave background radiation, etc. The usual practice is to reduce the importance of this noise by measuring the background and subtracting it from the main signal. Often the source and its background are observed in quick succession (chopping, see flicker noise above and Section 3.2). Alternatively, there may only be measurements of the background at the beginning and end of an observing run. In either case, some noise will remain, due to fluctuations of the background signal about its mean level. This noise source also reduces as the resolution of the telescope improves. If the resolution is 1″, then a point source has to have an energy equal to that coming from a square second of arc of the background sky in order to have a signal-to-noise ratio of unity. But if the resolution were to be 0.1″, then the same point source would have a signal-to-noise ratio of 100 since it is only competing with 0.01 square seconds of arc of the background. Since the light grasp of a telescope increases as the diameter is squared, for diffraction-limited telescopes, the signal-to-noise ratio for point sources thus improves as D^4 (see Section 1.1.21).

Noise also arises from the quantum nature of light. At low signal levels, photons arrive at the detector sporadically. A Poisson distribution gives the probability of arrival, and this has a standard deviation of $\sqrt{n}$ (where n

is the mean number of photons per unit time). Thus, the signal will fluctuate about its mean value. To reduce the fluctuations to less than x%, the signal must be integrated for $10^4/(nx^2)$ times the unit time. At high photon densities, photons tend to cluster more than a Poisson distribution would suggest because they are subject to Bose–Einstein statistics. This latter noise source may dominate at radio wavelengths (Section 1.2), but is not normally of importance over the optical region.

1.1.16.3 Digitization

Signals are digitized in two ways, signal strength and time. The effect of the first is obvious; there is an uncertainty (i.e., noise) in the measurement corresponding to plus or minus half the measurement resolution. The effect of sampling a time-varying signal is more complex. The well-known sampling theorem (Section 2.1.1) states that the highest frequency in a signal that can be determined is half the measurement frequency. Thus, if a signal is bandwidth-limited to some frequency, f, then it may be completely determined by sampling at $2f$ or higher frequencies. In fact, sampling at higher than twice the limiting (or Nyquist) frequency is a waste of effort; no further information will be gained. However, in a non-bandwidth-limited signal, or a signal containing components above the Nyquist frequency, errors or noise will be introduced into the measurements through the effect of those higher frequencies. In this latter effect, or aliasing (see Section 1.2) as it is known, the beat frequencies between the sampling rate and the higher frequency components of the signal appear as spurious low-frequency components of the signal.

1.1.17 Telescopes

The apologia for the inclusion of telescopes in the chapter titled detectors has already been given, but it is perhaps just worth repeating that for astronomers, the use of a detector in isolation is almost unknown. Some device to restrict the angular acceptance zone, and possibly to increase the radiation flux as well, is almost invariably a necessity. In the optical region this generally implies the use of a telescope, and so they are discussed here as a required ancillary to detectors.

1.1.17.1 Telescopes from the Beginning

The telescope may justly be regarded as the symbol of the astronomer, for notwithstanding the heroic work done with the naked eye in the past and the more recent advances in exotic regions of the spectrum; our picture of the universe is still to a large extent based upon optical observations made through telescopes.

The invention of the telescope is nowadays usually attributed to a German spectacle maker, Hans Lipperhey (or Lippershey, 1570?–1619), who settled during the late sixteenth century in Middelburg, the capital of Zeeland in the southwest of the Netherlands. Sometime during the summer of 1608, Lipperhey applied to the Zeeland government for a patent for "a certain device by means of which all things at a very great distance can be seen as if they were nearby." The Zeeland authorities sent Lipperhey to the States General (the national government) at the Hague to plead his case and provided him with a letter of support dated September 25, 1608 (Gregorian calendar). That letter is the first unequivocal record of a genuine and usable telescope. The application was discussed at the states general on October 2, 1608 but was eventually rejected partly because the device was too simple to be kept secret and partly because another Dutchman, Jacob Metius (1571?–1628) of Alkmaar applied for a patent on a telescope design just a few weeks after Lipperhey. Later it would emerge that another Middelburg spectacle maker, Zacharias Janssen (ca. 1580–1638) also had a design for or even an actual telescope at about the same time as Lipperhey's patent application.

Lipperhey's telescopes have not survived although he was commissioned to produce several telescopes, including binocular versions, by the Stadholder, Prince Maurits of Orange, and others. Almost certainly, the telescopes were of the design that we now call, rather unfairly, the Galilean refractor. This design uses a converging lens as the objective and a diverging lens as the eyepiece and produces an upright image. Lipperhey's telescopes had magnifications of around three or four times. Lipperhey's design and probably also those of Metius and Janssen were almost certainly the result of a serendipitous combination of actual lenses—indeed one, though probably apocryphal, tale has the invention occurring while Lipperhey's children played with some of his spare lenses. The following year, however, Galileo Galilei (1564–1642), then living in Padua in Italy, heard of what Lipperhey's invention could do, but not of how it worked. His studies of optics though enabled him to design a telescope from theoretical principles and then to manufacture the required lenses. He was soon constructing telescopes of the Galilean refractor design that magnified up to 30 times and these enabled him go on and make his astounding astronomical observational discoveries such as lunar craters and the four major satellites of Jupiter.

How is it that within a period of just a few months in 1608/1609 the telescope was effectively invented four times? Essentially by the start of the seventeenth century, the telescope was a device whose time had come and indeed it had possibly already been invented several more times during the preceding decades. If Lipperhey, Metius, Janssen, and Galileo had not produced their instruments when they did, then someone else would have done so soon afterward. The accidental discovery of the telescope requires a supply of lenses and/or curved mirrors with reasonable optical quality and a variety of focal lengths. The design of a telescope from theoretical principles requires a good understanding of the laws of optics. Both these requirements have been achieved by the middle of the fifteenth century.

The earliest known example of a lens, made from natural quartz (rock crystal) and dated at around 640 BC, was found during excavations at Nineveh. While the earliest written mention seems to be that of burning glasses (very short focal length lenses used to concentrate sunlight sufficiently to start fires) in Aristophanes' 424 BC play *The Clouds*. The Romans also knew of burning glasses and numerous examples made from quartz, possibly turned on pole lathes and dating from around the tenth century AD onward have been excavated from Viking graves in Gotland, Sweden. Convex lenses for use as magnifiers came into use in the thirteenth century and had been incorporated into frames to make the first reading spectacles by around 1350. Spectacles using concave lenses to correct short sight followed about a century later. On the theoretical side, the Arabian mathematician Ibn Sahl (ca. 940–1000) first used the law of refraction, now known as Snell's law, to design lenses in the late tenth century and wrote about the use of burning lenses and mirrors. While Ibn al-Haitham (965–1039) explained the formation of images within the eye soon afterwards. Galileo's own work on optics added significantly to the understanding of the principles of optical devices.

Thus, by 1450 or soon after, the telescope was potentially capable of being invented; that it took another one and a half centuries to be invented is the surprising aspect of the story, not that four people invented it quasi-simultaneously. The long delay may be due to suitable lenses not being available—the Galilean refractor, for example, requires a long focal length converging lens for the objective and a short focal length diverging lens for the eyepiece—or potential inventors may not have known the earlier Arabian writings. However, it seems more likely that the delay is illusory and that the telescope was indeed invented several times during the centuries before 1608, but the resulting devices did not become widely known and the inventions were soon lost. Thus, Bishop Robert Grosseteste (ca. 1175–1253) writing in *De Iride* (the Rainbow) says

> This part of optics, when well understood, shows us how we may make things a very long distance off appear as if placed very close, and large near things appear very small, and how we may make small things placed at a distance appear any size we want, so that it may be possible for us to read the smallest letters at incredible distances, or to count sand, or seed, or any sort or minute objects.

At least one reasonable interpretation of this text is that it is describing a telescope, used both normally and in reverse (looking through the objective) and perhaps also a compound microscope, the invention of which is usually attributed to Zacharias Janssen sometime between 1590 and 1609. In 1266, Roger Bacon (ca. 1214–1294) published his *Opus Majus* (Great Work) and writes

> The wonders of refracted vision are still greater; for it is easily shown...that very large objects can be made to appear small, and the reverse, and very distant objects will seem very close at hand, and conversely. For we can so shape transparent bodies, and arrange them in such a way with respect to our sight and objects of vision, that the rays will be refracted and bent in any direction that we desire, and under any angle we wish we shall see the object near or at a distance.

The *Opus Majus*, however, was a compendium of Bacon's knowledge rather than a report of his own work. Since Bacon certainly knew of Grosseteste's work, the similarity to Grosseteste's writing suggests that Bacon is reporting on that rather than describing a new invention of his own. A genuine independent invention of the telescope may be attributable to Leonard Digges (1520–1559). His son Thomas (1546–1595) writes about his father in 1571 in *Pantometrica* that

> ...sundry times hath by proportional glasses duly situate in convenient angles, not only discovered things far off, read letters, numbered pieces of money with the very coin and subscription thereof, cast by some of his friends of purpose upon downs in open fields, but also seven miles off declared what hath been done at that instant in private places.

One plausible interpretation of this (if it is not a gross exaggeration by Thomas) is as the description of some type of telescope, perhaps using a concave mirror as the eyepiece. Yet another possible inventor is Giambattista della Porta (ca. 1535–1615). He is credited with the invention of the camera obscura and claimed also to have invented the telescope, but he died before he could give any account of his invention and his claim is now regarded with some skepticism.

Using the modern practice of attributing the credit for a discovery to the first person to publish a clear account of it within the public domain, Lipperhey is thus correctly identified as the inventor of the telescope (but only by the skin of his teeth and perhaps he was not genuinely the first person to make and use a telescope).

The Galilean design of telescope is difficult to manufacture with high magnifications. Since magnification in a simple telescope is just the focal length of the objective divided by the focal length of the eyepiece (Equation 1.64), high magnification requires a long focal length for the objective and/or a short focal length for the eyepiece. A long focal length for the objective makes the whole telescope long and unwieldy and the field of view (Equation 1.66) very small, so that it becomes difficult to find and then follow objects as they move across the sky. The negative lens required for the eyepiece must have very deep curves for its surfaces if it is to have a short focal length and these were hard to create with the techniques available four centuries ago. A telescope design that was easier to construct was needed and just three years after Lipperhey's invention, Johannes Kepler (1571–1630) provided it, publishing in his 1611 book *Dioptrice* a design that used two converging lenses. The Keplerian telescope, or as it now more commonly known, the astronomical refractor, uses a long focal length converging lens for the objective and a short focal length converging lens for the eyepiece. The eyepiece is placed after the focal point of the objective, resulting in an inverted image. The inverted image is a major drawback if the telescope is to be used for everyday purposes and had such an instrument been accidentally put together, the way that Lipperhey's design may have been, then it would probably have been discarded as being of no use. However, it matters little whether we see the stars and planets same way that they appear in the sky to the unaided eye, or inverted, so Kepler's design, being far easier to produce with high magnifications, eventually replaced the Galilean telescope for astronomical observing. The first to employ an astronomical refractor to observe the sky was probably Christoph Scheiner (1573–1650) sometime around 1613, although its use does not seem to have spread widely until after the publication of Scheiner's *Rosa Ursina* in 1630. The addition of a third lens, the relay or erector lens, to an astronomical refractor enables it to produce an upright image, resulting in the terrestrial telescope design, which for nonastronomical uses also quickly replaced the Galilean design. The Galilean design survives today only in the form of opera glasses, where high magnifications are not needed and the shorter length of the instrument compared with that of a comparable terrestrial telescope makes it more convenient to use.

Once invented, the astronomical refractor followed a pattern of development that has been repeated several times since opticians successively manufacture bigger and better telescopes following the design until some flaw in or difficulty with the design makes further improvement impractical or too expensive. Additional progress then has to await a new design or technique, and when that appears, the same pattern of development recurs. The optical telescope has passed through four of these major phases of development, each of which has caused a quantum jump in astronomical knowledge. We are now at the start of the fifth phase when very large diffraction-limited telescopes can be formed from smaller mirrors or segments with active control of their alignment, etc. We may hope for a similar rapid development of our knowledge as these instruments come on stream in the next few decades.

The development of the astronomical refractor was limited by the optical aberrations of the simple lenses then in use, especially chromatic aberration (Figure 1.30). The effects of the aberrations could be reduced somewhat by minimizing the curvatures of the surfaces of the lenses—but this meant long focal lengths, even for the eyepiece lens. Useful levels of magnification (100 times or so) thus required very long focal lengths for the objective. By 1656, Christiaan Huygens (1629–1695) had produced a telescope 7 m long and Johannes Hevelius (1611–1687) had a 43 m long telescope by the early 1670s. Such lengthy instruments were known as aerial telescopes and dispensed with the telescope tube. The lenses were simply mounted on the ends of a long pole, one end of which was attached to a tower and which could be moved up and down while the observer moved the other end (with the eyepiece) around the tower to acquire and follow whatever was being observed. For the very longest aerial telescopes even the pole was dispensed with and replaced by a cord linking the objective and the eyepiece. The observer then also had to keep the string in tension to align the two lenses while simultaneously performing his (at that time there were few, if any, female observers) ballet around the tower to find and track the object. As may be imagined, observations with such instruments were exceedingly difficult and the slightest breath of wind made them impossible. Nonetheless, Giovanni Domenico Cassini (1625–1712) made several significant discoveries using aerial telescopes, including in 1675 that of "his" division (or gap) in Saturn's rings.

Most observers, however, did not have Cassini's skill or patience so it was fortunate that as the astronomical refractor in the form of the aerial telescope was reaching the practical limit of the first phase of its development, several new designs appeared. These new designs all employed a mirror for the telescope objective, so eliminating at one stroke the effects of chromatic aberration, since mirrors reflect all wavelengths equally (although lenses continued to be used for the eyepieces, so chromatic aberration did not disappear completely). The mirror forming the objective reflects the light back toward the direction from which it came, so the eyepiece cannot simply be placed behind the objective's focal point because the observer's head would then block the incoming light. Mirror-based (or reflecting) telescope designs thus have to use a second mirror to reflect the light from the objective (primary) mirror to an accessible point. The secondary mirror is much smaller than the primary and so only obstructs a small and acceptable proportion of the light gathered by the primary.

Chronologically, the first mirror-based (or reflecting) telescope design was devised by the Scottish mathematician, James Gregory (1638–1675), and published in his *Optica Promota* in 1663. His design used a concave paraboloidal primary mirror (the paraboloidal shape eliminates spherical aberration, Figure 1.69) and a concave ellipsoidal mirror placed beyond the primary's focal point as the secondary mirror. The secondary mirror reflected the light through a hole in the center of the primary mirror to the eyepiece placed behind that mirror. Gregory had a prototype instrument constructed but it never seems to have worked satisfactorily. The failure of the telescope is perhaps not surprising since there was at that time no way of ensuring that the two mirrors were of the correct shape—another century would pass before paraboloidal mirrors could be made reliably. The first person therefore to design and build a functioning reflecting telescope was Sir Isaac Newton (1642–1727) who showed a working instrument to the Royal Society in 1671. It had a 2 in. (50 mm) primary mirror, was some 6 in. (150 mm) in length, and magnified by about 25 times. It was more powerful, as Newton boasted, than a refracting telescope 10 or 12 times larger (large at this time meaning the length).

Newton had actually first constructed a reflecting telescope in 1668 and he avoided the necessity for a second curved mirror and a hole in the primary mirror required for the Gregorian telescope by making the secondary mirror flat and angled at 45° to the incoming light. The light reflected from the primary mirror was thus reflected to the side of the telescope by the secondary and thence into the eyepiece (Figure 1.45). The simplicity of the design of the Newtonian telescope made it very popular and it is still widely used today for small- to medium-sized instruments, especially by amateur astronomers. In 1672, yet a third design appeared. It is very similar to the Gregorian telescope except that the secondary mirror is convex and hyperboloidal and positioned before the primary mirror's focus (Figure 1.41). The design is known as the Cassegrain telescope after its inventor although very little is known about him beyond his surname. His nationality was French and his first name is variously suggested to be Laurent, Guillaume, Jacques, or Nicolas and his profession either a Catholic priest and teacher or a sculptor and metal founder. He may have been born in 1625 or 1629 and died in 1693 or 1712—or quite possibly during none of these years. The optician, James Short (1710–1768) who also produced Gregorian and Newtonian telescopes, probably made the first working telescopes to Cassegrain's design during mid-1700s. A surviving example of his work is a 9 in. (225 mm) diameter reflector manufactured around 1767/68 that has several secondary mirrors enabling it to be used in all three modes at magnifications up to 400 times. It is a great pity that so little is known of Cassegrain since his design, far more so than Newton's, dominates modern observational astronomy. The Cassegrain design and its variants and extensions such as the Ritchey–Chretién, the Coudé (Figure 1.44), and the Nasmyth (see later in this section) are used for every major and many smaller telescopes. Even the very popular small Schmidt–Cassegrain and Maksutov (Figure 1.50) telescopes, mostly sold for amateur use, are just a Cassegrain design with the addition of a thin correcting lens. Large telescopes (over 2 m or so in diameter) are often used at prime focus (i.e., without a secondary mirror) but in almost all cases can also be converted to a Cassegrain mode of operation. The popularity of the Cassegrain design arises primarily from the way in which the secondary mirror expands the cone of light from the primary mirror, enabling the effective focal length (and so the magnification or image scale) to be several times the instrument's actual physical size. This size reduction reduces the construction costs of both the telescope and its dome by as much as a factor of 10 or more.

Apart from the difficulty in manufacturing the required ellipsoidal, paraboloidal, or hyperboloidal surfaces for the mirrors, the early reflecting telescope suffered from another problem. The mirrors had to be made from speculum metal, an alloy with around three parts of copper to one part of tin. Even when freshly polished, speculum metal has a reflectivity of only around 60%. Thus, after the two reflections required in the Cassegrain, Newtonian, and Gregorian designs, only a little better than a third of the light entering the telescope will be delivered to the eye. Furthermore, speculum metal tarnishes and requires regular repolishing but that process could and very likely did change the shape of the surface of the mirror so that the quality of the images deteriorated each time unless care was taken to ensure that the correct surface shapes were maintained.

Despite these problems, the early reflecting telescopes underwent the same "telescope race" experienced by the early refractors, becoming larger, longer, and with higher and higher magnifications. By 1789, Sir William Herschel (1738–1822) had produced a 1.2 m (4 ft) reflector with a primary mirror weighing nearly 1000 kg (1 t). Herschel's telescopes though did not use any of the designs so far mentioned, but to avoid the loss of light when the secondary mirror was used, he tilted the primary mirror so that he could look directly at the image from the top of the telescope without his head getting in the way. Unfortunately, unless an off-axis-paraboloid shape can be produced for the primary mirror (as happens today with many TV satellite dish receivers), the quality of the image seen in this manner is severely degraded by aberrations. Herschel, however, accepted this drawback for the sake of the improved brightness of the images. Reflecting telescopes using speculum metal mirrors reached their pinnacle with the construction of the Leviathan of Parsonstown in 1845. William Parsons (Lord Rosse, 1800–1867) had the telescope constructed within the grounds of his castle in central Ireland with a primary mirror 1.8 m (6 ft) in diameter, weighing around 4000 kg (4 t). Supported between substantial parallel walls of masonry, the telescope made one major discovery—some of the nebulous objects seen in the sky were spiral in shape—the first hint of the existence of galaxies outside the Milky Way.

At about the same time as the speculum metal reflecting telescopes were being developed, a lens that was far less affected by chromatic aberration was invented by Chester Moor Hall (1703–1771). In 1729, Hall suggested that combining two lenses made from different types of glass would reduce the color problems arising from simple lenses. The achromatic lens (Figure 1.31) was first manufactured commercially by John Dollond (1706–1761) in 1758 and reduced the chromatic aberration compared with that of a simple lens by about a factor of 20. Dollond went on to construct many refractors using achromatic lenses for their objectives and other opticians soon followed. The achromatic refractor could be of a reasonable length and before long was widely used by the general public as well as by astronomers, as it still is. The achromatic refractor followed the development pattern of other types of telescopes culminating in 1897 with the 1 m (40 in.) Yerkes telescope. Refractors larger than this were never built successfully because lenses can only be supported at their edges and their weight leads to distortions, ruining their optical quality (mirrors can be supported on their backs as well as around their edges and so their distortions can be kept within bounds).

By the end of the nineteenth century, both achromatic refractors and speculum metal mirror reflectors had thus reached the limits of practical development. Fortunately, a technological breakthrough came to the rescue. Glass mirrors with a reflective backing of tin and mercury amalgam had been produced for domestic use in Venice from the early sixteenth century onward. The chemical process required to deposit a thin layer of silver onto glass was discovered by Justus von Liebig (1803–1873) in 1835 and is known within chemistry as the silver mirror test. In 1857, Léon Foucault started manufacturing telescope mirrors from glass and used Liebig's process to put a reflective coating of silver onto their front surfaces. The silver coating was a far better reflector than speculum metal, but like the latter it quickly tarnished. However, the great advantage of silvered mirrors was that the silver could be simply removed chemically without changing the shape of the surface of the mirror, as repolishing a speculum metal mirror was liable to do. Furthermore, a fresh, untarnished coating of silver could be placed onto the glass in an hour or two, and the mirror would be as good as new. Thus, the era of metal-on-glass reflecting telescopes began. Today aluminum is normally used as the reflecting metal and it is deposited onto a low-expansion ceramic or quartz substrate by evaporation inside a vacuum chamber, but the principle is unchanged. Casting large blanks of material to make such mirrors is not easy but throughout the twentieth century opticians entered the telescope race yet again with a will. The practical limit of development has probably been reached for monolithic mirrors with ESO's VLT and its four 8.2 m diameter primary mirrors, the two Gemini 8.1 m telescopes, and the large binocular telescope's (LBT's) two 8.4 m mirrors.

Writing during the first years of the twenty-first century, we are at the start of the fifth telescope race. Just when monolithic mirrors have reached their limits, technology and especially computers have developed to the point where larger mirrors can be built by putting many smaller mirrors together. Multi-mirror telescopes like the Keck instruments with their 10 m mirrors each made up from thirty-six 1.8 m hexagonal smaller mirrors and the 91 component mirrors of the 11 × 10 m Hobby–Eberly telescope represent the starting points for the next great leap forward. Since these telescopes and related developments such as real-time atmospheric compensation are covered in Section 1.1.21, their further discussion will be left until then except to ask where will the race end this time? With plans for 30 m and larger multi-mirror instruments already in the initial funding stages and with more speculative suggestions for instruments up to 100 m in diameter, we can note that the largest fully steerable radio dishes are only 100 m in size. Their sizes are limited by gravitational and wind stresses and of course by costs. Such considerations would apply even more to 100 m optical telescopes, so they may be the end of the line for multi-mirror instruments. Anything even more powerful will have to await further, as yet unknown, developments in designs and technology.

1.1.17.2 Optical Theory

Before looking further at the designs of telescopes and their properties, it is necessary to cover some of the optical principles upon which they are based. It is assumed that the reader is familiar with the basics of the optics of lenses and mirrors such terminology as focal ratio, magnification, optical axis, etc.; with laws such as Snell's law, the law of reflection, and so on; and with equations such as the lens and mirror formulae and the lens maker's formula. If required, however, any basic optics book or reasonably comprehensive physics book will suffice to provide this background.

We start by considering the resolution of a lens (the same considerations apply to mirrors, but the light paths, etc. are more easily pictured for lenses). That this is limited at all is due to the wave nature of light. As light passes through any opening, it is diffracted and the wavefronts spread out in a shape given by the envelope of the Huygens' secondary wavelets (Figure 1.24). Huygens' secondary wavelets radiate spherically outward from all points on a wavefront with the velocity of light in the medium concerned. Three examples are shown in Figure 1.24. Imaging the wavefront after passing through a slit-shaped aperture produces an image whose structure is shown in Figure 1.25. The variation in intensity

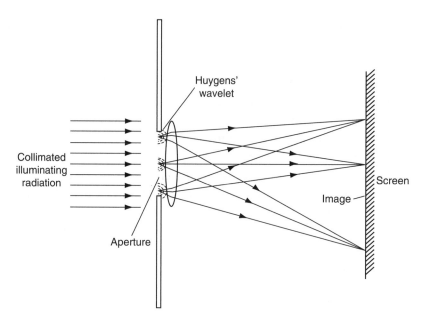

FIGURE 1.24 Fraunhofer diffraction at an aperture.

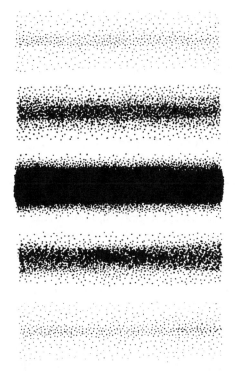

FIGURE 1.25 Image of a narrow slit (negative image).

is due to interference between waves originating from different parts of the aperture. The paths taken by such waves to arrive at a given point will differ and so also the distances that they have traveled. The waves will be out of step with each other to a greater or lesser extent depending upon the magnitude of this path difference. When the path difference is a half wavelength or an integer number of wavelengths plus a half wavelength, then the waves will be 180° out of phase with each other and so will cancel out. When the path difference is a whole number of wavelengths, the waves will be in step and will reinforce each other. Other path differences will cause intermediate degrees of cancellation or reinforcement. The central maximum arises from the reinforcement of many waves. The first minimum occurs when the path difference for waves originating at opposite edges of the aperture is one wavelength, then for every point in one-half of the aperture there is a point in the other half such that their path difference is half a wavelength, and all the waves cancel out completely. The intensity at a point within the image of a narrow slit may be obtained from the result that the diffraction pattern of an aperture is the power spectrum of the Fourier transform of its shape, and is given by

$$I_\theta = I_0 \frac{\sin^2 (\pi d \sin \theta / \lambda)}{(\pi d \sin \theta / \lambda)^2} \tag{1.6}$$

where
θ is the angle to the normal from the slit
d is the slit width
I_0 and I_θ are the intensities within the image on the normal and at an angle θ to the normal from the slit, respectively

With the image focused onto a screen at distance F from the lens, the image structure is as shown in Figure 1.26, where d is assumed to be large when compared with the wavelength λ. For a rectangular aperture with dimensions $d \times l$, the image intensity is similarly

$$I(\theta, \phi) = I_0 \frac{\sin^2 (\pi d \sin \theta / \lambda)}{(\pi d \sin \theta / \lambda)^2} \frac{\sin^2 (\pi l \sin \phi / \lambda)}{(\pi l \sin \phi / \lambda)^2} \tag{1.7}$$

where ϕ is the angle to the normal from the slit measured in the plane containing the side of length l. To obtain the image structure for a circular

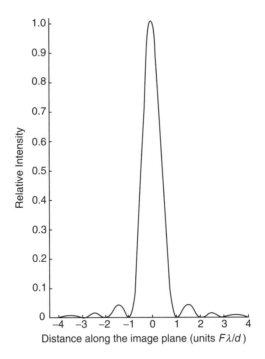

FIGURE 1.26 Cross section through the image of a narrow slit.

aperture, which is the case normally of interest in astronomy, we must integrate over the surface of the aperture. The image is then circular with concentric light and dark fringes. The central maximum is known as Airy's disk after the Astronomer Royal who first succeeded in completing the integration. The major difference from the previous case occurs in the position of the fringes.

Consider a circular aperture of radius r, illuminated by a beam of light normal to its plane (Figure 1.27), and consider also the light which is diffracted at an angle θ to the normal to the aperture from a point P, whose cylindrical coordinates with respect to the center of the aperture C, and the line AB which is drawn through C parallel to the plane containing the incident and diffracted rays, are (ϕ, ρ). The path difference, Δ, between diffracted rays from P and A is then

$$\Delta = (r - \rho \cos \phi) \sin \theta \qquad (1.8)$$

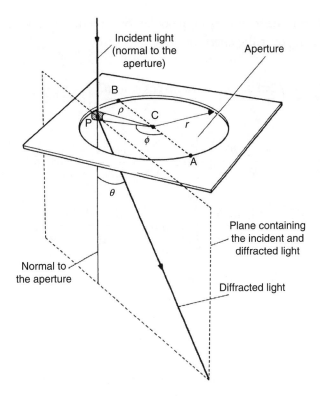

FIGURE 1.27 Diffraction by a circular aperture.

and the phase difference is

$$\frac{2\pi\Delta}{\lambda} = \frac{2\pi}{\lambda}(r - \rho\cos\phi)\sin\theta \tag{1.9}$$

The elemental area at P is just

$$dA = \rho\,d\phi\,d\rho \tag{1.10}$$

So the contribution to the electric vector of the radiation in the image plane by the elemental area around P for an angle θ to the normal is proportional to

$$\sin[\omega t + (2\pi/\lambda)(r - \rho\cos\phi)\sin\theta]\rho\,d\phi\,d\rho \tag{1.11}$$

where $\omega/2\pi$ is the frequency of the radiation, and the net effect is obtained by integrating over the aperture

$$\int_0^{2\pi}\int_0^r \sin\left[\omega t + \left(\frac{2\pi r\sin\theta}{\lambda}\right) - \left(\frac{2\pi\rho\cos\phi\sin\theta}{\lambda}\right)\right]\rho\,d\rho\,d\phi$$

$$= \sin\left(\omega t + \frac{2\pi r\sin\theta}{\lambda}\right)\int_0^{2\pi}\int_0^r \rho\cos\left(\frac{2\pi\rho\cos\phi\sin\theta}{\lambda}\right)d\rho\,d\phi$$

$$- \cos\left(\omega t + \frac{2\pi r\sin\theta}{\lambda}\right)\int_0^{2\pi}\int_0^r \rho\sin\left(\frac{2\pi\rho\cos\phi\sin\theta}{\lambda}\right)d\rho\,d\phi \qquad (1.12)$$

The second integral on the right-hand side is zero, which we may see by substituting

$$s = \frac{2\pi\rho\cos\phi\sin\theta}{\lambda} \qquad (1.13)$$

so that

$$\cos\left(\omega t + \frac{2\pi r\sin\theta}{\lambda}\right)\int_0^{2\pi}\int_0^r \rho\sin\left(\frac{2\pi\rho\cos\phi\sin\theta}{\lambda}\right)d\rho\,d\phi$$

$$= \cos\left(\omega t + \frac{2\pi r\sin\theta}{\lambda}\right)\int_0^r \rho\int_{s=2\pi\rho\sin\theta/\lambda}^{s=2\pi\rho\sin\theta/\lambda}\frac{-\sin s}{[(4\pi^2\rho^2\sin^2\theta/\lambda^2)-s^2]^{1/2}}ds\,d\phi$$

$$= 0 \qquad (1.14)$$

since the upper and lower limits of the integration with respect to s are identical. Thus, we have the image intensity $I(\theta)$ in the direction θ to the normal

$$I(\theta) \propto \left[\sin\left(\omega t + \frac{2\pi r\sin\theta}{\lambda}\right)\int_0^{2\pi}\int_0^r \rho\cos\left(\frac{2\pi\rho\cos\phi\sin\theta}{\lambda}\right)d\rho\,d\phi\right]^2$$

$$\qquad (1.15)$$

$$\propto \left(r^2 \int_0^{2\pi} \frac{\sin(2m\cos\phi)}{2m\cos\phi}\,d\phi - \frac{1}{2}r^2 \int_0^{2\pi} \frac{\sin^2(m\cos\phi)}{(m\cos\phi)^2}\,d\phi \right)^2 \quad (1.16)$$

where

$$m = \frac{\pi r \sin\theta}{\lambda} \quad (1.17)$$

Now

$$\frac{\sin(2m\cos\phi)}{2m\cos\phi} = 1 - \frac{(2m\cos\phi)^2}{3!} + \frac{(2m\cos\phi)^4}{5!} - \cdots \quad (1.18)$$

and

$$\frac{\sin^2(m\cos\phi)}{(m\cos\phi)^2} = 1 - \frac{2^3(m\cos\phi)^2}{4!} + \frac{2^5(m\cos\phi)^4}{6!} - \cdots \quad (1.19)$$

so that

$$I(\theta) \propto \left(r^2 \sum_0^\infty (-1)^n \int_0^{2\pi} \frac{(2m\cos\phi)^{2n}}{(2n+1)!}\,d\phi - \frac{1}{2}r^2 \sum_0^\infty (-1)^n \int_0^{2\pi} \frac{2^{2n+1}(m\cos\phi)^{2n}}{(2n+2)!}\,d\phi \right)^2$$

$$(1.20)$$

Now

$$\int_0^{2\pi} \cos^{2n}\phi\,d\phi = \frac{(2n)!}{2(n!)^2}\pi \quad (1.21)$$

and so

$$I(\theta) \propto \pi^2 r^4 \left[\sum_0^\infty (-1)^n \frac{1}{n+1}\left(\frac{m^n}{n!}\right)^2 \right]^2 \quad (1.22)$$

$$\propto \frac{\pi^2 r^4}{m^2}[J_1(2m)]^2 \quad (1.23)$$

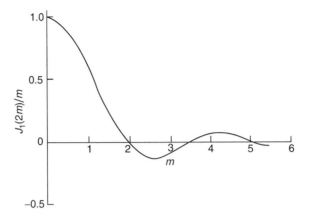

FIGURE 1.28 Variation of $J_1(2m)/m$ with m.

where $J_1(2m)$ is the Bessel function of the first kind of order unity. Thus, $I(\theta)$ is maximum or zero accordingly as $J_1(2m)/m$ reaches an extremum or is zero. Its variation is shown in Figure 1.28. The first few zeros occur for values of m of

$$m = 1.916, 3.508, 5.087, \ldots \tag{1.24}$$

and so in the image, the dark fringes occur at values of θ given by

$$\sin\theta = \frac{1.916\lambda}{\pi r}, \frac{3.508\lambda}{\pi r}, \frac{5.087\lambda}{\pi r}, \ldots \tag{1.25}$$

or for small values of θ

$$\theta \approx \frac{1.220\lambda}{d}, \frac{2.233\lambda}{d}, \frac{3.238\lambda}{d}, \ldots \tag{1.26}$$

where d is now the diameter of the aperture. The image structure along a central cross section is therefore similar to that of Figure 1.26, but with the minima expanded outward the points given in Equation 1.26 and the fringes, of course, are circular.

If we now consider two distant point sources separated by an angle α, then two such images will be produced and will be superimposed. There will not be any interference effects between the two images since the sources are mutually incoherent, and so their intensities will simply add together. The combined image will have an appearance akin to that shown

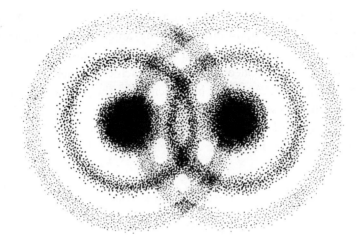

FIGURE 1.29 Image of two distant point sources through a circular aperture (negative image).

in Figure 1.29. When the center of the Airy disk of one image is superimposed upon the first minimum of the other image (and vice versa), then we have Rayleigh's criterion for the resolution of a lens (or mirror). This is the normally accepted measure of the theoretical resolution of a lens. It is given in radians by (from Equation 1.26)

$$\alpha = \frac{1.220\lambda}{d} \tag{1.27}$$

It is a convenient measure, but it is quite arbitrary. For sources of equal brightness, the image will appear noncircular for separations of about one-third of α, while for sources differing appreciably in brightness, the separation may need to be an order of magnitude larger than α for them to be resolved. For the eye, the effective wavelength is 510 nm for faint images so that the resolution of a telescope used visually is given by

$$R = \frac{0.128}{d} \tag{1.28}$$

where
d is the objective's diameter in meters
R is the angular resolution in seconds of arc

An empirical expression for the resolution of a telescope of $0.116''/d$, known as Dawes' limit, is often used by amateur astronomers since the slightly better value it gives for the resolution takes some account of the abilities of a skilled observer. To achieve either resolution in practice, the magnification must be sufficient for the angular separation of the images in the eyepiece to exceed the resolution of the eye (see the earlier discussion). Taking this to be an angle, β, we have the minimum magnification required to realize the Rayleigh limit of a telescope, M_m

$$M_m = \frac{\beta d}{1.220\lambda} \tag{1.29}$$

so that for β equal to $3'$, which is about its average value,

$$M_m = 1300d \tag{1.30}$$

where d is again measured in meters. Of course, most astronomical telescopes are actually limited in their resolution by the atmosphere. A 1 m telescope might reach its Rayleigh resolution on one night a year from a good observing site. On an average night, scintillation will spread stellar images to about $2''$ so that only telescopes smaller than about 0.07 m can regularly attain their diffraction limit. Since telescopes are rarely used visually for serious work, such high magnifications as are implied by Equation 1.30 are hardly ever encountered today. However, some of William Herschel's eyepieces still exist and these, if used on his 1.2 m telescope, would have given magnifications of up to 8000 times.

The theoretical considerations of resolution that we have just seen are only applicable if the lens or mirror is of sufficient optical quality and that the image is not already degraded beyond this limit. There are many effects that will blur the image and these are known as aberrations. With one exception, they can all affect the images produced by either lenses or mirrors. The universal or monochromatic aberrations are known as the Seidel aberrations after Ludwig von Seidel who first analyzed them. The exception is chromatic aberration and its related second-order effects of transverse chromatic aberration and secondary color, and these affect only lenses.

Chromatic aberration arises through the change in the refractive index of glass or other optical material with the wavelength of the illuminating radiation. Some typical values of the refractive index of some commonly used optical glasses are tabulated in Table 1.5.

TABLE 1.5   Some Typical Values of the Refractive Index of Some Commonly
Used Optical Glasses

Glass Type	Refractive Index at the Specified Wavelengths				
	361 nm	486 nm	589 nm	656 nm	768 nm
Crown	1.539	1.523	1.517	1.514	1.511
High dispersion crown	1.546	1.527	1.520	1.517	1.514
Light flint	1.614	1.585	1.575	1.571	1.567
Dense flint	1.705	1.664	1.650	1.644	1.638

The degree to which the refractive index varies with wavelength is
called the dispersion, and is measured by the constringence, ν

$$\nu = \frac{\mu_{589} - 1}{\mu_{486} - \mu_{656}} \tag{1.31}$$

where μ_λ is the refractive index at wavelength λ. The three wavelengths
that are chosen for the definition of ν are those of strong Fraunhofer lines:
486 nm, the F line (Hβ); 589 nm, the D lines (Na); and 656 nm, the C line
(Hα). Thus, for the glasses listed earlier, the constringence varies from 57
for the crown glass to 33 for the dense flint (note that the higher the value of
the constringence, the less that rays of different wavelengths diverge from
each other). The effect of the dispersion upon an image is to string it out
into a series of different colored images along the optical axis (Figure 1.30).

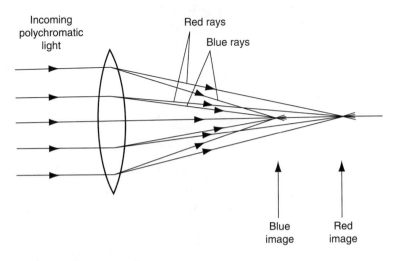

FIGURE 1.30 Chromatic aberration.

Looking at this sequence of images with an eyepiece, then at a particular point along the optical axis, the observed image will consist of a sharp image in the light of one wavelength surrounded by blurred images of varying sizes in the light of all the remaining wavelengths. To the eye, the best image occurs when yellow light is focused since it is less sensitive to the red and blue light. The image size at this point is called the circle of least confusion. The spread of colors along the optical axis is called the longitudinal chromatic aberration, while that along the image plane containing the circle of least confusion is called the transverse chromatic aberration.

Two lenses of different glasses may be combined to reduce the effect of chromatic aberration. Commonly in astronomical refractors, a biconvex crown glass lens is allied to a planoconcave flint glass lens to produce an achromatic doublet. In the infrared, achromats can be formed using barium, lithium, and strontium fluorides; zinc sulfate, or selenide and infrared-transmitting glasses. In the submillimeter region (i.e., wavelengths of several hundred microns), crystal quartz and germanium can be used. The lenses are either cemented together or separated by only a small distance (Figure 1.31). Despite its name, there is still some chromatic

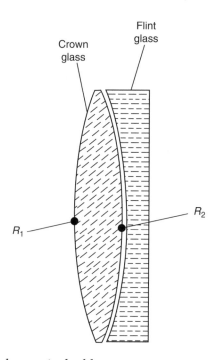

FIGURE 1.31 An achromatic doublet.

aberration remaining in this design of lens since it can only bring two wavelengths to a common focus. If the radii of the curved surfaces are all equal, then the condition for two given wavelengths, λ_1 and λ_2, to have coincident images is

$$2\Delta\mu_C = \Delta\mu_F \tag{1.32}$$

where $\Delta\mu_C$ and $\Delta\mu_F$ are the differences between the refractive indices at λ_1 and λ_2 for the crown glass and the flint glass, respectively. More flexibility in design can be attained if the two surfaces of the converging lens have differing radii. The condition for achromatism is then

$$\frac{|R_1| + |R_2|}{|R_1|}\Delta\mu_C = \Delta\mu_F \tag{1.33}$$

where
R_2 is the radius of the surface of the crown glass lens that is in contact with the flint lens (Note: the radius of the flint lens surface is almost invariably R_2 as well, in order to facilitate alignment and cementing)
R_1 is the radius of the other surface of the crown glass lens

By a careful selection of λ_1 and λ_2 an achromatic doublet can be constructed to give tolerable images. For example, by achromatizing at 486 and 656 nm, the longitudinal chromatic aberration is reduced when compared with a simple lens of the same focal length by a factor of about 30. Nevertheless, since chromatic aberration varies as the square of the diameter of the objective and inversely with its focal length, refractors larger than about 0.25 m still have obtrusively colored images. More seriously, if filters are used then the focal position will vary with the filter. Similarly, the scale on photographic plates will alter with the wavelengths of their sensitive regions. Further lenses may be added to produce apochromats that have three corrected wavelengths and superapochromats with four corrected wavelengths. But such designs are impossibly expensive for telescope objectives of any size, although eyepieces and camera lenses may have 8 or 10 components and achieve very high levels of correction.

A common and severe aberration of both lenses and mirrors is spherical aberration. In this effect, annuli of the lens or mirror that are of different radii have different focal lengths. It is illustrated in Figure 1.32 for

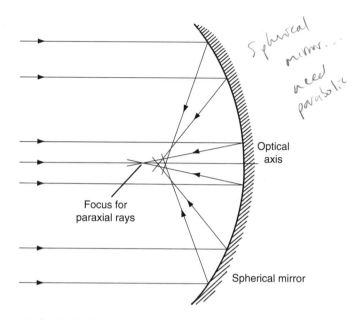

Spherical mirror...

Need parabolic

FIGURE 1.32 Spherical aberration.

a spherical mirror. For rays parallel to the optical axis, it can be eliminated completely by deepening the sphere to a paraboloidal surface for the mirror. It cannot be eliminated from a simple lens without using aspheric surfaces, but for a given focal length it may be minimized. The shape of a simple lens is measured by the shape factor, q

$$q = \frac{R_2 + R_1}{R_2 - R_1} \tag{1.34}$$

where R_1 is the radius of the first surface of the lens and R_2 is the radius of the second surface of the lens. The spherical aberration of a thin lens then varies with q as shown in Figure 1.33, with a minimum at $q = +0.6$. The lens is then biconvex with the radius of the surface nearer to the image three times the radius of the other surface. Judicious choice of surface radii in an achromatic doublet can lead to some correction of spherical aberration while still retaining the color correction. Spherical aberration in lenses may also be reduced by using high refractive index glass since the curvatures required for the lens' surfaces are lessened, but this is likely to increase chromatic aberration. Spherical aberration increases as the cube of the aperture.

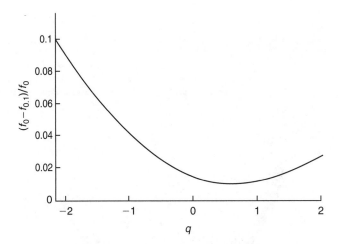

FIGURE 1.33 Spherical aberration in thin lenses. f_x is the focal length for rays parallel to the optical axis, and at a distance x times the paraxial focal length away from it.

The deepening of a spherical mirror to a paraboloidal one to correct for spherical aberration unfortunately introduces a new aberration called coma. This also afflicts mirrors of other shapes and lenses. It causes the images for objects away from the optical axis to consist of a series of circles that correspond to the various annular zones of the lens or mirror and which are progressively shifted toward or away from the optical axis (Figure 1.34). The severity of coma is proportional to the square of the aperture. It is zero in a system that obeys Abbe's sine condition

$$\frac{\sin \theta}{\sin \phi} = \frac{\theta_p}{\phi_p} = \text{constant} \qquad (1.35)$$

where the angles are defined in Figure 1.35. A doublet lens can be simultaneously corrected for chromatic and spherical aberrations, and coma within acceptable limits, if the two lenses can be separated. Such a system is called an aplanatic lens. A parabolic mirror can be corrected for coma by adding thin correcting lenses before or after the mirror, as discussed in more detail in Section 1.1.18. The severity of the coma at a given angular distance from the optical axis is inversely proportional to the square of the focal ratio. Hence, using as large a focal ratio as possible can also reduce its effect. In Newtonian reflectors, a focal ratio of $f8$ or larger gives acceptable coma for most purposes. At $f3$, coma will limit the

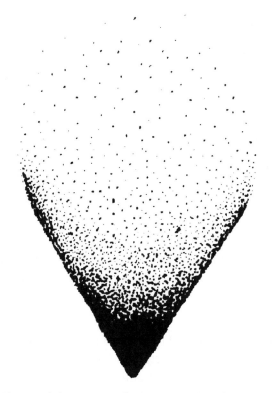

FIGURE 1.34 Shape of the image of a point source due to coma (negative image).

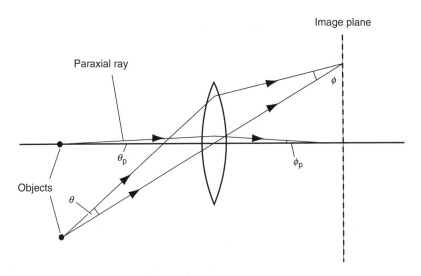

FIGURE 1.35 Parameters for Abbe's sine condition.

useful field of view to about 1 minute of arc, so that prime focus imaging almost always requires the use of a correcting lens to give reasonable fields of view.

Astigmatism is an effect whereby the focal length differs for rays in the plane containing an off-axis object and the optical axis (the tangential plane), in comparison with rays in the plane at right angles to this (the sagittal plane). It decreases more slowly with focal ratio than coma so that it may become the dominant effect for large focal ratios. It is possible to correct astigmatism, but only at the expense of introducing yet another aberration, field curvature. This is simply that the surface containing the sharply focused images is no longer a flat plane but is curved. A system in which a flat image plane is retained and astigmatism is corrected for at least two radii is termed an anastigmatic system.

The final aberration is distortion, and this is a variation in the magnification over the image plane. An optical system will be free of distortion only if the condition

$$\frac{\tan \theta}{\tan \phi} = \text{constant} \tag{1.36}$$

holds for all values of θ (see Figure 1.36 for the definition of the angles). Failure of this condition to hold results in pincushion or barrel distortion (Figure 1.37) accordingly as the magnification increases or decreases with distance from the optical axis. A simple lens is very little affected by distortion and it can frequently be reduced in more complex systems by the judicious placing of stops within the system.

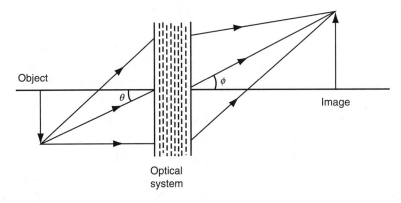

FIGURE 1.36 Terminology for distortion.

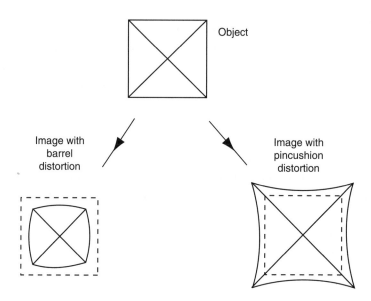

FIGURE 1.37 Distortion.

A fault of optical instruments as a whole is vignetting. This is not an attribute of a lens or mirror and so is not included among the aberrations. It arises from the uneven illumination of the image plane, usually due to obstruction of the light path by parts of the instrument. Normally it can be avoided by careful design, but it may become important if stops are used in a system to reduce other aberrations.

This long catalog of faults of optical systems may well have led the reader to wonder if the Rayleigh limit can ever be reached in practice. However, optical designers have a wide variety of variables to play with— refractive indices, dispersion, focal length, mirror surfaces in the form of various conic sections, spacings of the elements, number of elements, and so on—so that it is usually possible to produce a system which will give an adequate image for a particular purpose. Other criteria such as cost, weight, production difficulties, etc. may well prevent the realization of the system in practice even though it is theoretically possible. Multipurpose systems can usually only be designed to give lower quality results than single-purpose systems. Thus, the practical telescope designs that are discussed later in this section are optimized for objects at infinity and if they are used for nearby objects, their image quality will deteriorate. The approach to designing an optical system is largely an empirical one. There are two important methods in current use. The older approach requires an analytical expression for each of the aberrations. For example, the third-order approximation for spherical aberration of a lens is

$$
\frac{1}{f_x} - \frac{1}{f_p} = \frac{x^2}{8f^3\mu(\mu-1)} \left[\frac{\mu+2}{\mu-1} q^2 + 4(\mu+1)\left(\frac{2f}{v}-1\right)q \right.
$$
$$
\left. + (3\mu+2)(\mu-1)\left(\frac{2f}{v}-1\right)^2 + \frac{\mu^2}{\mu-1} \right] \tag{1.37}
$$

where

f_x is the focal distance for rays passing through the lens at a distance x from the optical axis

f_p is the focal distance for paraxial rays from the object (paraxial rays are rays that are always close to the optical axis and which are inclined to it by only small angles)

f is the focal length for paraxial rays that are initially parallel to the optical axis

v is the object distance

For precise work, it may be necessary to involve fifth-order approximations of the aberrations, and so the calculations rapidly become very cumbersome. The alternative and more modern approach is via ray tracing. The concepts involved are much simpler since the method consists simply of accurately following the path of a selected ray from the object through the system and finding its arrival point on the image plane. Only the basic formulas are required:

Snell's law for lenses

$$
\sin i = \frac{\mu_1}{\mu_2} \sin r \tag{1.38}
$$

and the law of reflection for mirrors

$$
i = r \tag{1.39}
$$

where

i is the angle of incidence

r is the angle of refraction or reflection as appropriate

μ_1 and μ_2 are the refractive indices of the materials on either side of the interface

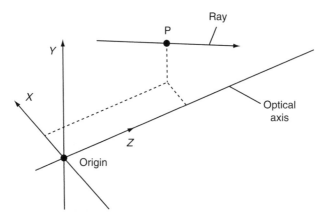

FIGURE 1.38 Ray tracing coordinate system.

The calculation of i and r for a general ray requires knowledge of the ray's position and direction in space, and the specification of this is rather more cumbersome. Consider first of all a ray passing through a point P, which is within an optical system (Figure 1.38). We may completely describe the ray by the spatial coordinates of the point P, together with the angles that the ray makes with the coordinate system axes (Figure 1.39). We may

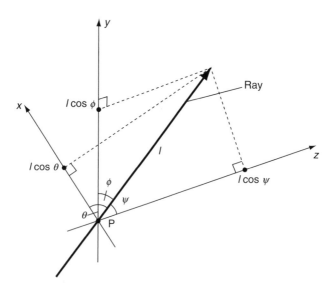

FIGURE 1.39 Ray tracing angular coordinate system.

without any loss of generality set the length of the ray, l, equal to unity, and therefore write

$$\gamma = \cos \theta \qquad (1.40)$$

$$\delta = \cos \phi \qquad (1.41)$$

$$\varepsilon = \cos \psi \qquad (1.42)$$

The angular direction of the ray is thus specified by the vector

$$\mathbf{v} = (\gamma, \delta, \varepsilon) \qquad (1.43)$$

The quantities γ, δ, and ε are commonly referred to as the direction cosines of the ray. If we now consider the point P as a part of an optical surface, then we require the angle of the ray with the normal to that surface in order to obtain i, and thereafter, r. Now we may similarly specify the normal at P by its direction cosines, forming the vector, $\mathbf{v'}$,

$$\mathbf{v'} = (\gamma', \delta', \varepsilon') \qquad (1.44)$$

and we then have

$$\cos i = \frac{\mathbf{v} \cdot \mathbf{v'}}{\|\mathbf{v}\| \, \|\mathbf{v'}\|} \qquad (1.45)$$

or

$$i = \cos^{-1} (\gamma\gamma' + \delta\delta' + \varepsilon\varepsilon') \qquad (1.46)$$

The value of r can now be obtained from Equation 1.38 or 1.39. We may again specify its track by direction cosines (the vector $\mathbf{v''}$)

$$\mathbf{v''} = (\gamma'', \delta'', \varepsilon'') \qquad (1.47)$$

and we may find the values of the components as follows. Three simultaneous equations can be obtained; the first from the angle between the incident ray and the refracted or reflected ray, the second from the angle between the normal and the reflected or refracted ray, and the third by requiring the incident ray, the normal, and the refracted or reflected ray to be coplanar, and these are

$$\gamma\gamma'' + \delta\delta'' + \varepsilon\varepsilon'' = \cos 2r \text{ (reflection)} \tag{1.48}$$

$$\gamma\gamma'' + \delta\delta'' + \varepsilon\varepsilon'' = \cos(i - r) \text{ (refraction)} \tag{1.49}$$

$$\gamma'\gamma'' + \delta'\delta'' + \varepsilon'\varepsilon'' = \cos r \tag{1.50}$$

$$(\varepsilon\delta' - \varepsilon'\delta)\gamma'' + (\gamma\varepsilon' - \gamma'\varepsilon)\delta'' + (\delta\gamma' - \delta'\gamma)\varepsilon'' = \cos r \tag{1.51}$$

After a considerable amount of manipulation, one obtains

$$\varepsilon'' = \frac{\{[(\varepsilon\delta' - \varepsilon'\delta)\delta' - (\gamma\varepsilon' - \gamma'\varepsilon)\gamma'](\gamma'\cos\alpha - \gamma\cos r) - (\gamma'\delta - \gamma\delta')(\varepsilon\delta' - \varepsilon'\delta)\cos r\}}{\{[(\varepsilon\delta' - \varepsilon'\delta)\delta' - (\gamma\varepsilon' - \gamma'\varepsilon)\gamma'](\gamma'\varepsilon - \gamma\varepsilon') - [(\varepsilon\delta' - \varepsilon'\delta)\varepsilon' - (\delta\gamma' - \delta'\gamma)\gamma'](\gamma'\delta - \gamma\delta')\}} \tag{1.52}$$

$$\delta'' = \frac{\gamma'\cos\alpha - \gamma\cos r - (\gamma'\varepsilon - \gamma\varepsilon')\varepsilon''}{(\gamma'\delta - \gamma\delta')} \tag{1.53}$$

$$\gamma'' = \frac{\cos\alpha - \delta\delta'' - \varepsilon\varepsilon''}{\gamma} \tag{1.54}$$

where α is equal to $2r$ for reflection, and $(i - r)$ for refraction.

The direction cosines of the normal to the surface are easy to obtain when its center of curvature is on the optical axis

$$\gamma' = \frac{x_1}{R} \tag{1.55}$$

$$\delta' = \frac{y_1}{R} \tag{1.56}$$

$$\varepsilon' = \frac{(z_1 - z_R)}{R} \tag{1.57}$$

where (x_1, y_1, z_1) is the position of P and $(0, 0, z_R)$ is the position of the center of curvature. If the next surface is at a distance s from the surface under consideration, then the ray will arrive on it at a point (x_2, y_2, z_2) still with the direction cosines γ'', δ'', and ε'', where

$$x_2 = \frac{\gamma''s}{\varepsilon''} + x_1 \tag{1.58}$$

$$y_2 = \frac{\delta''s}{\varepsilon''} + y_1 \tag{1.59}$$

$$z_2 = s + z_1 \tag{1.60}$$

We may now repeat the calculation again for this surface, and so on. Ray tracing has the advantage that all the aberrations are automatically included by it, but it has the disadvantage that many rays have to be followed to build up the structure of the image of a point source at any given place on the image plane, and many images have to be calculated to assess the overall performance of the system. Ray tracing, however, is eminently suitable for programming onto a computer, while the analytical approach is not, since it requires frequent value judgments. The approach of a designer, however, to a design problem is similar irrespective of the method used to assess the system. The initial prototype is set up purely on the basis of what the designer's experience suggests that may fulfill the major specifications such as cost, size, weight, resolution, etc. A general rule of thumb is that the number of optical surfaces needed will be at least as many as the number of aberrations to be corrected. The performance is then assessed either analytically or by ray tracing. In most cases, it will not be good enough, so a slight alteration is made with the intention of improving the performance, and it is reassessed. This process continues until the original prototype has been optimized for its purpose. If the optimum solution is within the specifications, then there is no further problem. If it is outside the specifications, even after optimization, the whole procedure is repeated starting from a different prototype. The performances of some of the optical systems favored by astronomers are considered in the next subsection.

Even after a design has been perfected, there remains the not inconsiderable task of physically producing the optical components to within the accuracies specified by the design. The manufacturing steps for both lenses and mirrors are broadly similar, although the details may vary. The surface is roughly shaped by moulding or by diamond milling. It is then matched to another surface formed in the same material whose shape is its inverse, called the tool. The two surfaces are ground together with coarse carborundum or other grinding powder between them until the required surface begins to approach its specifications. The pits left behind by this coarse grinding stage are removed by a second grinding stage in which a finer powder is used. The pits left by this stage are then removed in turn by a third stage using still finer powder, and so on. As many as 8 or 10 such stages may be necessary. When the grinding pits are reduced to a micron or so in size, the surface may be polished. This process employs a softer powder such as iron oxide or cerium oxide that is embedded in a soft matrix such as pitch. Once the surface has been polished, it can be tested for the accuracy of its fit to its specifications. Since, in general, it will not be within the specifications after the initial polishing, a final stage, which is termed figuring, is necessary. This is simply additional polishing to adjust the surface's shape until it is correct. The magnitude of the changes involved during this stage is only about a micron or two, so that if the alteration that is needed is larger than this, it may be necessary to return to a previous stage in the grinding to obtain a better approximation. There are a number of tests that can determine the shape of the mirrors surface to a precision of ± 50 nm or better, such as the Foucault, Ronchi, Hartmann, and Null tests. The details of these tests are beyond the scope of this book but may be found in books on optics and telescope making.

Small mirrors can be produced in this way by hand, and many an amateur astronomer has acquired his or her telescope at very low cost by making their own optics. Larger mirrors require machines that move the tool (now smaller than the mirror) in an epicyclic fashion. The motion of the tool is similar to that of the planets under the Ptolemaic model of the solar system, and so such machines have become known as planetary polishers. The epicyclic motion can be produced by a mechanical arrangement, but commercial production of large mirrors now relies on computer-controlled planetary polishers. The deepening of the surface to a parabolic, hyperbolic, or other shape is accomplished by preferential polishing and sometimes by the use of a flexible tool whose shape can be adjusted. The mirror segments for large instruments like the 10 m Keck telescopes are clearly small, off-axis parts of the total hyperbolic shape. These, and also the complex shapes required for Schmidt telescope corrector lenses, have been produced using stressed polishing. The blank is deformed by carefully designed forces from a warping harness, and then polished to a spherical shape (or flat for Schmidt corrector plates). When the deforming forces are released, the blank springs into the required nonspherical asymmetric shape. For the Keck and Hobby–Eberly mirror segments, the final polishing was undertaken using ion beams. The ion beam is an accelerated stream of ions, such as argon, that wears away the surface of an atom at a time. Since it applies almost no force to the mirror, the process can be used to correct defects left by normal polishing techniques such as deformations at the edge and print-through of the honeycomb back on lightweight mirrors.

Recently, the requirements for non-axisymmetric mirrors for segmented-mirror telescopes (see later discussion) and for glancing incidence x-ray telescopes (Section 1.3) have led to the development of numerically controlled diamond milling machines, which can produce the required shaped and polished surface directly to an accuracy of 10 nm or better.

The defects in an image that are due to surface imperfections on a mirror will not exceed the Rayleigh limit if those imperfections are less than about one-eighth of the wavelength of the radiation for which the mirror is intended. Thus, we have the commonly quoted $\lambda/8$ requirement for the maximum allowable deviation of a surface from its specifications. Sometimes the resulting deviations of the wavefront are specified rather than those of the optical surface. This is twice the value for the surface, that is, a limit of $\lambda/4$. The restriction on lens surfaces is about twice as large as those of mirror surfaces since the ray deflection is distributed over the two lens faces. However, as we have seen, the Rayleigh limit is an arbitrary one, and for some purposes, the fit must be several times better than this limit. This is particularly important when viewing extended objects such as planets, and improvements in the contrast can continue to be obtained by figuring surfaces to $\lambda/20$ or better.

The surface must normally receive its reflection coating after its production. The vast majority of astronomical mirror surfaces have a thin layer of aluminum evaporated onto them by heating aluminum wires suspended over the mirror inside a vacuum chamber. Other materials, such as silicon carbide are occasionally used, especially for ultraviolet work, since the reflectivity of aluminum falls off below 300 nm. The initial reflectivity of an aluminum coating in the visual region is around 90%. This however can fall to 75% or less within a few months as the coating ages. Mirrors therefore have to be re-aluminized at regular intervals. The intervals between re-aluminizing can be lengthened by gently cleaning the mirror every month or so. The currently favored methods of cleaning are rinsing with de-ionized water and drifting carbon dioxide snow across the mirror surface. Mirrors coated with a suitably protected silver layer can achieve 99.5% reflectivity in the visible, and are becoming more and more required since some modern telescope designs can have four or five reflections. With ancillary instrumentation, the total number of reflections can then reach 10 or more. A silver coating also lowers the infrared emission from the mirror itself from around 4% at a wavelength of 10 μm for an aluminum coating to around 2%. For this reason, the 8.1 m Gemini telescopes use protected silver coatings that are applied to the surfaces of the mirrors by evaporation inside a vacuum chamber.

Lens surfaces also normally receive a coating after their manufacture, but in this case the purpose is to reduce reflection. Uncoated lenses reflect about 5% of the incident light from each surface so that a system containing, say, five uncoated lenses could lose 40% of its available light through this process. To reduce the reflection losses, a thin layer of material covers the lens, for which

$$\mu' = \sqrt{\mu} \tag{1.61}$$

where
μ' is the refractive index of the coating
μ is the refractive index of the lens material

The thickness, t, should be

$$t = \frac{\lambda}{4\mu'} \tag{1.62}$$

This gives almost total elimination of reflection at the selected wavelength, but some will still remain at other wavelengths. Lithium fluoride and silicon dioxide are commonly used materials for the coatings. Recently, it has become possible to produce antireflection coatings that are effective simultaneously over a number of different wavelength regions through the use of several interfering layers.

Mirrors only require the glass as a support for their reflecting film. Thus, there is no requirement for it to be transparent; on the other hand, it is essential for its thermal expansion coefficient to be low. Today's large telescope mirrors are therefore made from materials other than glass. The coefficient of thermal expansion of glass is about 9×10^{-6} K^{-1}, that of Pyrex about 3×10^{-6} K^{-1}, and for fused quartz it is about 4×10^{-7} K^{-1}. Among the largest telescopes, almost the only glass mirror is the one for the 2.5 m Hooker telescope on Mount Wilson. After its construction, Pyrex became the favored material, until the last three decades, when quartz or artificial materials with a similar low coefficient of expansion such as CerVit and Zerodur, etc. but which are easier to manufacture have been used. These low-expansion materials have a ceramic glass structure that is partially crystallized. The crystals are around 50 nm across and contract as the temperature rises, while the surrounding matrix of amorphous material expands. An appropriate ratio of crystals to matrix provides very small expansion (10^{-7} K^{-1}) over a wide range of temperatures. Ultra-low-expansion (ULE) fused silica has an even smaller coefficient of expansion (3×10^{-8} K^{-1}). A radically different alternative is to use a material with a very high thermal conductivity, such as silicon carbide, graphite epoxy, steel, beryllium or aluminum. The mirror is then always at a uniform temperature and its surface has no stress distortion. Most metallic mirrors, however, have a relatively coarse crystalline surface that cannot be polished adequately. They must therefore be coated with a thin layer of nickel before polishing, and there is always some risk of this being worn through if the polishing process extends for too long.

Provided that they are mounted properly, small solid mirrors can be made sufficiently thick that once ground and polished, their shape is maintained simply by their mechanical rigidity. However, the thickness required for such rigidity scales as the cube of the size so that the weight of a solid rigid mirror scales as D^5. Increasing solid mirror blanks to larger sizes thus rapidly becomes expensive and impractical. Most mirrors larger than about 0.5–1 m therefore have to have their weight reduced in some way. There are two main approaches for reducing the mirror weight: thin mirrors and honeycomb mirrors. The thin mirrors are also subdivided into monolithic and segmented mirrors. In both cases, active support of the mirror is required to retain its correct optical shape. The 8.2 m mirrors of ESO's VLT are an example of thin monolithic mirrors, they are 178 mm thick Zerodur, with a weight of 23 t each, but need 150 actuators to maintain their shape. The 10 m Keck telescopes use thin, segmented Zerodur mirrors. There are 36 individual hexagonal segments in each main mirror, with each segment 1.8 m across and about 70 mm thick, giving a total mirror weight of just 14.4 t (compared with 14.8 t for the 5 m Hale telescope mirror).

Honeycomb mirrors are essentially thick solid blanks that have had a lot of the material behind the reflecting surface removed, leaving only thin struts to support that surface. The struts often surround hexagonal cells of removed material giving the appearance of honeycomb, but other cell shapes such as square, circular, or triangular can be used. The mould for a honeycomb mirror blank has shaped cores protruding from its base that produce the empty cells, with the channels between the cores filling with the molten material to produce the supporting ribs. Once the blank has solidified and been annealed, it may be ground and polished as though it were solid. However, if the original surface of the blank is flat, then considerable amounts of material will remain near the edge after grinding, adding to the final weight of the mirror. Roger Angel of the University of Arizona has therefore pioneered the technique of spin-casting honeycomb mirror blanks. The whole furnace containing the mould and molten material is rotated so that the surface forms a parabola (see liquid mirrors below) close to the finally required shape. The surface shape is preserved as the furnace cools and the material solidifies. Potentially honeycomb mirrors up to 8 m in diameter may be produced by this method. Howsoever it may have been produced, thinning the ribs and the underside of the surface may further reduce the weight of the blank. The thinning is accomplished by milling or etching with hydrogen fluoride. In this way, the 1 m diameter Zerodur secondary mirrors produced for the 8.1 m Gemini telescopes have supporting ribs just 3 mm wide, and weigh less than 50 kg.

Whatever approach is taken in reducing the mirror's weight, it will be aided if the material used for the blank has a highly intrinsic stiffness (resistance to bending). Beryllium has already been used to produce the 1.1 m secondary mirrors for the VLT, whose weights are only 42 kg. Silicon carbide and graphite epoxy are other high stiffness materials that may be used in the future.

A quite different approach to producing mirrors that, perhaps surprisingly gives low weights, is to use a rotating bath of mercury. Isaac Newton was the first to realize that the surface of a steadily rotating liquid would take up a paraboloidal shape under the combined influences of gravity and centrifugal acceleration. If that liquid reflects light, like mercury, gallium, gallium-indium alloy, or an oil suffused with reflecting particles, then it can act as a telescope's primary mirror. Of course the mirror has to remain accurately horizontal, so that it always points toward the zenith, but with suitable, perhaps active, correcting secondary optics the detector can be moved around the image plane to give a field of view currently tens of minutes of arc wide, and conceivably in the future up to 8° across. Moving the electrons across the CCD detector at the same speed as the image movement enables short time exposures to be obtained (TDI). The earth's rotation enables such a zenith telescope to sample up to 7% of the sky for less than 7% of the cost of an equivalent fully steerable telescope. The "bath" containing the mercury is in fact a lightweight structure whose surface is parabolic to within 0.1 mm. Only a thin layer of mercury is thus needed to produce the accurate mirror surface, and so the mirror overall is a lightweight one despite the high density of mercury. The bath is rotated smoothly around an accurately vertical axis at a constant angular velocity using a large air bearing. Mercury is toxic, so that suitable precautions have to be taken to protect the operators. Also, its reflectivity is less than

80% in the visible, but since all the other optics in the instrument can be conventional mirrors or lenses, this penalty is not too serious. Until recently, NASA operated a 3 m liquid-mirror telescope in New Mexico to track debris orbiting the earth (NASA Orbital Debris Observatory or NODO), and a 2.7 m liquid-mirror telescope has operated since 1995 in British Columbia. The large zenith telescope (LZT), also in British Columbia, saw first light in 2003 using a 6 m diameter liquid mirror.

The large-aperture mirror array (LAMA), a proposal for an array of eighteen 10 m liquid-mirror telescopes, would give the equivalent collecting power of a single 42 m dish. In space, a liquid mirror of potentially almost any size could be formed from a ferromagnetic liquid confined by electromagnetic forces.

An approach to producing mirrors whose shape can be rapidly adjusted is to use a thin membrane with a reflection coating. The membrane forms one side of a pressure chamber, and its shape can be altered by changing the gas pressure inside the chamber. $\lambda/2$ surface quality or better can currently be achieved over mirror diameters up to 5 mm, and such mirrors have found application in optical aperture synthesis systems (Section 2.5.5).

1.1.18 Telescope Designs

1.1.18.1 Background

Most serious work with telescopes uses equipment placed directly at the focus of the telescope. But for visual work such as finding and guiding on objects, an eyepiece is necessary. Often it matters little whether the image produced by the eyepiece is of a high quality or not. Ideally, however, the eyepiece should not degrade the image noticeably more than the main optical system. There are an extremely large number of eyepiece designs, whose individual properties can vary widely. For example, one of the earliest eyepiece designs of reasonable quality is the Kellner. This combines an achromat and a simple lens and typically has a field of view of 40°–50°. The Plössl uses two achromats and has a slightly wider field of view. More recently, the Erfle design employs six or seven components and gives fields of view of 60°–70°, while the current state of the art is represented by designs such as the Nagler with eight or more components and fields of view up to 85°. Details of these and other designs may generally be found in books on general astronomy or on optics or from the manufacturers. For small telescopes used visually, a single low magnification wide-angle eyepiece may be worth purchasing for the magnificent views of large objects like the Orion Nebula that it will provide. There is little point in the higher power eyepieces being wide-angle ones (which are very expensive), since these will normally be used to look at angularly small objects. The use of a Barlow lens provides an adjustable magnification for any eyepiece. The Barlow lens is an achromatic negative lens placed just before the telescope's focus. It increases the telescope's effective focal length, and so the magnification.

For our purposes, only four aspects of eyepieces are of concern; light losses, eye relief, exit pupil, and angular field of view. Light loss occurs through vignetting when parts of the light beam fail to be intercepted by the eyepiece optics, or are obstructed by a part of the structure of the eyepiece, and also through reflection, scattering and absorption by the optical components. The first of these can generally be avoided by careful eyepiece design and selection, while the latter effects can be minimized by antireflection coatings and by keeping the eyepieces clean.

The exit pupil is the image of the objective produced by the eyepiece (Figure 1.40). All the rays from the object pass through the exit pupil, so that it must be smaller than the pupil of the human eye if all of the light gathered by the objective is to be utilized. Its diameter, E, is given by

$$E = \frac{F_e D}{F_o} \tag{1.63}$$

where
D is the objective's diameter
F_e is the focal length of the eyepiece
F_o is the focal length of the objective

Since magnification is given by

$$M = \frac{F_o}{F_e} \tag{1.64}$$

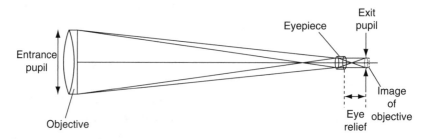

FIGURE 1.40 Exit pupil and eye relief.

and the diameter of the pupil of the dark-adapted eye is 6 or 7 mm, we must therefore have

$$M \geq \sim 170D \tag{1.65}$$

where D is in meters, if the whole of the light from the telescope is to pass into the eye. In practice, for small telescopes being used visually, the maximum useable magnification is around $\times 2000D$ (i.e., $\times 300$ for a 0.15 m telescope).

The eye relief is the distance from the final lens of the eyepiece to the exit pupil. It should be about 6–10 mm for comfortable viewing. If you wear spectacles, however, then the eye relief may need to be up to 20 mm.

The angular field of view is defined by the acceptance angle of the eyepiece, θ'. Usually this is about 40°–60°, but it may be up to 90° for wide-angle eyepieces. The angular diameter of the area of sky, which is visible when the eye is positioned at the exit pupil and which is known as the angular field of view, θ, is then just

$$\theta = \frac{\theta'}{M} \tag{1.66}$$

The brightness of an image viewed through a telescope is generally expected to be greater than when it is viewed directly. However, this is not the case for extended objects. The naked-eye brightness of a source is proportional to the eye's pupil diameter squared, while that of the image in a telescope is proportional to the objective diameter squared. If the eye looks at that image, then its angular size is increased by the telescope's magnification. Hence, the increased brightness of the image is spread over a greater area. Thus, we have

$$R = \frac{\text{brightness through a telescope}}{\text{brightness to the naked eye}} = \frac{D^2}{M^2 P^2} \tag{1.67}$$

where
D is the objective's diameter
P is the diameter of the pupil of the eye
M is the magnification

But from Equations 1.63 and 1.64 we have

$$R = 1 \tag{1.68}$$

when the exit pupil diameter is equal to the diameter of the eye's pupil, and

$$R < 1 \qquad (1.69)$$

when it is smaller than the eye's pupil.

If the magnification is less than $170D$ times, then the exit pupil is larger than the pupil of the eye and some of the light gathered by the telescope will be lost. Since the telescope will generally have other light losses due to scattering, imperfect reflection, and absorption, the surface brightness of an extended source is always fainter when viewed through a telescope than when viewed directly with the naked eye. This result is in fact a simple consequence of the second law of thermodynamics; for if it were not the case, then one could have a net energy flow from a cooler to a hotter object. The apparent increase in image brightness when using a telescope arises partly from the increased angular size of the object so that the image on the retina must fall to some extent onto the regions containing more rods, even when looked at directly, and partly from the increased contrast resulting from the exclusion of extraneous light by the optical system. Even with the naked eye, looking through a long cardboard tube enables faint extended sources such as M31 to be seen more easily.

The analysis that we have just seen does not apply to images that are physically smaller than the detecting element. For this situation, the image brightness is proportional to D^2. Again, however, there is an upper limit to the increase in brightness that is imposed when the energy density at the image is equal to that at the source. This limit is never approached in practice since it would require 4π sr for the angular field of view. Thus, stars that are fainter than those visible to the naked eye may be seen through a telescope by a factor called the light grasp. The light grasp is simply given by D^2/P^2. Taking $+6^m$ as the magnitude of the faintest star visible to the naked eye (see Section 3.1.1.2); the faintest star that may be seen through a telescope has a magnitude, m_l, which is the limiting magnitude for that telescope

$$m_l = 17 + 5\log_{10} D \qquad (1.70)$$

where D is in meters. If the stellar image is magnified to the point where it spreads over more than one detecting element, then we must return to the analysis for the extended sources. For an average eye, this upper limit to the magnification is given by

$$M \approx 850D \qquad (1.71)$$

where D is again in meters.

1.1.18.2 Designs

Probably the commonest format for large telescopes is the Cassegrain system, although most large telescopes can usually be used in several alternative different modes by interchanging their secondary mirrors. The Cassegrain system is based upon a paraboloidal primary mirror and a convex hyperboloidal secondary mirror (Figure 1.41). The nearer focus of the conic section, which forms the surface of the secondary, is coincident with the focus of the primary, and the Cassegrain focus is then at the more distant focus of the secondary mirror's surface. The major advantage of the Cassegrain system lies in its telephoto characteristic; the secondary mirror serves to expand the beam from the primary mirror so that the effective focal length of the whole system is several times that of the primary mirror. A compact and hence rigid and relatively cheap mounting can thus be used to hold the optical components while retaining the advantages of long focal length and large image scale. The Cassegrain

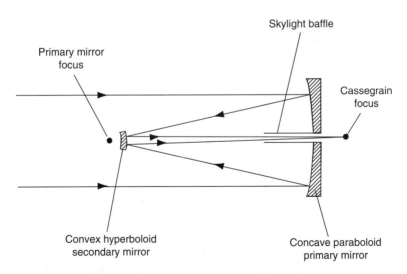

FIGURE 1.41 Cassegrain telescope optical system.

design is afflicted with coma and spherical aberration to about the same degree as an equivalent Newtonian telescope, or indeed to just a single parabolic mirror with a focal length equal to the effective focal length of the Cassegrain. The beam expanding effect of the secondary mirror means that Cassegrain telescopes normally work at focal ratios between 12 and 30, even though their primary mirror may be $f3$ or $f4$ (or even $f1$ as in the 4 m VISTA telescope). Thus, the images remain tolerable over a field of view that may be several tenths of a degree across (Figure 1.42). Astigmatism and field curvature are stronger than in an equivalent Newtonian system, however. Focusing of the final image in a Cassegrain system is customarily accomplished by moving the secondary mirror along the optical axis. The amplification due to the secondary means that it only has to be moved a short distance away from its optimum position to move the focal plane considerably. The movement away from the optimum position, however, introduces severe spherical aberration. For the 0.25 m $f4/f16$ system whose images are shown in Figure 1.42, the secondary mirror can only be moved by 6 mm either side of the optimum position before even the on-axis images without diffraction broadening become

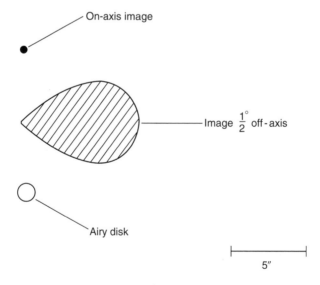

On-axis image

Image $\frac{1}{2}°$ off-axis

Airy disk

5″

FIGURE 1.42 Images in a 0.25 m $f4/f16$ Cassegrain telescope. (The images were obtained by ray tracing. Since this does not allow for the effects of diffraction, the on-axis image in practice will be the same as the Airy disk, and the 0.5° off-axis image will be blurred even further.)

comparable in size with the Airy disk. Critical work with Cassegrain telescopes should therefore always be undertaken with the secondary mirror at or very near to its optimum position.

A very great improvement to the quality of the images may be obtained if the Cassegrain design is altered slightly to the Ritchey–Chrétien system. The optical arrangement is identical with that shown in Figure 1.41 except that the primary mirror is deepened to an hyperboloid and a stronger hyperboloid is used for the secondary. With such a design both coma and spherical aberration can be corrected and we have an aplanatic system. The improvement in the images can be seen by comparing Figure 1.43 with Figure 1.42. It should be noted, however, that the improvement is in fact considerably more spectacular since we have a 0.5 m Ritchey–Chrétien and a 0.25 m Cassegrain with the same effective focal lengths. A 0.5 m Cassegrain telescope would have its off-axis image twice the size of that shown in Figure 1.42 and its Airy disk half the size shown there.

Another variant on the Cassegrain system is the Dall–Kirkham telescope that has a concave ellipsoidal main mirror and spherical convex secondary mirror. The design is popular with amateur telescope makers since the curves on the mirrors are easier to produce. However, it suffers badly from aberrations and its field of sharp focus is only about a third that of an equivalent conventional Cassegrain design.

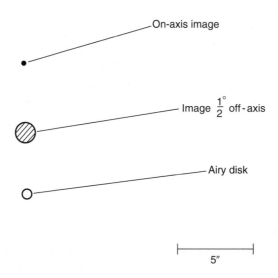

FIGURE 1.43 Ray tracing images in a 0.5 m $f3/f8$ Ritchey–Chrétien telescope.

Alternatively, a Cassegrain or Ritchey–Chrétien system can be improved by the addition of correctors just before the focus. The correctors are low- or zero-power lenses whose aberrations oppose those of the main system. There are numerous successful designs for correctors although many of them require aspheric surfaces or the use of exotic materials such as fused quartz. Images can be reduced to less than the size of the seeing disk over fields of view of up to 1°, or sometimes over even larger angles. The 3.9 m Anglo-Australian telescope (AAT), for example, uses correcting lenses to provide an unvignetted 2° field of view (2° field or 2dF), over which 400 optical fibers can be positioned to feed a spectroscope. The 4 m VISTA telescope uses a modified Ritchey–Chrétien design that leaves some significant aberrations in images produced by the mirrors alone. However, using three correcting lenses produces a fully corrected image over 2° across in the visible and 1.6° across in the infrared. Reflective corrective optics are also possible and a four mirror system including aspheric active optics has been proposed for the overwhelmingly large (OWL) telescope project (see below). The corrective optics may be combined with a focal reducer to enable the increased field of view to be covered by the detector array. A focal reducer is the inverse of a Barlow lens and is a positive lens, usually an apochromatic triplet, placed just before the focal point of the telescope that decreases the effective focal length and so gives a smaller image scale.

Another telescope design that is again very closely related to the Cassegrain is termed the Coudé system. It is in effect a very long focal length Cassegrain or Ritchey–Chrétien whose light beam is folded and guided by additional flat mirrors to give a focus whose position is fixed in space irrespective of the telescope position. One way of accomplishing this (there are many others) is shown in Figure 1.44. After reflection from the secondary, the light is reflected down the hollow declination axis by a diagonal flat mirror, and then down the hollow polar axis by a second diagonal. The light beam then always emerges from the end of the polar axis, whatever part of the sky the telescope may be inspecting. Designs with similar properties can be devised for most other types of mountings although additional flat mirrors may be needed in some cases. With alt–az mountings, the light beam can be directed along the altitude axis to one of the two Nasmyth foci on the side of the mounting. These foci still rotate as the telescope changes its azimuth, but this poses far fewer problems than the changing altitude and attitude of a conventional Cassegrain focus. On large modern telescopes, platforms of considerable size are often constructed at the Nasmyth foci allowing large ancillary instruments to be

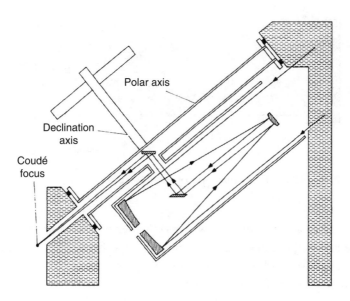

FIGURE 1.44 Coudé system for a modified English mounting.

used. The fixed or semi-fixed foci of the Coudé and Nasmyth systems are a very great advantage when bulky items of equipment, such as high dispersion spectrographs, are to be used, since these can be permanently mounted in a nearby separate laboratory and the light brought to them, rather than have to have the equipment mounted on the telescope. The design also has several disadvantages; the field of view rotates as the telescope tracks an object across the sky, and is very tiny due to the large effective focal ratio ($f25$ to $f40$) which is generally required to bring the focus through the axes, finally the additional reflections will cause loss of light.

The simplest of all designs for a telescope is a mirror used at its prime focus. That is, the primary mirror is used directly to produce the images and the detector is placed at the top end of the telescope. The largest telescopes have a platform or cage that replaces the secondary mirror and which is large enough for the observer to ride in while he or she operates and guides the telescope from the prime focus position. With smaller instruments, too much light would be blocked, so that they must be guided using a separate guide telescope or use a separate detector monitoring another star (the guide star) within the main telescope's field of view. The image quality at the prime focus is usually poor even a few tens of seconds of arc away from the optical axis, because the primary mirror's focal ratio may be as short as $f3$ or less to reduce the length of the

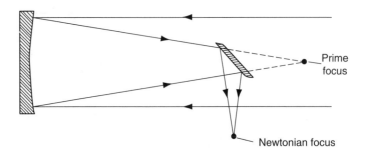

FIGURE 1.45 Newtonian telescope optical system.

instrument to a minimum. Thus, correcting lenses are invariably essential to give acceptable images and reasonably large fields of view. These are similar to those used for correcting Cassegrain telescopes and are placed immediately before the prime focus.

A system that is almost identical to the use of the telescope at prime focus and which was the first design to be made into a working reflecting telescope is that due to Newton and hence is called the Newtonian telescope. A secondary mirror is used which is a flat diagonal placed just before the prime focus. This reflects the light beam to the side of the telescope from where access to it is relatively easy (Figure 1.45). The simplicity and cheapness of the design make it very popular as a small telescope for the amateur market, but it is rarely encountered in telescopes larger than about 1 m. There are several reasons for its lack of popularity for large telescope designs; the main ones being that it has no advantage over the prime focus position for large telescopes since the equipment used to detect the image blocks out no more light than the secondary mirror, the secondary mirror introduces additional light losses, and the position of the equipment high on the side of the telescope tube causes difficulties of access and counterbalancing. The images in a Newtonian system and at prime focus are very similar and are of poor quality away from the optical axis as shown in Figure 1.46.

A very great variety of other catoptric (reflecting objective) telescope designs abound, but most have few or no advantages over the two groups of designs discussed above. Some find specialist applications, for example, the Gregorian design is similar to the Cassegrain except that the secondary mirror is a concave ellipsoid and is placed after the prime focus. It was used for the stratoscope II telescope, and the two 8.4 m instruments of the LBT on Mount Graham include Gregorian secondary mirrors for infrared work.

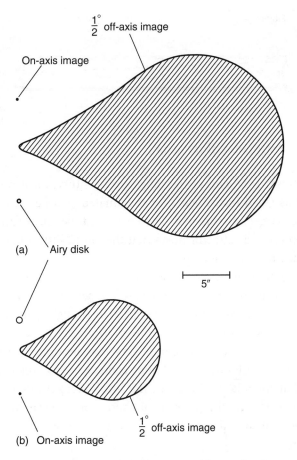

FIGURE 1.46 Images in Newtonian telescopes. (a) 1 m *f*4 telescope; (b) 0.5 m *f*8 telescope.

The Pfund-type telescope uses a siderostat to feed a fixed paraboloidal main mirror that reflects light through a hole in the flat mirror of the siderostat to a focus behind it. The infrared spatial interferometer (ISI) (Section 2.5.2) used Pfund-type telescopes but other examples are uncommon. A few such specialized designs may also be built as small telescopes for amateur use because of minor advantages in their production processes, or for space-saving reasons as in the folded schiefspieglers, but most such designs will be encountered very rarely.

Of the dioptric (refracting objective) telescopes, only the basic refractor using an achromatic doublet, or very occasionally a triplet, as its objective is in any general use (Figure 1.47). Apart from the large refractors that

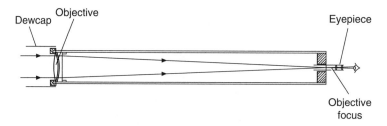

FIGURE 1.47 Astronomical refractor.

were built toward the end of the nineteenth century, most refractors are now found as small instruments for the amateur market or as the guide telescopes of larger instruments. The enclosed tube and relatively firmly mounted optics of a refractor mean that they need little adjustment once aligned with the main telescope.

The one remaining class of optical telescopes is the catadioptric group, of which the Schmidt camera is probably the best known. A catadioptric system uses both lenses and mirrors in its primary light-gathering section. Very high degrees of correction of the aberrations can be achieved because of the wide range of variable parameters that become available to the designer in such systems. The Schmidt camera uses a spherical primary mirror so that coma is eliminated by not introducing it into the system in the first place. The resulting spherical aberration is eliminated by a thin correcting lens at the mirror's radius of curvature (Figure 1.48). The only

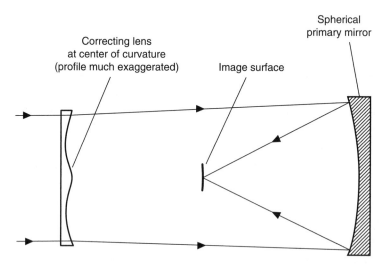

FIGURE 1.48 Schmidt camera optical system.

major remaining aberration is field curvature and the effect of this is eliminated through the use of additional correcting lenses (field flatteners) for CCDs and other array detectors or, if still used, by moulding the photographic plate to the shape of the image surface. The correcting lens can introduce small amounts of coma and chromatic aberration, but is usually so thin that these aberrations are negligible. Diffraction-limited performance over fields of view of several degrees with focal ratios as fast as $f1.5$ or $f2$ is possible. As larger CCDs are becoming available, they are being used on Schmidt cameras. They are still smaller than photographic plates, and so fields of view are restricted to only a degree or so. However, the high detecting efficiency, the linear response, and the machine-readable output of the CCD means that they have replaced the photographic plate for most purposes. The CCD can also be used over wider fields of view, usually for spectroscopy, by using fiber-optic feeds to pick-off images of individual stars or galaxies—see, for example, the 6dF system for the U.K. Schmidt camera (Section 4.2).

The need to use a lens in the Schmidt design limits the sizes possible for them. Larger sizes can be achieved using a correcting mirror. The largest such instrument is the 4 m large sky area multi-object fiber spectroscopic telescope (LAMOST) at the Xinglong Station of the Chinese National Astronomical Observatory. It operates using a horizontally fixed primary mirror fed by a coelostat (see below). The coelostat mirror is actively controlled and also acts as the corrector. With up to 4000 optical fibers, the instrument is one of the most rapid spectroscopic survey tools in the world and it started observations in May 2007. Other large Schmidt cameras include the 1.35 m Tautenburg Schmidt and the Oschin Schmidt at Mount Palomar and the U.K. Schmidt camera at the AAT both of which have entrance apertures 1.2 m across.

The Schmidt design suffers from having a tube length at least twice its focal length. A number of alternative designs to provide high-quality images over fields of view of 5°–10° have therefore been proposed, though none with any significant sizes have been built. For example, the Willstrop three-mirror telescope (Figure 1.49) is completely achromatic and can potentially be made as large as the largest conventional optical telescope. The primary mirror is close to parabolic, the corrector (secondary mirror) is close to spherical with a turned-down edge, while the tertiary is also close to spherical in shape. It has a 5° field and a tube length one-third to one-half that of the equivalent Schmidt camera. But its focal surface is curved like that of the Schmidt, and all except the central 2° of the field of

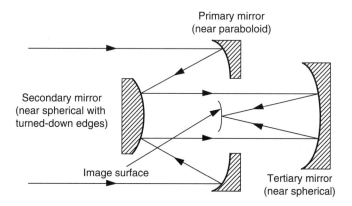

FIGURE 1.49 Willstrop three-mirror camera optical system.

view suffers from vignetting. A 0.5 m version of the design has been successfully built and operated at Cambridge (United Kingdom).

The Schmidt camera cannot be used visually since its focus is inaccessible. There are several modifications of its design, however, that produce an external focus while retaining most of the beneficial properties of the Schmidt. One of the best of these is the Maksutov (Figure 1.50) that originally also had an inaccessible focus, but which is now a kind of Schmidt–Cassegrain hybrid. All the optical surfaces are spherical, and

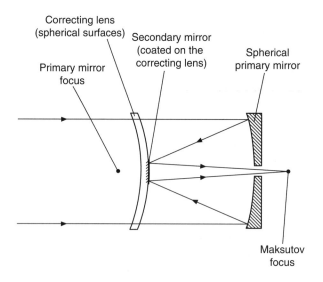

FIGURE 1.50 Maksutov optical system.

spherical aberration, astigmatism, and coma are almost eliminated, while the chromatic aberration is almost negligible. A similar system is the Schmidt–Cassegrain telescope itself. This uses a thin correcting lens like the Schmidt camera and has a separate secondary mirror. The focus is accessible at the rear of the primary mirror as in the Maksutov system. Schmidt–Cassegrain telescopes are now available commercially in sizes up to 0.4 m diameter. They are produced in large numbers by several firms, at very reasonable costs, for the amateur and education markets. They are also finding increasing use in specific applications, such as site testing, for professional astronomy.

Although it is not a telescope, there is one further system that deserves mention and that is the coelostat. This comprises two flat mirrors that are driven so that a beam of light from any part of the sky is directed toward a fixed direction. They are particularly used in conjunction with solar telescopes whose extremely long focal lengths make them impossible to move. One mirror of the coelostat is mounted on a polar axis and driven at half sidereal rate. The second mirror is mounted and driven to reflect the light into the fixed telescope. Various arrangements and adjustments of the mirrors enable differing declinations to be observed. A related system is the siderostat which uses just a single flat mirror to feed a fixed telescope. The flat mirror has to be driven at variable speeds in altitude and azimuth to track an object across the sky but even a small computer can undertake the calculations needed for this purpose and control the driving motors.

No major further direct developments in the optical design of large telescopes seem likely to lead to improved resolution, since such instruments are already limited by the atmosphere. Better resolution therefore requires that the telescope be lifted above the earth's atmosphere, or that the distortions that the atmosphere introduces be overcome by a more subtle approach. Telescopes mounted on rockets, satellites, and balloons are discussed later in this section, while the four ways of improving the resolution of earth-based telescopes—interferometry, speckle interferometry, occultations, and real-time compensation—are discussed in Sections 2.5 through 2.7 and 1.1.21, respectively. A fairly common technique for improving resolution, deconvolution, is discussed in Section 2.1.1. This requires that the precise nature of the degradations of the image be known so that they can be removed. Since the atmosphere's effects are changing in a fairly random manner on a timescale of milliseconds, it is not an appropriate technique for application here, although it should be noted that one approach to speckle interferometry is in fact related to this method and deconvolution may be needed on the compensated image.

A technique termed apodization (literally meaning removal of the feet, see also Sections 2.5 and 4.1) gives an apparent increase in resolution. It is actually, however, a method of enabling the telescope to achieve its normal resolution, and sometimes even reduces that resolution slightly. For optical telescopes, apodization is usually accomplished through the use of masks over the entrance aperture of the telescope. These have the effect of changing the diffraction pattern (see earlier discussion about the Airy disk, and also Section 2.1). The nominal resolution given by the size of the central disk may then worsen, but the outer fringes disappear, or become of a more convenient shape for the purpose in mind. A couple of examples will illustrate the process. Firstly, when searching for a faint object next to a bright one, such as Sirius B next to Sirius A, or for an extra-solar planet next to a star, a square aperture with fuzzy edges, or a four-armed star aperture will produce fringes in the form of a cross, and suppress the normal circular fringes. If the aperture is rotated so that the faint object lies in one of the regions between the arms of the cross, then it will have a much higher signal-to-noise ratio than if superimposed upon a bright fringe from a circular aperture, and so be more easily detected. Secondly, for observing low-contrast features on planets, a mask that is a neutral density filter varying in opacity in a Gaussian fashion from clear at the center to nearly opaque at the edge of the objective may be used. This has the effect of doubling the size of the Airy disk, but also of eliminating the outer fringes completely. The resulting image will be significantly clearer since the low-contrast features on extended objects are normally swamped by the outer fringes of the point spread function (PSF, see Section 2.1). A professionally made neutral density filter of this type would be extremely expensive, but a good substitute may be made using a number of masks formed out of a mesh. Each mask is an annulus with an outer diameter equal to that of the objective and an inner diameter that varies appropriately in size. When they are all superimposed on the objective, and provided that the meshes are not aligned, then they will obstruct an increasingly large proportion of the incoming light toward the edge of the objective. The overall effect of the mesh masks will thus be similar to the Gaussian neutral density filter.

Now although resolution may not be improved in any straightforward manner (but see discussion in Section 1.1.21), the position is less hopeless for the betterment of the other main function of a telescope, which is light gathering. The upper limit to the size of an individual mirror is probably being approached with the 8 m telescopes presently in existence. A diameter of 10–15 m for a metal-on-glass mirror would probably be about the ultimate possible with present-day techniques and their foreseeable developments such as a segmented honeycomb construction for the mirror blank or a thin mirror on active supports that compensate for its flexure. The ESO's VLT, for example, has four main telescopes each with thin actively supported 8.2 m monolithic mirrors. While the LBT on Mount Graham, Arizona, which is currently nearing completion (first light using just one of the mirrors occurred in October 2005), will have two 8.4 m diameter monolithic honeycomb mirrors on a single mounting. The mirrors will have very short focal ratios ($f1.14$), and the telescopes will be of Gregorian design.

An additional practical constraint on mirror size is that imposed by needing to transport the mirror from the manufacturer to the telescope's observing site. Even with 8 m mirrors, roads may need widening, obstructions leveling, bends have to have their curvature reduced, etc. and this can add very significantly to the cost of the project.

Greater light-gathering power may therefore be acquired only by the use of several mirrors, by the use of a single mirror formed from separate segments, by aperture synthesis (Section 2.5.5) or perhaps in the future by using liquid mirrors (see above). Several telescopes are currently operating or are under construction using single large mirrors formed from several independent segments. For example, there are the two 10 m Keck telescopes that are each constructed from thirty-six 1.8 m hexagonal segments (see Figure 1.51) and the Hobby–Eberly spectroscopic telescope and the very similar Southern African large telescope (SALT) which each have ninety-one 1.0 m spherical segments forming 11×10 m mirrors. The latter telescopes use correcting optics to produce an adequate image and are on fixed mountings (see later discussion). The gran telescopio Canarias (GTC), sited on La Palma in the Canary Isles, is a Spanish-led project to build a 10.4 m segmented-mirror telescope. It has 36 segments that can be actively positioned and deformed to maintain their shape. It saw first light in July 2007. These types of designs require that each segment of the mirror be independently mounted, and that their positions be continuously monitored and adjusted to keep them within a tenth of a wavelength of their correct positions (active optics—see Section 1.1.21).

Alternatively, the secondary mirror may be segmented as well, and those segments moved to compensate for the variations in the primary

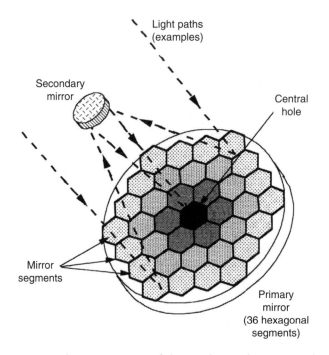

FIGURE 1.51 Optical arrangement of the Keck I and II 10 m telescopes.

mirror segments. The monitoring and control of the segments would be very similar to those required for the remaining option: the multi-mirror telescope. The first such telescope was built at Mount Hopkins about three decades ago. This telescope was based upon six fairly conventional 1.8 m Cassegrain systems, with mirrors of honeycomb construction, and arranged in a hexagonal array 6.9 m across. They were supported on a single mounting and fed a common focus. The system was equivalent to a 4.4 m telescope in area, but only cost about one-third as much. Since the cost of a telescope varies as about the cube of its diameter, far larger savings would be possible with the use of more mirrors. The Mount Hopkins telescope, however, experienced problems, especially with the mirror position monitoring system. It has therefore now been refitted with a single 6.5 m mirror.

The accuracy and stability of surfaces, alignments, etc. required for both the multi-mirror and the segmented-mirror systems are a function of the operating wavelength. For infrared and short-microwave work, far less technically demanding limits are therefore possible and their telescopes are correspondingly easier to construct. Specialist infrared telescopes do, however, have other requirements that add to their complexity. First, it is often necessary to chop between the source and the background to subtract the latter. Some purpose-built infrared telescopes such as the 3.8 m UKIRT oscillate the secondary mirror through a small angle to achieve this (a mass equal to that of the chopping mirror may need to be moved in the opposite direction to avoid vibrating the telescope). The UKIRT secondary mirror can also be moved to correct for flexure of the telescope tube and mounting and for buffeting by the wind. For NIR observations, the telescope must be sited above as much of the atmospheric water vapor as possible, limiting the sites to places like Mauna Kea on Hawaii, the Chilean altiplano, or the Antarctic plateau. For MIR and some FIR works, it may be necessary also to cool the telescope, or parts of it, to reduce its thermal emission, and such telescopes have to be lifted above most of the atmosphere by balloon or aircraft or flown on a spacecraft. SOFIA, for example, is scheduled to start observing at infrared wavelengths in 2009 although its full capability will not be reached until 2014. It will be a 2.5 m telescope flown on board a Boeing 747 aircraft that will fly at altitudes up to 13,700 m and will operate at 250 K and be able to cover wavelengths from 0.3 to 1600 μm.

The atmosphere is completely opaque to radiation shorter than about 320 nm. Telescopes designed for ultraviolet observations therefore have to be launched on rockets or carried onboard spacecraft. Except in the extreme ultraviolet (EUV: ~6 to ~90 nm), where glancing optics are needed (see Section 1.3), conventional telescope designs are used. Aluminum suffices as the mirror coating down to 100 nm. At shorter wavelengths, silicon carbide may be used.

The limits may also be relaxed if the system is required as a light bucket rather than for imaging. Limits of 10 times the operating wavelength may still give acceptable results when a photometer or spectroscope is used on the telescope, providing that close double stars of similar magnitudes and extended sources are avoided.

1.1.19 Telescopes in Space

The most direct way to improve the resolution of a telescope that is limited by atmospheric degradation is to lift it above the atmosphere, or at least above the lower, denser parts of the atmosphere. The three main methods of accomplishing this are to place the telescope on an aircraft, a balloon-borne platform, or an artificial satellite. The Kuiper Airborne Observatory (KAO—a 0.9 m telescope on board a C141 aircraft that operated until 1995) and SOFIA (see above) are examples of the first method, but aircraft are limited to maximum altitudes of about 15 km. Balloon-borne telescopes, of which the 0.9 m Stratoscope II is the best-known but far from the only example, have the advantages of relative cheapness, and that their images may be recorded by fairly conventional means. Their disadvantages include low maximum altitude (40 km) and short flight-duration (a few days). Recently a 1 m Gregorian solar telescope has been flown by balloon to a height of 36 km during a test flight for project Sunrise. This project is expected to start science missions in 2009/2010 with flights of 1–2 weeks duration close to the north and south poles.

Many small telescopes have already been launched on satellites. The telescope aboard the *International Ultraviolet Explorer* (*IUE*) spacecraft was a fairly typical example. This spacecraft used a 0.45 m Ritchey–Chrétien telescope that fed images to a pair of ultraviolet spectrometers. In the infrared, the entire telescope and its ancillary instrumentation may be cooled. The *Infrared Space Observatory* (*ISO*) spacecraft, for example, had a 0.6 m telescope cooled to 3 K with the detectors held at 1.8 K, using 2300 L of liquid helium. While the *Spitzer Observatory* carries a Ritchey–Chrétien telescope with a 0.85 m beryllium mirror cooled to 5.5 K by some 360 L of liquid helium. The use of unconventional telescope designs is not uncommon for this type of work, since they may have minor advantages in their weight or volume which become vital when the restrictions imposed by the balloon's lift or launcher's power are taken into account. For example, off-axis systems, Gregorian telescopes (in Stratoscope II), and the Wynne camera, which is a sort of inverse Cassegrain system, have all been used recently. Glancing incidence systems, that are an extreme form of the off-axis format, are of great importance for short-wave ultraviolet and x-ray observations and are discussed in Section 1.3. The HST is a 2.4 m $f2.3/f24$ Ritchey–Chrétien telescope designed to operate from 115 nm to 2.5 μm, with a variety of instrumentation at its focus. With its corrective optics to compensate for the deficiencies of the primary mirror, and updated ancillary instruments, it is now regularly providing diffraction-limited images. Its replacement, now scheduled for launch in 2013, the James Webb space telescope (JWST), is planned to have a 6.5 m beryllium mirror. The mirror will have eighteen 1.3 m segments and be folded for launch and during the spacecraft's journey to its working position at Earth–Sun L2* point. It will be able to observe

---

* The outer Lagrangian point is 1.5 million km farther from the Sun than the earth on the Sun–earth line.

from 600 nm to 10 μm or longer wavelengths and will operate at a temperature of 50 K. Before the JWST, ESA's *Herschel** space observatory is due for launch in 2008 and will carry a 3.5 m silicon carbide mirror. It will also operate at the L2 point, and will observe from 80 to 670 μm. The telescope will be passively cooled to 80 K, with active cooling for the instrumentation down to less than 3 K.

Suggestions for the future include arrays of orbiting optical telescopes with separations of 100 km or more, capable of directly imaging Earth-sized planets around other stars. The Exo-Earth Imager, for example, is a proposal for one hundred and fifty 3 m mirrors in orbit around the Sun and spread over an area of 8000 km^2 to simulate a sparsely filled 100 km aperture telescope. It seems unlikely that such proposals will be funded in the near future. More feasible proposals for observing exoplanets are perhaps the new worlds observer (NWO) and the new worlds imager (NWI). The NWO would be a coronagraph-type instrument (Section 5.3.5) using an external shield to block the light from the star. The shield could be used with an existing space telescope such as the JWST and comprises a 40 m diameter petal-shaped (to reduce diffraction effects) occulting disk placed some 30,000 km away from the spacecraft. The NWI would be two NWOs separated by some 1500 km and acting as an interferometer. At the time of writing, the NWO remains a possible mission with a launch date around 2014.

1.1.20 Mountings

The functions of a telescope mounting are simple—to hold the optical components in their correct mutual alignment, and to direct the optical axis toward the object to be observed. The problems in successfully accomplishing this to within the accuracy and stability required by astronomers are so great, however, that the cost of the mounting can be the major item in funding a telescope. We may consider the functions of a mounting under three separate aspects: supporting the optical components, preserving their correct spatial relationships, and acquiring and holding the object of interest in the field of view.

Mounting the optical components is largely a question of supporting them in a manner that ensures their surfaces maintain their correct shapes. Lenses may only be supported at their edges, and it is the impossibility of doing this adequately for large lenses that limits their practicable

---

* Previously named FIRST (FIR and submillimeter telescope).

sizes to a little over a meter. There is little difficulty with smaller lenses and mirrors since most types of mount may grip them firmly enough to hold them in place without at the same time straining them. However, large mirrors require very careful mounting. They are usually held on a number of mounting points to distribute their weight, and these mounting points, especially those around the edges, may need to be active so that their support changes to compensate for the different directions of the mirror's weight as the telescope moves. The active support can be arranged by systems of pivoted weights, especially on older telescopes, or more normally nowadays by computer control of the supports. As discussed earlier, some recently built telescopes and many of those planned for the future have active supports for the primary mirror that deliberately stress the mirror so that it retains its correct shape whatever the orientation of the telescope or the temperature of the mirror. Segmented-mirror telescopes additionally need the supports to maintain the segments in their correct mutual alignments. On the 10 m Keck telescopes, for example, the active supports adjust the mirror segment's positions twice a second to maintain their places correctly to within ±4 nm.

The optical components are held in their relative positions by the telescope tube. With most large telescopes, the "tube" is in fact an open structure, but the name is still retained. For small instruments, the tube may be made sufficiently rigid that its flexure is negligible. But this becomes impossible as the size increases. The solution then is to design the flexure so that it is identical for both the primary and secondary mirrors. The optical components then remain aligned on the optical axis, but are no longer symmetrically arranged within the mounting. The commonest structure in use that allows this equal degree of flexure and also maintains the parallelism of the optical components is the Serrurier truss (Figure 1.52). However, this design is not without its own problems. The lower trusses may need to be excessively thin or the upper ones excessively thick to provide the required amount of flexure. Even so, many of the large reflecting telescopes built in the last four decades have used Serrurier truss designs for their tubes. More recently, computer-aided design has enabled other truss systems to be developed. This, however, is mostly for economic reasons; the design objectives of such supports remain unchanged.

A widely used means of mounting the telescope tube so that it may be pointed at an object and then moved to follow the object's motion across the sky is the equatorial mounting. This is a two-axis mounting, with one

Secondary mirror cell

Serrurier trusses

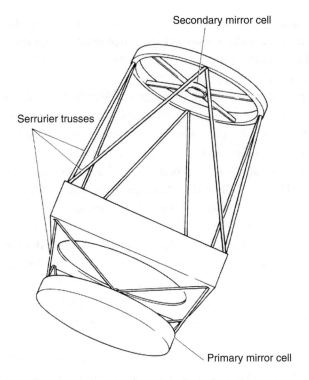

Primary mirror cell

FIGURE 1.52 Telescope tube construction based on Serrurier trusses.

axis, the polar axis, aligned parallel with the earth's rotational axis, and the other, the declination axis, perpendicular to the polar axis. The design has the enormous advantage that only a single constant velocity motor is required to rotate the mounting around the polar axis to track an object. It is also very convenient in that angular readouts on the two axes give the hour angle or right ascension and declination directly.

A large variety of different arrangements for the basic equatorial mounting exist, but these are not reviewed here. Books on telescopes and most general astronomy books list their details by the legion, and should be checked if the reader requires further information.

The alt–az mounting that has motions in altitude and azimuth is the other main two-axis mounting system. Structurally, it is a much simpler form than the equatorial and has therefore long been adopted for most large radio telescope dishes. Its drawbacks are that the field of view rotates with the telescope motion, and that it needs driving continuously in both axes and with variable speeds to track an object. For the last two decades, most new large optical telescopes have also used alt–az mountings, and the reduction in price and increase in capacity of small computers means that smaller telescopes even down to a few centimeters in size frequently have such a mounting as an option.

Telescopes up to 1 m in diameter constructed for or by amateur astronomers are often mounted on a Dobsonian alt–az mounting. This may be simply and cheaply constructed from sheets of plywood or similar material. The telescope is supported by two circles, which rest in two semicircular cut outs in three-sided cradle shaped like an inverted Greek letter "pi" (Π). These allow the telescope to move in altitude. The cradle then pivots at its center to give motion in azimuth. Telescopes on such mountings are usually un-driven, and are moved by hand to find and follow objects in the sky. Small telescopes on any alt–az design of mounting can be given a tracking motion for a short interval of time by placing them on a platform that is itself equatorially driven. There are several designs for such equatorial platforms using sloping planes or inclined bearings, and details may be found in modern books aimed at the amateur observer.*

Large telescopes, whether on equatorial or alt–az mountings, almost universally use hydrostatic bearings for the main moving parts. These bearings use a thin layer of pressurized oil between the moving and static parts, and the oil is continuously circulated by a small pump. The two solid surfaces are thus never in direct contact, and movement round the bearing is very smooth and of low friction.

Several telescopes have fixed positions. These include liquid-mirror telescopes that always point near the zenith (see above), and the 11 m Hobby–Eberly telescope and SALT (see Section 1.1.18.2). Tracking is then accomplished by moving the detector and any correcting optics to follow the motion of the image, or for short exposures, by moving the charges along the pixels of a CCD detector (TDI, see above). Zenith-pointing telescopes are generally limited to a few degrees either side of the zenith, but the Hobby-Eberly telescope can point to objects with declinations ranging from $-11°$ to $+71°$. It accomplishes this because it points to a fixed zenith angle of $35°$, but can be rotated in azimuth between observations. There are also a few instruments mounted on altitude–altitude mountings. These use a universal joint (such as a ball and socket) and

---

* See, for example, the author's *Telescopes and Techniques* 2nd Edition, 2003, Springer.

tip the instrument in two orthogonal directions to enable it to point anywhere in the sky.

With any type of mounting, acquisition and tracking accuracies of a second of arc or better are required, but these may not be achieved especially for the smaller earth-based telescopes. Thus, the observer may have to search for his or her object of interest over an area tens to hundreds of seconds of arc across after initial acquisition, and then guide, either himself or herself, or with an automatic system, to ensure that the telescope tracks the object sufficiently closely for his or her purposes. With balloon-borne and space telescopes such direct intervention is difficult or impossible. However, the reduction of external disturbances and possibly the loss of weight mean that the initial pointing accuracy is higher. For space telescopes, tracking is easy since the telescope will simply remain pointing in the required direction once it is fixed, apart from minor perturbations such as light pressure, the solar wind, gravitational anomalies, and internal disturbances from the spacecraft. Automatic control systems can therefore usually be relied upon to operate the telescope. Earlier, spacecraft such as the *IUE*, however, had facilities for transmitting a picture of the field of view after the initial acquisition to the observer on the ground, followed by corrections to the position to set onto the desired object.

With terrestrial telescopes, the guiding may be undertaken via a second telescope that is attached to the main telescope and aligned with it, or a small fraction of the light from the main telescope may be diverted for the purpose. The latter method may be accomplished by beam splitters, dichroic mirrors, or by intercepting a fraction of the main beam with a small mirror, depending upon the technique that is being used to study the main image. The whole telescope can then be moved using its slow-motion controls to keep the image fixed, or additional optical components can be incorporated into the light path whose movement counteracts the image drift, or the detector can be moved around the image plane to follow the image motion or the charges can be moved within a CCD chip (TDI). Since far less mass needs to be moved with the latter methods, their response times can be much faster than that of the first method. The natural development of the second method leads on to the active surface control, which is discussed later in this section and which can go some way toward eliminating the effects of scintillation.

If guiding is attempted by the observer then there is little further to add, except to advise a plentiful supply of black coffee in order to avoid going to sleep through the boredom of the operation. Automatic guiding has two major problem areas. The first is sensing the image movement and generating the error signal, while the second is the design of the control system to compensate for this movement. Detecting the image movement has been attempted in two main ways: CCDs and quadrant detectors.

With a CCD guide, the guide image is read out at frequent intervals, and any movement of the object being tracked is detected via software that also generates the correction signals. The guide CCD is adjacent to the main detector (which may or may not be a CCD itself), and thus it guides on objects other than those being imaged. This is termed offset guiding and often has advantages over guiding directly on the object of interest. The offset guide star can be brighter than the object of interest, and if the latter is an extended object, then guiding on it directly will be difficult. However for comets, asteroids, and other moving objects, the object itself must be followed. In some small CCD systems aimed at the amateur market, it is the main image that is read out at frequent intervals and the resulting multiple images are then shifted into mutual alignment before being added together. This enables poor tracking to be corrected provided that the image shift is small during any single exposure.

A quadrant detector is one whose detecting area is divided into four independent sectors, the signal from each of which can be separately accessed. If the image is centered on the intersection of the sectors, then their signals will all be equal. If the image moves, then at least two of the signals will become unbalanced and the required error signal can then be generated. A variant of this system uses a pyramidal "prism." The image is normally placed on the vertex, and so is divided into four segments each of which can then be separately detected. When guiding on stars by any of these automatic methods it may be advantageous to increase the image size slightly by operating slightly out of focus, scintillation "jitter" then becomes less important.

The problem of the design of the control system for the telescope to compensate for the image movement should be handed to a competent control engineer. If it is to work adequately, then proper damping, feedback, rates of motion, and so on must be calculated so that the correct response to an error signal occurs without "hunting" or excessive delays.

1.1.21 Real-Time Atmospheric Compensation

The resolution of ground-based telescopes of more than a fraction of a meter in diameter is limited by the turbulence in the atmosphere. The maximum diameter of a telescope before it becomes seriously affected by atmospheric turbulence is given by Fried's coherence length, r_0

$$r_0 \approx 0.114 \left(\frac{\lambda \cos z}{550} \right)^{0.6} \text{ m} \qquad (1.72)$$

where
λ is the operating wavelength in nm
z is the zenith angle

Fried's coherence length, r_0, is the distance over which the phase difference is one radian. Thus, for visual work telescopes of more than about 11.5 cm (4.5 in.) diameter will always have their images degraded by atmospheric turbulence.

In an effort to reduce the effects of this turbulence (the "seeing", "scintillation" or "twinkling"), many large telescopes are sited at high altitudes, or placed on board high-flying aircraft or balloons. The ultimate, though very expensive solution, of course, is to orbit the telescope beyond the earth's atmosphere out in space. Recently significant reductions in atmospheric blurring have been achieved through lucky imaging. In this process numerous short exposure images are obtained of the object (cf. Section 2.6). Only the sharpest are then combined to produce the final image. In this way, the final image can approach the diffraction limit for 2 or 3 m telescopes at visible wavelengths. However, up to 99% of the images may need to be discarded, making the process expensive in telescope time.

An alternative approach to obtain diffraction-limited performance for large telescopes, especially in the NIR and MIR, that is relatively inexpensive and widely applicable is to correct the distortions in the incoming light beam produced by the atmosphere. This atmospheric compensation is achieved through the use of adaptive optics. In such systems, one or more of the optical components can be changed rapidly and in such a manner that the undesired distortions in the light beam are reduced or eliminated. Although a relatively recent development in its application to large telescopes, such as the VLT, Gemini, Keck, William Herschel, and Subaru telescopes, adaptive optics is actually a very ancient technique familiar to us all. That is because the eye operates via an adaptive optic system to keep objects in focus, with the lens being stretched or compressed by the ciliary muscle (Figure 1.1).

The efficiency of an adaptive optics system is measured by the Strehl ratio. This quantity is the ratio of the intensity at the center of the corrected image to that at the center of a perfect diffraction-limited image of the same source. The normalized Strehl ratio is also used. This is the Strehl ratio of the corrected image divided by that for the uncorrected image. Strehl ratios up to 0.6 are currently being achieved in the NIR, and may reach 0.8 in the near future. While a ratio of 0.98 has been realized by the MMT using a deformable secondary mirror at MIR wavelengths (see below).

Adaptive optics is not a complete substitute for spacecraft-based telescopes, however, because in the visual and NIR, the correction only extends over a very small area (the isoplanatic patch, see below). Thus, for example, if applied to producing improved visual images of Jupiter, only a small percentage of the planet's surface would be seen sharply. Also, of course, ground-based telescopes are still limited in their wavelength coverage by atmospheric absorption.

There is often some confusion in the literature between adaptive optics and active optics. However, the most widespread usage of the two terms is that an adaptive optics system is a fast closed-loop system, and an active optics system a more slowly operating open- or closed-loop system. The division is made at a response time of a few seconds. Thus, the tracking of a star by the telescope drive system can be considered as an active optics system that is open loop if no guiding is used and closed-loop if guiding is used. Large thin mirror optical telescopes and radio telescopes may suffer distortion due to buffeting by the wind at a frequency of a tenth of a hertz or so, they may also distort under gravitational loadings or thermal stresses, and have residual errors in their surfaces from the manufacturing process. Correction of all of these effects would also be classified under active optics. There is additionally the term active support that refers to the mountings used for the optical components in either an adaptive or an active optics system.

An atmospheric compensation system contains three main components, a sampling system, a wavefront sensor, and a correcting system. We look at each of these in turn.

1.1.21.1 Sampling System

The sampling system provides the sensor with the distorted wavefront or an accurate simulacrum thereof. For astronomical adaptive optics systems, a beam splitter is commonly used. This is just a partially reflecting mirror that typically diverts about 10% of the radiation to the sensor, while allowing the other 90% to continue on to form the image. A dichroic mirror can also be used which allows all the light at the desired wavelength to pass into the image while diverting light of a different wavelength to the sensor. However, atmospheric effects change with wavelength, and so this latter approach may not give an accurate reproduction of the distortions unless the operating and sampling wavelengths are close together.

Since astronomers go to great lengths to gather photons as efficiently as possible, the loss of even 10% to the sensor is to be regretted. Many adaptive optics systems therefore use a guide star rather than the object of interest to determine the wavefront distortions. This becomes essential when the object of interest is a large extended object, since most sensors need to operate on point or near-point images. The guide star must be very close in the sky to the object of interest, or its wavefront will have undergone different atmospheric distortion (Figure 1.53). For a solar work, small sunspots or pores can be used as the guide object. The region of the sky over which images have been similarly affected by the atmosphere is called the isoplanatic area or patch. It is defined by the distance over which the Strehl ratio improvement due to the adaptive optics halves. In the visible it is about 15″ across. The size of the isoplanatic patch scales as $\lambda^{1.2}$, so it is larger in the infrared, reaching 80″ at 2.2 μm (K band—see Section 3.1).

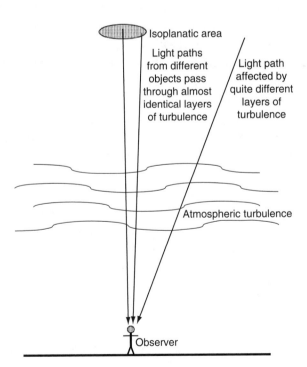

FIGURE 1.53 Isoplanatic area.

The small size of the isoplanatic area means that few objects have suitable guide stars. Less than 1% of the sky can be covered using real stars as guides, even in the infrared. Recently, therefore, artificial guide stars have been produced. This is accomplished using a powerful laser* pointed skyward. The laser is tuned to one of the sodium D-line frequencies and excites the free sodium atoms in the atmosphere at a height of about 90 km. The glowing atoms appear as a star-like patch that can be placed as near in the sky to the object of interest as required.

Guide stars at lower altitudes and at other wavelengths can be produced through back scattering by air molecules of a laser beam. Since these latter guide stars are produced via Rayleigh scattering, they are sometimes called Rayleigh stars. They can be of any wavelength, but a laser operating in the green is often chosen. The laser light can be sent out through the main telescope or more usually using an auxiliary telescope mounted on the main telescope. The 10 m Keck telescopes started operating with a single laser guide star in 2004 and now can achieve near diffraction-limited resolution in the NIR. Other large instruments such as the VLT, WHT, LBT, and the Gemini telescope have also recently started using laser guide stars for their adaptive optics systems or are in the process of developing them. While the 8.2 m Subaru telescope uses a deformable mirror with 188 actuators in combination with a laser guide star and a second-generation stellar coronagraph (high-contrast instrument for the Subaru next generation adaptive optics [HiCIAO]—see Section 5.3.5) in an attempt to image extra-solar planets directly.

Laser-produced guide stars have two problems that limit their usefulness. Firstly, for larger telescopes, the relatively low height of the guide star means that the light path from it to the telescope differs significantly from that for the object being observed (the cone problem—Figure 1.54). At 1 μm, the gain in Strehl ratio is halved through this effect, and in the visible it results in almost no improvement at all to the images. Secondly, the outgoing laser beam is affected by atmospheric turbulence, and therefore the guide star moves with respect to the object. This limits the correction of the overall inclination (usually known as the tip–tilt) of the wavefront resulting in a blurred image on longer exposures. A real star

---

* Continuous wave lasers with powers of up to 10 W may be used. These radiate sufficient energy to cause skin burns and retinal damage. Care therefore has to be taken to ensure that the beam does not intercept an aircraft. The VLT, for example, uses a pair of cameras and an automatic detection system to close a shutter over the laser beam should an aircraft approach.

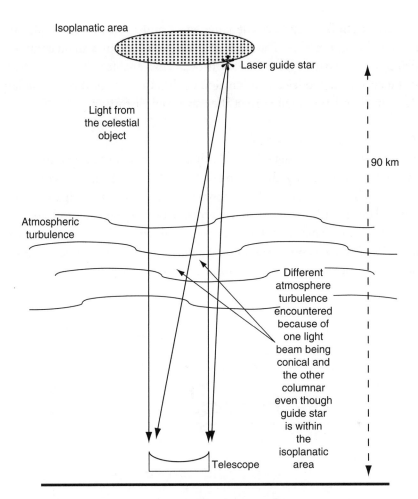

FIGURE 1.54 Light paths from a celestial object and a laser guide star to the telescope.

can, however, be used to determine the tip–tilt of the wavefront separately. The wavefront sensor simply needs to detect the motion of the star's image and since the whole telescope aperture can be utilized for this purpose, very faint stars can be observed. So most objects have a suitable star sufficiently nearby to act as a tip–tilt reference.

An adaptive optics system using two or more (up to nine) guide stars, such as those planned for the Gemini telescope and VLT and the proposed thirty meter telescope (TMT*), and known as MCAO (multi-conjugate

---

* Thirty meter telescope, previously known as the California extremely large telescope (CELT).

adaptive optics), eliminates the cone effect and produces an isoplanatic patch up to $120''$ across. The guide stars are separated by a small angle and detected by separate wavefront sensors. This enables the atmospheric turbulence to be modeled as a function of altitude. Two or three subsidiary deformable mirrors then correct the wavefront distortion.

1.1.21.2 Wavefront Sensing

The wavefront sensor detects the residual and changing distortions in the wavefront provided by the sampler after reflection from the correcting mirror. The Hartmann (also known as the Hartmann–Shack or the Shack–Hartmann) sensor is in widespread use in astronomical adaptive optics systems. This uses a two-dimensional array of small lenses (Figure 1.55). Each lens produces an image that is sensed by an array detector. In the absence of wavefront distortion, each image will be centered on

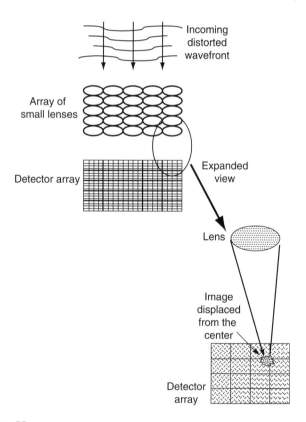

FIGURE 1.55 Hartmann sensor.

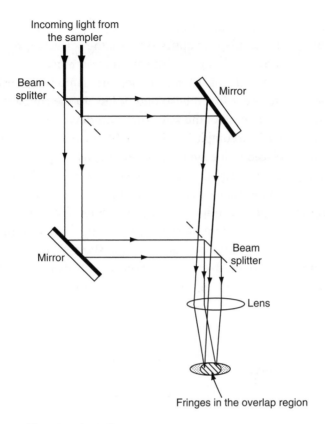

FIGURE 1.56 Shearing interferometer.

each detector. Distortion will displace the images from the centers of the detectors, and the degree of displacement and its direction is used to generate the error signal. An alternative sensor is based upon the shearing interferometer (Figure 1.56). This is a standard interferometer but with the mirrors marginally turned so that the two beams are slightly displaced with respect to each other when they are recombined. The deformations of the fringes in the overlap region then provide the slopes of the distortions in the incoming wavefront. The shearing interferometer was widely used initially for adaptive optics systems, but has now largely been replaced by the Hartman sensor. A new sensor, known as the curvature sensor, has been developed recently. It detects the wavefront distortions by comparing the illumination variations across slightly defocused images just inside and outside the focal point. A vibrating mirror is used to change the focus at kilo-hertz frequencies.

1.1.21.3 Wavefront Correction

In most astronomical adaptive optics systems, the correction of the wavefront is achieved by distorting a subsidiary mirror. Since the atmosphere changes on a timescale of 10 ms or so, the sampling, sensing, and correction have to occur in a millisecond or less. In the simplest systems, only the overall tip and tilt of the wavefront introduced by the atmosphere is corrected. That is accomplished by suitably inclining a plane or segmented mirror placed in the light beam from the telescope in the opposite direction (Figure 1.57). Tip–tilt correction systems for small telescopes are now available commercially at about 20% of the cost of a 0.2 m Schmidt–Cassegrain telescope and improve image sharpness by about a factor of two. An equivalent procedure, since the overall tilt of the wavefront causes the image to move, is shift and add. Multiple short exposure images are shifted until their brightest points are aligned, and then added together. Even this simple correction, however, can result in a considerable improvement of the images.

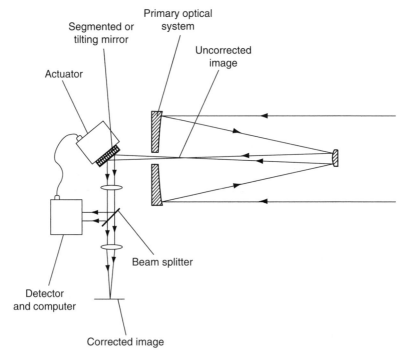

FIGURE 1.57 Schematic optical system for real-time atmospheric compensation.

More sophisticated approaches provide better corrections; either just of the relative displacements within the distorted wavefront, or of both displacement and fine-scale tilt. In some systems, the overall tilt of the wavefront is corrected by a separate system using a flat mirror whose angle can be changed. Displacement correction would typically use a thin mirror capable of being distorted by up to 100 piezo-electric or other actuators placed underneath it. The error signal from the sensor is used to distort the mirror in the opposite manner to the distortions in the incoming wavefront. The reflected wavefront is therefore almost flat. The noncorrection of the fine-scale tilt, however, does leave some small imperfections in the reflected wavefront. Nonetheless, currently operating systems using this approach can achieve diffraction-limited performance in the NIR for telescopes of 3 or 4 m diameter (i.e., about $0.2''$ at 2 μm wavelength). The 6.5 m MMT uses an adaptive secondary mirror 0.64 m across. It is a 2 mm thick Zerodur plate attached to a thick base plate by 336 actuators. At MIR wavelengths, a Strehl ratio of 0.98 can be achieved, compared with about 0.6 if the adaptive optics system is not used. It is planned to provide one of the VLT instruments with an adaptive secondary using four artificial guide stars by about 2012.

A recent proposal to improve the correcting mirrors is for a liquid mirror based upon reflective particles floating on a thin layer of oil. The oil contains magnetic grains and its surface can be shaped by the use of small electromagnets, but this has yet to be applied to astronomical image correction. At visual wavelengths, reductions of the uncorrected image size by about a factor of 10 are currently being reached. Further improvements can sometimes be achieved by applying blind or myopic deconvolution (Section 2.1.1) to the corrected images after they have been obtained.

Correction of fine-scale tilt within the distorted wavefront as well as displacement is now being investigated in the laboratory. It requires an array of small mirrors rather than a single bendy mirror. Each mirror in the array is mounted on four actuators so that it can be tilted in any direction as well as being moved linearly. At the time of writing, it is not clear whether such systems will be applied to many large astronomical telescopes, because at visual wavelengths few of them have optical surfaces good enough to take advantage of the improvements such a system would bring. Plans for future 50 and 100 m telescopes, however, include adaptive secondary or tertiary mirrors up to 8 m in diameter, requiring up to 500,000 actuators.

1.1.22 Future Developments

As a general guide to the future we may look to the past. Figure 1.58 shows the way in which the collecting area of optical telescopes has increased with time. Simple extrapolation of the trends shown there suggests collecting areas of 800 m^2 by the year 2050 and of 4500 m^2 by the year 2100 (diameters of 32 and 75 m, respectively, for filled circular apertures). Alternative historical analyses suggest a doubling of aperture every 30 years, leading to 100 m optical telescopes by the year 2100. However, there appear to be no fundamental technical differences between constructing 10 m segmented mirrors like those of the Keck telescopes and 30 m or even 100 m segmented mirrors. It is just a case of doing the same thing more times

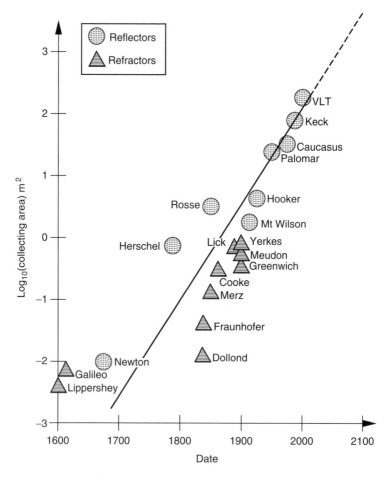

FIGURE 1.58 Optical telescope collecting area as a function of time.

and finding the increased funding required. Genuinely serious proposals for 20–40 m class telescopes (see below) suggest that such instruments will be working within the next two decades or less. As discussed earlier, the fact that the largest fully steerable radio telescopes are about 100 m in diameter suggests that this sort of size is likely to be the upper limit for individual optical telescopes as well. Of course, interferometers and aperture synthesis systems (Section 2.5) conceivably could be made thousands of kilometers across like the radio very long base-line interferometers.

Traditionally, the cost of a telescope is expected to rise as $D^{2.6}$ or thereabouts. This would suggest a cost of £10,000–£30,000 million for a 100 m class telescope. However, the use of lightweight mirrors, numerous small identical segments, short focal ratios, etc. means that not only is the cost of the optics reduced but also the mounting and enclosure are much cheaper to produce as well. Other expensive items of ancillary equipment such as re-aluminizing vacuum chambers can be reduced in size, and so on. Thus, current estimates are around £200–£300 million for a 30 m telescope and £600–£800 million for a 100 m telescope. These are still large sums, but not beyond the realms of possibility. The JWST, for example, is expected to cost around £500 million. The Apollo space program in today's money cost around £30,000 million, and the *HST* has so far cost some £1500 million.

There are several telescopes significantly larger than 10 m either now in operation or under construction. The VLT is currently starting to act as an aperture synthesis system (Section 2.5.5) with a sensitivity equal to that of a 16 m telescope and an unfilled aperture diameter of 100 m. The Keck telescopes will soon be able to act as an aperture synthesis system with a sensitivity equal to that of a 14 m telescope and an unfilled aperture diameter of 85 m. The giant Magellan telescope (GMT) has started construction with a planned completion date of 2016. It will be sited on Cerro Las Campanas in Chile and will comprise seven 8 m monolithic mirrors in a close packed array 25.4 m across.

There are also at least two detailed design studies underway for even larger telescopes. The European extremely large telescope (E-ELT*) is to be a 40 m class instrument due for completion around 2018/2020, while the United States and Canada are pursuing the TMT[†] project that is

---

* Now superseding the preliminary studies for the 100 m overwhelmingly large (OWL) and 50 m Euro50 telescopes.

[†] Now superseding the preliminary studies for the very large optical telescope (VLOT), the giant segmented-mirror telescope (GSMT), and the CELT.

expected to have a primary mirror formed from 492 hexagonal segments and perhaps to start operations in 2015. Less certain of completion is the already mentioned LAMA with a possible 42 m equivalent aperture and based upon liquid mirrors.

With MCAO correction of the images (which might require 100,000 active elements), a 40 m class telescope would have a resolution of 1 mas (milliarcsecond) in the visible region. The limiting magnitude for such a telescope might be 35^m in the visible, not just because of its increased light grasp, but because a 1 mas stellar image only has to equal the brightness of a 1 mas^2 area of the background sky in order to be detectable. This would enable the telescope to observe Jupiter-sized planets directly out to a distance of about 100 pc.

The multi-mirror concept may be extended or made easier by the development of fiber optics. Fused silica fibers can now be produced whose light losses and focal ratio degradation are acceptable for lengths of tens of meters. Thus, mirrors on independent mounts with a total area of hundreds of square meters, which all feed a common focus, will become technically feasible within a decade. Recently, the two Keck telescopes were linked by two 300 m fluoride glass cables thus potentially providing an alternative to the existing Keck interferometer (Section 2.5). A quite different future use of fiber optics may be the suppression of atmospheric emission lines in the NIR mentioned earlier.

At the other end of the scale, there is still a use for small telescopes—some of them very small indeed. Super wide-angle search for planets (SuperWASP), for example, uses four 0.11 m wide-angle lenses coupled to CCD cameras to search for exo-planets via transits. Likewise, the Trans-Atlantic Exoplanet Survey (TrES) comprises three 0.1 m instruments (stellar astrophysics and research on exoplanets (STARE) sited on Tenerife, Sleuth on Mount Palomar and planet search survey telescope (PSST) at the Lowell observatory, Arizona) monitoring thousands of stars at a time. Yet another exoplanet transit finder is Vulcan at the Lick Observatory which also has a 0.1 m telescope. A fifth exoplanet finder though takes the record of being the baby among all research instruments. The kilodegree extremely little telescope (KELT) uses a 0.05 m camera lens—the same size as the telescope that Newton showed to the Royal Society in 1761.

A technique that is in occasional use today but which may well gain in popularity is daytime observation of stars. Suitable filters and large focal ratios (that diminish the sky-background brightness, but not that of the star, see the earlier discussion) enable useful observations to be made of stars as faint as the seventh magnitude. The technique is particularly applicable in the infrared where the scattered solar radiation is minimal. If diffraction-limited 100 m telescopes are built, then with 10 nm bandwidths they would be able to observe stars (or other submas sources, but not extended objects) down to a visible magnitude of 23^m during the day.

Lenses seem unlikely to make a comeback as primary light-gathering components. However, they are still used extensively in eyepieces and in ancillary equipment. Three major developments seem likely to affect these applications, principally by making lens systems simpler and cheaper. The first is the use of plastic to form the lens. High-quality lenses then can be cheaply produced in quantity by moulding. The second advance is related to the first and it is the use of aspherical surfaces. Such surfaces have been used when they were essential in the past, but their production was expensive and time consuming. The investment in an aspherical mould, however, would be small if that mould were then to produce thousands of lenses. Hence, cheap aspherical lenses could become available. The final possibility is for lenses whose refractive index varies across their diameter or with depth. Several techniques exist for introducing such variability to the refractive index of a material such as the diffusion of silver into glass and the growing of crystals from a melt whose composition varies with time. Lenses made from liquid crystal are now being made in small sizes. The refractive index in this material depends upon an externally applied electric field, and can thus be changed rapidly. Such lenses may find application to wavefront correction in adaptive optics systems in due course. Thus, very highly corrected and relatively cheap lens systems may become available within a few years through the use of some or all of these techniques.

Space telescopes seem likely to follow the developments of terrestrially based telescopes toward increasing sizes, though if terrestrial diffraction-limited telescopes with diameters of several tens of meters or more become available, the space telescopes will only have advantages for spectral regions outside the atmospheric windows. Other developments may be along the lines of laser interferometer space antenna (LISA) (Section 1.6.2.2) with two or more separately orbiting telescopes forming very high resolution interferometers.

1.1.23 Observing Domes, Enclosures, and Sites

Any permanently mounted optical telescope requires protection from the elements. Traditionally this has taken the form of a hemi-spherical dome with an aperture through which the telescope can observe. The dome can be rotated to enable the telescope to observe any part of the sky, and the aperture may be closed to protect the telescope during the day and during inclement weather. Many recently built large telescopes have used cylindrical or other shapes for the moving parts of the enclosure for economic reasons; however, such structures are still clearly related to the conventional hemisphere. The dome, however, is an expensive item and it can amount to a third of the cost of the entire observatory, including the telescope. Domes and other enclosures also cause problems through heating-up during the day so inducing convection currents at night and through the generation of eddies as the wind blows across the aperture. Future VLTs may therefore operate without any enclosure at all, just having a movable shelter to protect them when not in use. This will expose the telescopes to wind buffeting, but active optics can now compensate for that at any wind speed for which it is safe to operate the telescope.

The selection of a site for a major telescope is at least as important for the future usefulness of that telescope as the quality of its optics. The main aim of site selection is to minimize the effects of the earth's atmosphere, of which the most important are usually scattering, absorption, and scintillation. It is a well-known phenomenon to astronomers that the best observing conditions at any site occur immediately before the final decision to place a telescope there. Thus, not all sites have fulfilled their expectations.

Scattering by dust and molecules causes the sky background to have a certain intrinsic brightness, and it is this that imposes a limit upon the faintest detectable object through the telescope. The main source of the scattered light is artificial light, and most especially street lighting. Thus, a first requirement of a site is that it should be as far as possible from built-up areas. If an existing site is deteriorating due to encroaching suburbs, etc. then some improvement may be possible for some types of observation by the use of a light pollution rejection (LPR) filter, which absorbs in the regions of the most intense sodium and mercury emission lines. Scattering can be worsened by the presence of industrial areas upwind of the site or by proximity to deserts, both of which inject dust into the atmosphere.

Absorption is due mostly to the molecular absorption bands of the gases forming the atmosphere. The two well-known windows in the spectrum, wherein radiation passes through the atmosphere relatively unabsorbed, extend from about 360 nm to 100 μm and from 10 mm to

100 m. But even in these regions there is some absorption, so that visible light is decreased in its intensity by 10%–20% for vertical incidence. The infrared region is badly affected by water vapor, OH, and other molecules to the extent that portions of it are completely obscured. Thus, the second requirement of a site is that it be as high an altitude as possible to reduce the air paths to a minimum, and that the water content be as low as possible. We have already seen how balloons and spacecraft are used as a logical extension of this requirement, and a few high-flying aircraft are also used for this purpose. The possibility of using a fiber-optic Bragg grating to reduce the sky background in the NIR by a factor of up to 50 has been mentioned earlier.

Scintillation is the change in the wavefront due to the varying densities of the atmospheric layers. It is sometimes divided into seeing, due to low-level turbulence, and scintillation proper, arising at higher levels. It causes the image to shimmer when viewed through a telescope, changing its size, shape, position, and brightness at frequencies from one to a thousand hertz. It is the primary cause of the low resolution of large telescopes, since the image of a point source is rarely less than half a second of arc across due to this blurring. Thus, the third requirement for a good site is a steady atmosphere. The ground-layer effects of the structures and landscape in the telescope's vicinity may worsen scintillation. A rough texture to the ground around the dome, such as low-growing bushes, seems to reduce scintillation when compared with that found for smooth (e.g., paved) and very rough (e.g., tree-covered) surfaces. Great care also needs to be taken to match the dome, telescope, and mirror temperatures to the ambient temperature or the resulting convection currents can worsen the scintillation by an order of magnitude or more. Cooling the telescope during the day to the predicted night-time temperature is helpful, provided that the weather forecast is correct. Forced air circulation by fans and thermally insulating and siting any ancillary equipment that generates heat as far from the telescope as possible, or away from the dome completely, can also reduce convection currents.

These requirements for an observing site restrict the choice very considerably and lead to a clustering of telescopes on the comparatively few optimum choices. Most are now found at high altitudes and on oceanic islands, or with the prevailing wind from the ocean. Built-up areas tend to be small in such places and the water vapor is usually trapped near the sea level by an inversion layer. The long passage over the ocean by the winds tends to minimize dust and industrial pollution. The remaining problem is that there is quite a correlation of good observing sites and noted tourist spots so that it is somewhat difficult to persuade nonastronomical colleagues that one is going to, say, the Canary isles or Hawaii, to work. This latter problem disappears for infrared and submillimeter wave observers, since their optimum site is the Antarctic plateau with its cold, dry, and stable atmospheric conditions. The Antarctic plateau is also optimum for radio studies of the aurora, etc. with, for example, the EISCAT (European Incoherent Scatter) fixed and movable dishes operating at wavelengths of a few hundreds of millimeters.

The recent trend in the reduction of the real cost of data transmission lines, whether these are via cables or satellites, is likely to lead to the greatly increased use of remote control of telescopes. In a decade or two, most major observatories are likely to have only a few permanent staff physically present at the telescope wherever that might be in the world, and the astronomer would only need to travel to a relatively nearby control center for his or her observing shifts. There are also already a few observatories with completely robotic telescopes, in which all the operations are computer controlled, with no staff on site at all during their use. Many of these robotic instruments are currently used for long-term photometric monitoring programs or for teaching purposes. However, their extension to more exacting observations such as spectroscopy is now starting to occur. The Liverpool telescope on La Palma and the two Faulkes telescopes in Hawaii and Australia are 2 m robotic telescopes that can be linked to respond to γ-ray burst (GRB) alerts in under 5 min. Five percent of the observing time on the Liverpool telescope and most of the time on the Faulkes telescopes is devoted to educational purposes with students able to control the instruments via the Internet. Other examples of robotic telescopes include the rapid eye mount (REM) a 0.6 m instrument designed for GRB observing and sited at La Silla, the panchromatic robotic optical monitoring and polarimetry telescopes (PROMPT) that use six 0.4 m telescopes covering the violet to NIR regions and designed for GRB and supernova observations as well as for education, and Peters automated infrared imaging telescope (PAIRITEL) located on Mount Hopkins that utilizes the 1.3 m telescope constructed for the 2MASS and is also for GRB observing.

Exercise 1.1

Calculate the effective theoretical resolutions of the normal and the dark-adapted eye, taking their diameters to be 2 and 6 mm, respectively.

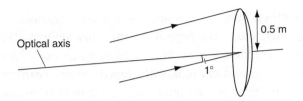

FIGURE 1.59 Optical arrangement for Exercise 1.3.

Exercise 1.2

Calculate the focal lengths of the lenses of a cemented achromatic doublet, whose overall focal length is 4 m. Assume the contact surfaces are of equal radius, the second surface of the diverging lens is flat, and that the components are made from crown and dense flint glasses. Correct it for wavelengths of 486 and 589 nm.

Exercise 1.3

By ray tracing calculate the physical separation in the focal plane of two rays that are incident onto a parabolic mirror of focal length 4 m. The rays are parallel before the reflection and at an angle of 1° to the optical axis. The first ray intersects the mirror at its center, while the second intersects it 0.5 m from the optical axis, and in the plane containing the rays and the optical axis (Figure 1.59).

Exercise 1.4

Calculate the maximum usable eyepiece focal length for the 5 m $f3.3/f16$ Mount Palomar telescope at its Cassegrain focus, and hence its lowest magnification when used visually.

Exercise 1.5

Taking Newton's reflecting telescope to have a diameter of 0.05 m and the dark-adapted eye to respond to an effective wavelength of 510 nm, what would be the maximum zenith distance that could Newton have observed objects in the sky without being limited by atmospheric turbulence?

1.2   RADIO AND MICROWAVE DETECTION

1.2.1   Introduction

Radio astronomy is the oldest of the "new" astronomies since it is now three-quarters of a century since Karl Jansky first observed radio waves from the Milky Way galaxy. It has passed beyond the developmental stage wherein the other "new" astronomies—infrared, ultraviolet, x-ray, neutrino astronomy, etc.—still remain. It thus has quite well-established instruments and techniques that are not likely to change overmuch.

The reason why radio astronomy developed relatively early is that radio radiation penetrates to ground level. For wavelengths from about 10 mm to 10 m, the atmosphere is almost completely transparent. The absorption becomes almost total at about 0.5 mm wavelength. Between 0.5 and 10 mm there are a number of absorption bands that are mainly due to oxygen and water vapor, with more or less transparent windows between the bands. The scale height for water vapor in the atmosphere is about 2000 m, so that observing from high altitudes reduces the short-wave absorption considerably. Radiation with wavelengths longer than about 50 m again fails to penetrate to ground level, but this time the cause is reflection by the ionosphere. Thus, this section is concerned with the detection of radiation with wavelengths longer than 0.1 mm. That is, frequencies less than 3×10^{12} Hz, or photon energies less than 2×10^{-21} J (0.01 eV). The detection of radiation in the 0.1 to a few mm wavelength region using bolometers is covered in Section 1.1.13.

The unit of intensity that is commonly used at radio wavelengths is the jansky (Jy)

$$1 \text{ Jy} = 10^{-26} \text{ W m}^{-2} \text{ Hz}^{-1} \tag{1.73}$$

and detectable radio sources vary from about 10^{-3} to 10^6 Jy. Most radio sources of interest to astronomers generate their radio flux as thermal radiation, when the Rayleigh–Jeans law gives their spectrum

$$F_\nu = \frac{2\pi k}{c^2} T \nu^2 \tag{1.74}$$

or as synchrotron radiation from energetic electrons spiraling around magnetic fields, when the spectrum is of the form

$$F_\nu \propto \nu^{-\alpha} \tag{1.75}$$

where

F_ν is the flux per unit frequency interval at frequency ν

α is called the spectral index of the source, and is related to the energy distribution of the electrons

For many sources $0.2 \le \alpha \le 1.2$.

1.2.2 Detectors and Receivers

The detection of radio signals is a two-stage process in which the sensor produces an electrical signal that then has to be processed until it is in a directly usable form. Coherent detectors, which preserve the phase information of the signal, are available for use over the whole of the radio spectrum, in contrast to the optical and infrared detectors discussed in Section 1.1.13 that respond only to the total power of the signal. The detection of submillimeter and millimeter radiation (FIR) via the use of bolometers is covered in Section 1.1. In the MHz radio region, the sensor is normally a dipole placed directly at the focus of the telescope, such as the half-wave dipole shown in Figure 1.60, although this can only be optimized for one wavelength and has very restricted bandwidths. The two halves of such a dipole are each a quarter of a wavelength long. Connection to the remainder of the system is by coaxial cable.

In the GHz and higher frequency regions, a horn antenna is normally used to collect the radiation, usually with waveguides for the connection to the rest of the system, though plastic and quartz lenses may be used at very high frequencies. Recent developments in the design of horn antennas have enabled them to have much wider bandwidths than earlier designs. These developments include using a corrugated internal surface for the horn and reducing its diameter in a series of steps rather than smoothly, giving a bandwidth that covers a factor of two in terms of frequency. Bandwidths with factors of nearly eight in frequency are possible with

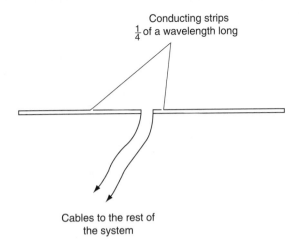

Conducting strips
$\frac{1}{4}$ of a wavelength long

Cables to the rest of
the system

FIGURE 1.60 Half-wave dipole.

dielectric-loaded horns. These are smooth-walled horns filled with an appropriate dielectric except for a small gap between the filling and the walls of the horn through which the wave propagates. At high frequencies, the feed horns may need cooling to cryogenic temperatures.

The sensor at the higher frequencies is nowadays normally a superconductor–insulator–superconductor (SIS) device. In an SIS detector, an electron in one superconducting film absorbs a photon, giving the electron enough energy to tunnel through the insulating barrier into the other superconducting film (cf. Section 1.1.11). This process, known as photon-assisted tunneling, produces one electron for every absorbed photon. Modern devices are based upon two niobium layers separated by an insulating region of aluminum oxide around 1 nm thick and the whole cooled to 4 K or less. Such SIS devices can operate up to about 700 GHz. SIS devices can be used up to 1.2 THz using niobium titanium nitride. In addition to SCUBA2 (Section 1.1.13.2) the 15 m JCMT uses a 4 × 4 array of SIS detectors, known as heterodyne array receiver program (HARP), to obtain spectra and images over the 325–375 GHz region. Even higher frequencies require the use of Schottky diodes, or more recently TES devices (Section 1.1.13.2) based upon niobium, and generally known among radio astronomers as hot electron bolometers (HEBs). The 12 m APEX telescope, for example, started observing with CONDOR (CO N + deuterium observation receiver) in 2005. This detector uses a niobium–titanium nitride HEB cooled to 4 K to convert the THz radiation to GHZ frequencies and thereby observe over the 200–240 μm (1.52–1.25 THz) region.

The signal from the sensor is carried to the receiver whose purpose is to convert the high-frequency electrical currents into a convenient form. The behavior of the receiver is governed by five parameters: sensitivity, amplification, bandwidth, receiver noise level, and integration time.

The sensitivity and the other parameters are very closely linked, for the minimum detectable brightness, B_{min}, is given by

$$B_{min} = \frac{2k\nu^2 KT_s}{c^2\sqrt{t\,\Delta\nu}} \tag{1.76}$$

where
T_s is the noise temperature of the system
t is the integration time
$\Delta\nu$ is the frequency bandwidth
K is a constant close to unity that is a function of the type of receiver

The bandwidth is usually measured between output frequencies whose signal strength is half the maximum when the input signal power is constant with frequency. The amplification and integration time are self-explanatory, so that only the receiver noise level remains to be explained. This noise originates as thermal noise within the electrical components of the receiver, and may also be called Johnson or Nyquist noise (see also Section 1.1.16.1). The noise is random in nature and is related to the temperature of the component. For a resistor, the RMS voltage of the noise per unit frequency interval, $\overline{V}$, is given by

$$\overline{V} = 2\sqrt{kTR} \tag{1.77}$$

where
R is the resistance
T is the temperature

The noise of the system is then characterized by the temperature T_s that would produce the same noise level for the impedance of the system. It is given by

$$T_s = T_1 + \frac{T_2}{G_1} + \frac{T_3}{G_1 G_2} + \cdots \frac{T_n}{G_1 G_2 \cdots G_{n-1}} \tag{1.78}$$

where
T_n is the noise temperature of the nth component of the system
G_n is the gain (or amplification) of the nth component of the system

It is usually necessary to cool the initial stages of the receiver with liquid helium to reduce T_s to an acceptable level. Other noise sources that may be significant include shot noise resulting from random electron emission, g–r noise due to a similar effect in semiconductors (Section 1.1.16.1), noise from the parabolic reflector or other collector that may be used to concentrate the signal onto the antenna, radiation from the atmosphere, and last but by no means least, spillover from radio taxis, microwave ovens, and other artificial sources.

Many types of receivers exist; the simplest is a development of the heterodyne system employed in the ubiquitous transistor radio. The basic layout of a heterodyne receiver is shown in a block form in Figure 1.61. The preamplifier operates at the signal frequency and will typically have a gain of 10–1000. It is often mounted close to the feed and cooled to near

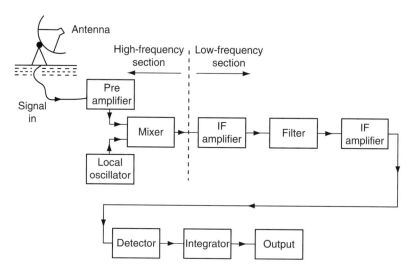

FIGURE 1.61 Block diagram of a basic heterodyne receiver.

absolute zero to minimize its contribution to the noise of the system (see above). The most widely used amplifiers today are based upon cooled gallium arsenide and indium phosphide HFETs (heterostructure field-effect transistors, also known as high electron mobility transistors [HEMTs]) in which the current-carrying electrons are physically separated from the donor atoms. The current-carrying electrons are restricted to a thin (10 nm) layer of undoped material producing a fast, low-noise device. Above 40 GHz, the mixer must precede the preamplifier to decrease the frequency before it can be amplified; a second lower frequency local oscillator is then employed to reduce the frequency even further.

The local oscillator produces a signal that is close to but different from the main signal in its frequency. Thus, when the mixer combines the main signal and the local oscillator signal, the beat frequency between them (intermediate frequency or IF) is at a much lower frequency than that of the original signal. The relationship is given by

$$\nu_{\text{SIGNAL}} = \nu_{\text{LO}} \pm \nu_{\text{IF}} \tag{1.79}$$

where

ν_{SIGNAL} is the frequency of the original signal (i.e., the operating frequency of the radio telescope)

ν_{LO} is the local oscillator frequency

ν_{IF} is the IF

Normally, at lower frequencies, only one of the two possible signal frequencies given by Equation 1.79 will be picked up by the feed antenna or passed by the preamplifier. At high frequencies, both components may contribute to the output.

The power of the IF emerging from the mixer is directly proportional to the power of the original signal. The IF amplifiers and filter determine the pre-detector bandwidth of the signal and further amplify it by a factor of 10^6 to 10^9. The detector is normally a square-law device; that is to say, the output voltage from the detector is proportional to the square of the input voltage. Thus, the output voltage from the detector is proportional to the input power. In the final stages of the receiver, the signal from the detector is integrated, usually for a few seconds, to reduce the noise level. Then it is fed to an output device, usually an analogue-to-digital input to a computer for further processing. Further advances in speed and noise reduction come from combining the sensor and much of the electronics onto MMICs (monolithic microwave integrated circuits). The devices are cheap and are used, for example, in SETI to construct low-noise amplifiers, filters, and local oscillators.

The basic heterodyne receiver has a high system temperature, and its gain is unstable. The temperature may be lowered by applying an equal and opposite voltage in the later stages of the receiver, and the stability of the gain may be greatly improved by switching rapidly from the antenna to a calibration noise source and back again, with a phase-sensitive detector (Section 5.2) to correlate the changes. Such a system is then sometimes called a Dicke radiometer. The radiometer works optimally if the calibration noise source level is the same as that of the signal, and so it may be further improved by continuously adjusting the noise source to maintain the balance, and it is then termed a null-balancing Dicke radiometer. Since the signal is only being detected half the time, the system is less efficient than the basic receiver, but using two alternately switched receivers will restore its efficiency. The value of T_s for receivers varies from 10 K at meter wavelengths to 10,000 K at millimeter wavelengths. The noise sources must therefore have a comparable range, and at long wavelengths are usually diodes, while at the shorter wavelengths a gas discharge tube inside the waveguide and inclined to it by an angle of about $10°$ is used.

Receivers are generally sky-background limited just like terrestrial optical telescopes. The earth's atmosphere radiates at 100 K and higher temperatures below a wavelength of about 3 mm. Only between 30 and

100 mm does its temperature fall as low as 2 K. Then, at longer wavelengths, the galactic emission becomes important, rising to temperatures of 10^5 K at wavelengths of 30 m.

Spectrographs at radio frequencies can be obtained in several different ways. In the past, the local oscillator has been tuned, producing a frequency-sweeping receiver, or the receiver has been a multichannel device so that it registered several different discrete frequencies simultaneously. For pulsar observations such filter banks may have several hundred channels over a bandwidth of a few megahertz. Today, most radio spectroscopy is carried out by autocorrelation, even at the highest frequencies. Successive delays are fed into the signal that is then cross-correlated with the original signal in a computer. The spectrum is obtained from the Fourier transform of the result. The polarization of the original signal is determined by separately detecting the orthogonal components and cross-correlating the electrical signals later within the receiver.

Alternatively, the radio signal may be converted into a different type of wave, and the variations of this secondary wave studied instead. This is the basis of the acousto-optical radio spectrometer (AOS). The radio signal is converted into an ultrasonic wave whose intensity varies with that of the radio signal and whose frequency is also a function of that of the radio signal. In the first such instruments, water was used for the medium in which the ultrasound propagated, and the wave was generated by a piezoelectric crystal driven either directly from the radio signal or by a frequency-reduced version of the signal. More recently, materials such as fused silica, lithium niobate, and lead molybdate have replaced water as the acoustic medium to improve the available spectral range. A laser illuminates the cell containing the acoustic medium and a part of the light beam is diffracted by the sound wave. The angle of diffraction depends upon the sound wave's frequency, while the intensity of the diffracted light depends on the sound wave's intensity. Thus, the output from the device is a fan beam of light, position within which ultimately depends upon the observed radio frequency, and whose intensity at that position ultimately depends upon the radio intensity. The fan beam may then simply be detected by a linear array of detectors or by scanning, and the spectrum inferred from the result. The AOS initially found application to the observation of solar radio bursts (Section 5.3.8) at meter wavelengths, but is now employed more in the submillimeter and infrared regions since autocorrelation techniques have replaced it at the longer wavelengths and are even starting to do so in the millimeter region.

A major problem at all frequencies in radio astronomy is interference from artificial noise sources. In theory, certain regions of the spectrum (Table 1.6) are reserved partially or exclusively for use by radio astronomers. But leakage from devices such as microwave ovens, incorrectly tuned receivers and illegal transmissions often overlap into these bands. The Russian satellite navigation system GLONASS, for example, overlapped into the band reserved for the interstellar OH lines at 1.61 GHz. The use of highly directional aerials (see Section 1.2.3) reduces the problem to some extent. But it is likely that radio astronomers will have to follow their optical colleagues to remote parts of the globe or place their aerials in space if their work is to continue in the future. Even the latter means of escape may be threatened by solar power satellites with their potentially enormous microwave transmission intensities.

The wave bands listed in Table 1.6 are those allocated by the European Science Foundation's Committee on Radio Astronomy Frequencies in 2006. Not all frequencies are kept free in all places; some allocations are shared with other users and the allocations are continually changing. The reader should consult an up-to-date local source for specific information relative to his or her requirements. These can often be found by searching the Internet.

The International Telecommunications Union in collaboration with Commission 40 (Radio Astronomy) of the International Astronomical Union has identified the following bands (Table 1.7) for the exclusive use of radio astronomy. Until now though the ITU has only allocated frequencies up to 275 GHz.

1.2.3 Radio Telescopes

Although the antenna and the receiver are the main active portions of a radio detecting system, they are far less physically impressive than the large structures that serve to gather and concentrate the radiation and to shield the antenna from unwanted sources. Before going on, however, to the consideration of these large structures that form most people's ideas of what comprises a radio telescope, we must look in a little more detail at the physical background of antennae.

The theoretical optics of light and radio radiation is identical, but different traditions in the two disciplines have led to differences in the mathematical and physical formulations of their behaviors. Thus, the image in an optical telescope is discussed in terms of its diffraction structure (e.g., Figure 1.26), while that of a radio telescope is discussed

TABLE 1.6   Radio Astronomy Reserved Frequencies:
European Science Foundation's Committee
on Radio Astronomy Frequencies

13.36–13.41 MHz	79–81 GHz
25.55–25.67 MHz	81–84 GHz
37.5–38.25 MHz	84–86 GHz
73.0–74.6 MHz	86–92 GHz
79.25–80.25 MHz	92–94 GHz
150.05–153 MHz	94–94.1 GHz
322.0–328.6 MHz	94.1–95 GHz
406.1–410.0 MHz	95–100 GHz
608.0–614.0 MHz	100–102 GHz
1.3300–1.4000 GHz	102–105 GHz
1.4000–1.4270 GHz	105–109.5 GHz
1.6106–1.6138 GHz	109.5–111.8 GHz
1.6600–1.6700 GHz	111.8–114.25 GHz
1.7188–1.7222 GHz	114.25–116 GHz
2.6550–2.6900 GHz	123–126 GHz
2.6900–2.7000 GHz	126–130 GHz
3.2600–3.2670 GHz	130–134 GHz
3.3320–3.3390 GHz	134–136 GHz
3.3458–3.3525 GHz	136–141 GHz
4.8000–4.9900 GHz	141–148.5 GHz
4.9900–5.0000 GHz	148.5–151.5 GHz
6.6500–6.6752 GHz	151.5–155.5 GHz
10.6–10.7 GHz	155.5–158.5 GHz
14.47–14.50 GHz	164–167 GHz
15.35–15.40 GHz	168–170 GHz
22.01–22.21 GHz	170–174.5 GHz
22.21–22.50 GHz	182–185 GHz
22.81–22.86 GHz	191.8–200 GHz
23.07–23.12 GHz	200–202 GHz
23.60–24.00 GHz	202–209 GHz
31.2–31.8 GHz	209–217 GHz
36.43–36.5 GHz	217–226 GHz
42.5–43.5 GHz	226–231.5 GHz
47.2–50.2 GHz	241–248 GHz
51.4–54.25 GHz	248–250 GHz
58.2–59.0 GHz	250–252 GHz
64.0–65.0 GHz	252–265 GHz
76–77.5 GHz	265–275 GHz
77.5–78 GHz	275–1000 GHz

TABLE 1.7   Radio Astronomy Exclusive Frequencies:
Allocated by the International
Telecommunications Union in collaboration
with Commission 40 (Radio Astronomy)
of the International Astronomical Union

1.400–1.427 GHz	50.2–50.4 GHz
2.690–2.700 GHz	52.6–54.25 GHz
3.260–3.267 GHz	86–92 GHz
3.332–3.339 GHz	100–102 GHz
3.3458–3.3525 GHz	109.5–111.8 GHz
10.68–10.70 GHz	114.25–116 GHz
13.35–15.4 GHz	148.5–151.5 GHz
22.01–22.21 GHz	164–167 GHz
23.6–24 GHz	182–185 GHz
31.3–31.5 GHz	200–202 GHz
31.5–31.8 GHz	226–231.5 GHz
48.94–49.04 GHz	250–252 GHz

in terms of its polar diagram. However, these are just two different approaches to the presentation of the same information. The polar diagram is a plot, in polar coordinates, of the sensitivity or voltage output of the telescope, with the angle of the source from the optical axis. (Note that the polar diagrams discussed herein are all far-field patterns, i.e., the response for a distant source, near-field patterns for the aerials may differ from those shown here.) The polar diagram may be physically realized by sweeping the telescope past a point source, or by using the telescope as a transmitter and measuring the signal strength around it.

The simplest antenna, the half-wave dipole (Figure 1.60), accepts radiation from most directions, and its polar diagram is shown in Figure 1.62. This is only a cross section through the beam pattern; the full three-dimensional polar diagram may be obtained by rotating the pattern shown in Figure 1.62 about the dipole's long axis, and thus has the appearance of a toroid that is filled-in to the center. The polar diagram, and hence the performance of the antenna, may be described by four parameters: the beam width at half-power points (BWHP), the beam width at first nulls (BWFN), the gain, and the effective area. The first nulls are the positions on either side of the optical axis where the sensitivity of the antenna first decreases to zero, and the BWFN is just the angle between them. Thus, the value of the BWFN for the half-wave dipole is 180°.

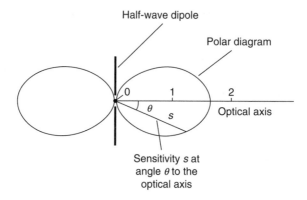

FIGURE 1.62 Polar diagram of a half-wave dipole.

The first nulls are the direct equivalent of the first fringe minima in the diffraction pattern of an optical image, and for a dish aerial type of radio telescope, their position is given by Equation 1.26 thus

$$\text{BWFN} = 2x\frac{1.22\lambda}{D} \tag{1.80}$$

The Rayleigh criterion of optical resolution may thus be similarly applied to radio telescopes; two point sources are resolvable when one is on the optical axis and the other is in the direction of a first null. The half-power points may be best understood by regarding the radio telescope as a transmitter, they are then the directions in which the broadcast power has fallen to one-half of its peak value. The BWHP is just the angular separation of these points. For a receiver they are the points at which the output voltage has fallen by a factor of the square root of two, and hence the output power has fallen by half. The maximum gain or directivity of the antenna is also best understood in terms of a transmitter. It is the ratio of the peak value of the output power to the average power. In a receiver it is a measure of the output from the system compared with that from a comparable (and hypothetical) isotropic receiver. The effective area of an antenna is the ratio of its output power to the strength of the incoming flux of the radiation that is correctly polarized to be detected by the antenna, that is,

$$A_e = \frac{P_\nu}{F_\nu} \tag{1.81}$$

where

A_e is the effective area

P_ν is the power output by the antenna at frequency ν

F_ν is the correctly polarized flux from the source at the antenna at frequency ν

The effective area and the maximum gain, g, are related by

$$g = \frac{4\pi}{c^2} \nu^2 A_e \tag{1.82}$$

For the half-wave dipole, the maximum gain is about 1.6, and so there is very little advantage over an isotropic receiver.

The performance of a simple dipole may be improved by combining the outputs from several dipoles that are arranged in an array. In a collinear array, the dipoles are lined up along their axes and spaced at intervals of half a wavelength (Figure 1.63). The arrangement is equivalent to a

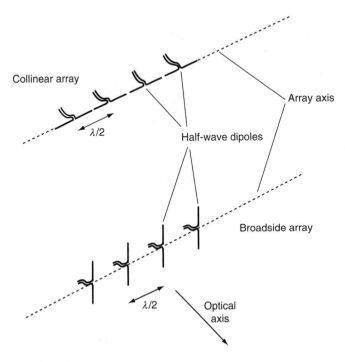

FIGURE 1.63 Dipole arrays.

diffraction grating and so the sensitivity at an angle θ to the long axis of the array, $s(\theta)$, is given by

$$s(\theta) = s_0 \left[\frac{\sin (n\pi \sin \theta)}{\sin (\pi \sin \theta)} \right] \tag{1.83}$$

where
n is the number of half-wave dipoles
s_0 is the maximum sensitivity (cf. Equation 4.8)

Figure 1.64 shows the polar diagrams for 1, 2, and 4 dipole arrays, their three-dimensional structure can be obtained by rotating these diagrams around a vertical axis so that they become lenticular toroids. The resolution along the axis of the array, measured to the first null is given by

$$\alpha = \sin^{-1}\left(\frac{1}{n}\right) \tag{1.84}$$

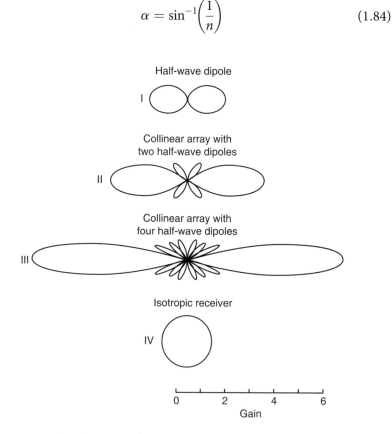

Half-wave dipole

Collinear array with two half-wave dipoles

Collinear array with four half-wave dipoles

Isotropic receiver

Gain

FIGURE 1.64 Polar diagrams for collinear arrays.

The structure of the polar diagrams in Figure 1.64 shows a new development. Apart from the main lobe whose gain and resolution increase with n as might be expected, a number of smaller side lobes have appeared. Thus, the array has sensitivity to sources that are at high angles of inclination to the optical axis. These side lobes correspond precisely to the fringes surrounding the Airy disk of an optical image (Figure 1.26, etc.). Although the resolution of an array is improved over that of a simple dipole along its optical axis, it will still accept radiation from any point perpendicular to the array axis. The use of a broadside array in which the dipoles are perpendicular to the array axis and spaced at half wavelength intervals (Figure 1.63) can limit this 360° acceptance angle somewhat. For a four-dipole broadside array, the polar diagram in the plane containing the optical and array axes is given by polar diagram number II in Figure 1.64, while in the plane containing the optical axis and the long axis of an individual dipole, the shape of the polar diagram is that of a single dipole (number I in Figure 1.64), but with a maximum gain to match that in the other plane. The three-dimensional shape of the polar diagram of a broadside array thus resembles a pair of squashed balloons placed end to end. The resolution of a broadside array is given by

$$\alpha = \sin^{-1}\left(\frac{2}{n}\right) \tag{1.85}$$

along the array axis, and is that of a single dipole, that is, 90°, perpendicular to this. Combinations of broadside and collinear arrays can be used to limit the beam width further if necessary.

With the arrays as shown, there is still a two-fold ambiguity in the direction of a source that has been detected, however narrow the main lobe may have been made, due to the forward and backward components of the main lobe. The backward component may easily be eliminated, however, by placing a reflector behind the dipole. This is simply a conducting rod about 5% longer than the dipole and unconnected electrically with it. It is placed parallel to the dipole and about one-eighth of a wavelength behind it. For an array, the reflector may be a similarly placed electrically conducting screen. The polar diagram of a four-element collinear array with a reflector is shown in Figure 1.65 to the same scale as the diagrams in Figure 1.64. It is apparent that not only has the reflector screened out the backward lobe but also has doubled the gain of the main lobe. Such a reflector is termed a parasitic element since it is not a part of

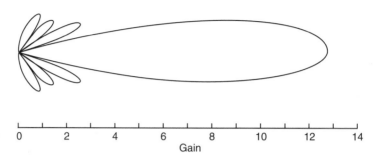

FIGURE 1.65 Polar diagram for a four-element collinear array with a mesh reflector.

the electrical circuit of the antenna. Similar parasitic elements may be added in front of the dipole to act as directors. These are about 5% shorter than the dipole. The precise lengths and spacings for the parasitic elements can only be found empirically, since the theory of the whole system is not completely understood. With a reflector and several directors we obtain the parasitic or Yagi antenna, familiar from its appearance on so many rooftops as a television aerial. A gain of up to 12 is possible with such an arrangement and the polar diagram is shown in Figure 1.66. The Rayleigh resolution is 45° and it has a bandwidth of about 3% of its operating frequency. The main use of parasitic antennae in radio astronomy is as the receiving element (sometimes called the feed) of a larger reflector such as a parabolic dish.

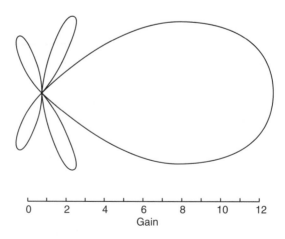

FIGURE 1.66 Polar diagram of a parasitic aerial.

The use of a single dipole or even several dipoles in an array is the radio astronomy equivalent of a naked-eye observation. The Netherlands-centered low-frequency array (LOFAR) though, which has just started making its first observations over the 10–250 MHz band, does use simple dipole antennas. Although currently operating with just 96 receivers, it is planned eventually to have 15,000 spread over much of Western Europe.

However, just as in optical astronomy, the observations may be greatly facilitated by the use of some means of concentrating the signal from over a wide area onto the antenna. The most familiar of these devices are the large parabolic dishes that are the popular conception of a radio telescope. These are directly equivalent to an optical reflecting telescope. They are usually used at the prime focus or at the Cassegrain focus (Section 1.1.18.2). The gain may roughly be found by substituting the dishes' area for the effective area in Equation 1.82. The Rayleigh resolution is given by Equation 1.27. The sizes of the dishes are very large because of the length of the wavelengths being observed; for example, to obtain a resolution of $1°$ at a wavelength of 0.1 m requires a dish 7 m across, which is larger than most optical reflectors for a resolution over 10^4 times poorer. The requirement on surface accuracy is the same as that for an optical telescope—deviations from the paraboloid to be less than $\lambda/8$ if the Rayleigh resolution is not to be degraded. Now, however, the longer wavelength helps since it means that the surface of a telescope working at 0.1 m, say, can deviate from perfection by over 10 mm without seriously affecting the performance. In practice, a limit of $\lambda/20$ is often used, for as we have seen the Rayleigh limit does not represent the ultimate limit of resolution. These less-stringent physical constraints on the surface of a radio telescope ease the construction problems greatly, more importantly, however, it also means that the surface need not be solid, as a wire mesh with spacings less than $\lambda/20$ will function equally well as a reflector. The weight and wind resistance of the reflector are thus reduced by very large factors. At the shorter radio wavelengths, a solid reflecting surface may be more convenient however, and at very short wavelengths (<1 mm) active surface control to retain the accuracy is used along somewhat similar lines to the methods discussed in Section 1.1 for optical telescopes. The shape of the mirror is monitored holographically. The dishes are usually of very small focal ratio, $f0.5$ is not uncommon, and the reason for this is that the dish acts as a screen against unwanted radiation as well as concentrating the desired radiation. Fully steerable dishes up to 100 m across have been built such as the Green Bank and Effelsberg telescopes, while the Arecibo telescope in Puerto Rico, which, at the time of writing, is threatened with closure by 2011, is a fixed dish 300 m across. This latter instrument acts as a transit telescope and has some limited ability to track sources and look at a range of declinations by moving the feed antenna around the image plane. Such fixed telescopes may have spherical surfaces to extend this facility and use a secondary reflector to correct the resulting spherical aberration. The feed antenna for such dishes may be made of such a size that it intercepts only the center lobe of the telescope response (i.e., the Airy disk in optical terms). The effects of the side lobes are then reduced or eliminated. This technique is known as tapering the antenna and is the same as the optical technique of apodization (Section 4.1.4.2). The tapering function may take several forms and may be used to reduce background noise as well as eliminating the side lobes. Tapering reduces the efficiency and resolution of the dish, but this is usually more than compensated for by the improvement in the shape of the response function. In the microwave region, the largest dishes are currently the 30 m IRAM instrument on Pico Veleta in Spain and the 45 m telescope at Nobeyama in Japan. A 50 m dish (the large millimeter telescope) is currently under construction near Mexico City and due to start science operations in 2008. The 25 m Cornell Caltech Atacama telescope (CCAT) is planned to begin operating in 2013 with its mirror segments under active control.

With a single feed, the radio telescope is a point-source detector only. Images have to be built up by scanning (Section 2.4) or by interferometry (Section 2.5). Scanning used to be accomplished by pointing the telescope at successive points along the raster pattern and observing each point for a set interval. This practice, however, suffers from fluctuations in the signal due to the atmosphere and the electronics. Current practice is therefore to scan continuously or "on-the-fly." The atmospheric or electronic variations are then on a longer timescale than the changes resulting from moving over a radio source and can be separated out.

True imaging can be achieved through the use of cluster or array feeds. These are simply multiple individual feeds arranged in a suitable array at the telescope's focus. Each feed is then the equivalent of a pixel in a CCD or other type of detector. The number of elements in such cluster feeds currently remains small compared with their optical equivalents. For example, the 64 m Parkes radio telescope uses a 13-beam receiver for 21 cm radiation, and the 14 m telescope at the Five College Radio Astronomy Observatory (FCRAO) in Massachusetts uses a 32-beam array operating at millimeter wavelengths (see also SCUBA2 and Bolometers in Section 1.1).

Very many other systems have been designed to fulfill the same function as a steerable paraboloid but which are easier to construct. The best known of these are the multiple arrays of mixed collinear and broadside type, or similar constructions based upon other aerial types. They are mounted onto a flat plane which is oriented east–west and which is tiltable in altitude to form a transit telescope. Another system such as the 600 m RATAN telescope uses an off-axis paraboloid that is fixed and which is illuminated by a tiltable flat reflector. Alternatively, the paraboloid may be cylindrical and tiltable itself around its long axis. For all such reflectors, some form of feed antenna is required. A parasitic antenna is a common choice at longer wavelengths, while a horn antenna, which is essentially a flared end to a waveguide, may be used at higher frequencies.

A quite different approach is used in the Mills cross type of telescope. These uses two collinear arrays oriented north–south and east–west. The first provides a narrow fan beam along the north–south meridian, while the second provides a similar beam in an east–west direction. Their intersection is a narrow vertical pencil beam, typically 1° across. The pencil beam may be isolated from the contributions of the remainders of the fan beams by comparing the outputs when the beams are added in phase with when they are added out of phase. The in-phase addition is simply accomplished by connecting the outputs of the two arrays directly together. The out-of-phase addition delays one of the outputs by half a wavelength before the addition, and this is most simply done by switching in an extra length of cable to one of the arrays. In the first case, radiation from objects within the pencil beam will interfere constructively, while in the second case, there will be destructive interference. The signals from objects not within the pencil beam will be mutually incoherent and so will simply add together in both cases. Thus, looking vertically down onto the Mills cross, the beam pattern will alternate between the two cases shown in Figure 1.67. Subtraction of the one from the other will then just leave the pencil beam. The pencil beam may be displaced by an angle θ from the vertical by introducing a phase shift between each dipole. Again the simplest method is to switch extra cable into the connections between each dipole, the lengths of the extra portions, L, given by

$$L = d \sin \theta \qquad (1.86)$$

where d is the dipole separation. The pencil beam may thus be directed around the sky as wished. In practice, the beam is only moved along the

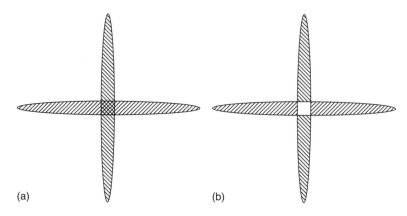

FIGURE 1.67 Beam patterns of a Mills cross radio telescope (a) with the beams added together (i.e., in phase); (b) with the beams subtracted from each other (i.e., 180° out of phase).

north–south plane, and the telescope is used as a transit telescope, since the alteration of the cable lengths between the dipoles is a lengthy procedure. The resolution of a Mills cross is the same as that of a parabolic dish whose diameter is equal to the array lengths. The sensitivity, however, is obviously much reduced from that of a dish, since only a tiny fraction of the aperture is filled by the dipoles. In addition to being cumbersome to operate, the Mills cross suffers from the disadvantage that it can only operate at a single wavelength unless its dipoles are all changed. Furthermore, confusion of sources may arise if a strong source is in one of the pencil beams formed by the side lobes, since it will have the same appearance as a weak source in the main beam. An alternative and related system uses a single fan beam and observes the object at many position angles. The structure of the source is then retrievable from the observations in a relatively unambiguous manner.

The phasing of two dipoles, as we have seen, can be used to alter the angle of the beam to their axis. By continuously varying the delay, the lobe can be swept across the sky, or accurately positioned on an object and then moved to track that object. This is an important technique for use with interferometers (Section 2.5), and for solar work. It also forms the basis of a phased array. This latter instrument is basically a Mills cross in which the number of dipoles has been increased until the aperture is filled. It provides great sensitivity since a large area may be covered at a relatively low cost.

A radically different approach for concentrating the radiation is based upon refraction. The Luneburg lens has yet to find much application to radio astronomy, but may be used in the square kilometer array (SKA; Section 2.5.5). The Luneburg lens is a solid sphere within which the refractive index increases linearly inward from unity at the surface. With a central refractive index of 2, the focus is on the surface of the lens. Since there is no axis of symmetry, the lens can be used to observe in many directions simultaneously, simply by having numerous feeds distributed around it. Many materials potentially can be used for the lens' construction, but to date high-density polystyrene is the one that has been used in practice.

A number of spacecraft carrying microwave detectors have been launched. These include *COBE* (*Cosmic Background Explorer Satellite*) that included Dicke radiometers operating at 31.5, 53, and 90 GHz, and the current *WMAP* (*Wilkinson Microwave Anisotropy Probe*) mission observing from 22 to 90 GHz. Few longer wave spacecraft-borne telescopes have been used, since the atmosphere is then transparent. However, the *Japanese Halca* spacecraft carried an 8 m dish operating at centimeter wavelengths as part of an aperture synthesis system (Section 2.5.5).

Great improvements in resolution and sensitivity of radio telescopes may be obtained through the use of interferometers and aperture synthesis, and these devices and their associated receivers and detectors are discussed in detail in Section 2.5.

1.2.3.1 Construction

The large dishes that we have been discussing pose major problems in their construction. Both the gravitational and wind loads on the structure can be very large, and shadowing of parts of the structure can lead to inhomogeneous heating and hence to expansion- and contraction-induced stresses.

The worst problem is due to wind, since its effect is highly variable. The force can be very large—1.5×10^6 N (150 t) for a 50 m dish facing directly into a gale-force wind, for example. A rough rule of thumb to allow scaling the degree of wind distortion between dishes of different sizes is that the deflection, Δe, is given by

$$\Delta e \propto \frac{D^3}{A} \tag{1.87}$$

where
D is the diameter of the dish
A is the cross-sectional area of its supporting struts

Thus, doubling the diameter of the design of a dish would require the supporting members' sizes to be increased by a factor of eight if it were to still work at the same wavelength. There are only two solutions to the wind problem; to enclose the dish, or to cease using it when the wind load becomes too great. Some smaller dishes, especially those working at short wavelengths where the physical constraints on the surface accuracy are most stringent, are enclosed in radomes, or space-enclosing structures built from nonconducting materials. But this is not usually practicable for the larger dishes. These must generally cease operating and be parked in their least wind-resistant mode once the wind speed rises above $10-15$ m s^{-1}.

The effects of gravity are easier to counteract. The problem arises from the varying directions and magnitudes of the loads placed upon the reflecting surface as the orientation of the telescope changes. Three approaches have been tried successfully for combating the effects of gravity. The first is the brute force approach, whereby the dish is made so rigid that its deformations remain within the surface accuracy limits. Except for very small dishes, this is an impossibly expensive option. The second approach is to compensate for the changing gravity loads as the telescope moves. The surface may be kept in adjustment by systems of weights, levers, guy ropes, springs, etc. or more recently by computer-controlled hydraulic jacks. The final approach is much more subtle and is termed the homological transformation system. The dish is allowed to deform, but its supports are designed so that its new shape is still a paraboloid, although in general an altered one, when the telescope is in its new position. The only active intervention that may be required is to move the feed antenna to keep it at the foci of the changing paraboloids.

There is little that can be done about inhomogeneous heating, other than painting all the surfaces white so that the absorption of the heat is minimized. Fortunately, it is not usually a serious problem in comparison with the first two.

The supporting framework of the dish is generally a complex, cross-braced skeletal structure, whose optimum design requires a computer for its calculation. This framework is then usually placed onto an alt–az mounting (Section 1.1), since this is generally the cheapest and simplest structure to build, and also because it restricts the gravitational load variations to a single plane, and so makes their compensation much easier. A few, usually quite small, radio telescopes do have equatorial mountings, though.

Exercise 1.6

Show that the BWHP of a collinear array with n dipoles is given by

$$\mathrm{BWHP} \approx 2\sin^{-1}\left(\frac{6n^2 - 3}{2n^4\pi^2 - \pi^2}\right)^{1/2}$$

when n is large.

Exercise 1.7

Calculate the dimensions of a Mills cross to observe at a wavelength of 0.3 m with a Rayleigh resolution of 0.25°.

Exercise 1.8

Show that the maximum number of separable sources using the Rayleigh criterion, for a 60 m dish working at a wavelength of 0.1 m, is about 3.8×10^5 over a complete hemisphere.

1.3   X-RAY AND γ-RAY DETECTION

1.3.1   Introduction

The electromagnetic spectrum comprises radiation of an infinite range of wavelengths, but the intrinsic nature of the radiation is unvarying. There is, however, a tendency to regard the differing wavelength regions from somewhat parochial points of view, and this tends to obscure the underlying unity of the processes that may be involved. The reasons for these attitudes are many; some are historical hangovers from the ways in which the detectors for the various regions were developed, others are more fundamental in that different physical mechanisms predominate in the radiative interactions at different wavelengths. Thus, high-energy γ rays may interact directly with nuclei, at longer wavelengths, we have resonant interactions with atoms and molecules producing electronic, vibrational, and rotational transitions, while in the radio region, currents are induced directly into conductors. But primarily the reason may be traced to the academic backgrounds of the workers involved in each of the regions. Because of the earlier reasons, workers involved in investigations in one spectral region will tend to have different bias in their backgrounds compared with workers in a different spectral region. Thus, there will be different traditions, approaches, systems of notation, etc. with a consequent tendency to isolationism and unnecessary failures in communication. The discussion of the point-source image in terms of the Airy disk and its fringes by optical astronomers and in terms of the polar diagram by radio astronomers as already mentioned is one good example of this process, and many more exist. It is impossible to break out from the straitjacket of tradition completely, but an attempt to move toward a more unified approach has been made in this work by dividing the spectrum much more broadly than is the normal case. We only consider three separate spectral regions, within each of which the detection techniques bear at least a familial resemblance to each other. The overlap regions are fairly diffuse with some of the techniques from each of the major regions being applicable. We have already discussed two of the major regions, radio and microwaves, and optical and infrared. The third region, x-rays and γ rays, is the high-energy end of the spectrum and is the most recent area to be explored. This region also overlaps to a very considerable extent, in the nature of its detection techniques, with the cosmic rays discussed in Section 1.4. None of the radiation discussed in this section penetrates down to ground level, so its study had to await the availability of observing platforms in space, or near the top of the earth's atmosphere. Thus, significant work on these photons has only been possible since the 1960s, and many of the detectors and observing techniques are still under development.

The high-energy spectrum is fairly arbitrarily divided into

- Extreme ultraviolet (EUV or XUV region): 10–100 nm wavelengths (12–120 eV photon energies)

- Soft x-rays: 1–10 nm (120–1200 eV)

- X-rays: 0.01–1 nm (1.2–120 keV)

- Soft γ rays: 0.001–0.01 nm (120–1200 keV)

- γ rays: less than 0.001 nm (greater than 1.2 MeV)

We shall be primarily concerned with the last four regions in this section, that is, with wavelengths less than 10 nm and photon energies greater than 120 eV. (Note that the electron volt, eV, is 1.6×10^{-19} J and is a convenient unit for use in this spectral region and also when discussing cosmic rays in the next section.) The main production mechanisms for high-energy radiation include electron synchrotron radiation, the inverse Compton effect, free–free radiation, and pion decay, while the sources include the Sun, supernova remnants, pulsars, bursters, binary systems, cosmic rays, the intergalactic medium, galaxies, Seyfert galaxies, and quasars. Absorption of the radiation can be by ionization with a fluorescence photon or an Auger electron produced in addition to the ion and electron, by Compton scattering, or in the presence of matter, by pair production. This latter process is the production of a particle and its antiparticle, and not the pair production process discussed in Section 1.1 which was simply the excitation of an electron from the valence band. The interstellar absorption in this spectral region varies roughly with the cube of the wavelength, so that the higher energy radiation can easily pass through the whole galaxy with little chance of being intercepted. At energies under about 2 keV, direct absorption by the heavier atoms and ions can be an important process. The flux of the radiation varies enormously with wavelength. The solar emission alone at the lower energies is sufficient to produce the ionosphere and thermosphere on the earth. At 1 nm wavelength, for example, the solar flux is 5×10^9 photons m^{-2} s^{-1}, while the total flux from all sources for energies above 10^9 eV is only a few photons per square meter per day.

1.3.2   Detectors
1.3.2.1   Geiger Counters
The earliest detection of high-energy radiation from a source other than the Sun took place in 1962 when soft x-rays from a source that later became known as Sco X-1 were detected by large area Geiger counters flown on a sounding rocket.

The principle of the Geiger counter is well known. Two electrodes inside an enclosure are held at such a potential difference that a discharge in the medium filling the enclosure is on the point of occurring. The entry of ionizing radiation triggers this discharge, resulting in a pulse of current between the electrodes that may then be amplified and detected. The electrodes are usually arranged as the outer wall of the enclosure containing the gas and as a central coaxial wire (Figure 1.68). The medium inside the tube is typically argon at a low pressure with a small amount of an organic gas, such as alcohol vapor, added. The electrons produced in the initial ionization are accelerated toward the central electrode by the applied potential; as these electrons gain energy, they cause further ionization, producing more electrons, which in turn are accelerated toward

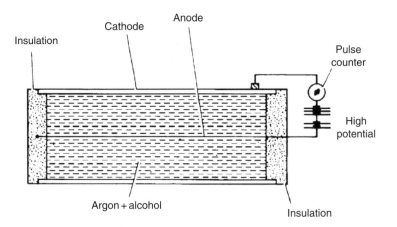

FIGURE 1.68 Typical arrangement for a Geiger counter.

the central electrode, and so on. The amplification factor can be as high as 10^8 electrons arriving at the central electrode for every one in the initial ionization trail. The avalanche of electrons rapidly saturates, so that the detected pulse is independent of the original energy of the photon. This is a serious disadvantage if the device is used as the primary detector and not as a trigger for a more sophisticated detector (Section 1.4), and so Geiger counters have been replaced by proportional counters (see Section 1.3.2.2) for this purpose. Another disadvantage of Geiger counters that also applies to many of the other detectors discussed in this and the following sections is that a response to one event leaves the detector inoperative for a short interval, known as the dead time. In the Geiger counter, the cause of the dead time is that a discharge lowers the potential between the electrodes, so that it is momentarily insufficient to cause a second avalanche of electrons should another x-ray enter the device. The length of the dead time is typically 200 μs.

1.3.2.2 Proportional Counters
These devices are very closely related to Geiger counters and are also known as gas-filled ionization detectors. They are in effect Geiger counters operated at less than the trigger voltage. By using a lower voltage, saturation of the pulse is avoided, and its strength is then proportional to the energy of the original interaction. The gain of the system operated in this way is reduced to about 10^4 or 10^5, but it is still sufficient for further detection by conventional amplifiers, etc. Provided that all the energy of

the ionizing radiation is absorbed within the detector, its original total energy may be obtained from the strength of the pulse, and we have a proportional counter. At low photon energies, a window must be provided for the radiation. These are typically made from thin sheets of plastic, mica, or beryllium, and absorption in the windows limits the detectors to energies above a few hundred electron volts. When a window is used, the gas in the detector has to be continuously replenished because of losses by diffusion through the window. At high photon energies, the detector is limited by the requirement that all the energy of the radiation be absorbed within the detector's enclosure. To this end, proportional counters for high-energy detection may have to be made very large. About 30 eV on average are required to produce one ion–electron pair, so that a 1 keV photon produces about 36 electrons and a 10 keV photon about 360 electrons. The spectral energy resolution to 2.5 standard deviations from the resulting statistical fluctuations of the electron numbers is thus about 40% at 1 keV and 12% at 10 keV. The quantum efficiencies of proportional counters approach 100% for energies up to 50 keV.

The position of the interaction of the x-ray along the axis of the counter may be obtained through the use of a resistive anode. The pulse is abstracted from both ends of the anode and a comparison of its strength and shape from the two ends then leads to a position of the discharge along the anode. The anode wires are very thin, typically 20 μm across, so that the electric field is most intense very near to the wire. The avalanche of electrons thus develops close to the wire, limiting its spread, and giving a precise position. The concept may easily be extended to a two-dimensional grid of anodes to allow genuine imaging. Spatial resolutions of about a tenth of a millimeter are possible. In this form, the detector is called a position-sensitive proportional counter. They are also known as multi-wire chambers, especially in the context of particle physics (see also drift chambers, Section 1.5.7.1).

Many gases can be used to fill the detector; argon, methane, xenon, carbon dioxide, and mixtures thereof at pressures near that of the atmosphere, are among the commonest ones. The inert gases are to be preferred since there is then no possibility of the loss of energy into the rotation or vibration of the molecules.

Position-sensitive proportional counters were widely used on x-ray space missions until the 1990s. The last major observatory to use them was *Röntgen Satellite (ROSAT)*, which carried three such detectors filled with argon, xenon, and methane and which operated until 1999. The *Rossi X-ray*

Timing Explorer (*RXTE*) spacecraft, launched in 1995 and still operating at the time of writing also uses an array of proportional counters.

1.3.2.3 Scintillation Detectors

The ionizing photons do not necessarily knock out only the outermost electrons from the atom or molecule with which they interact. Electrons in lower energy levels may also be removed. When this happens, a hole is left behind into which one of the higher electrons may drop, with a consequent emission of radiation. Should the medium be transparent to this radiation, the photons may be observed, and the medium used as an x-ray detector. Each interaction produces a flash or scintilla of light, from which the name of the device is obtained. There are many materials that are suitable for this application. Commonly used ones include sodium iodide doped with an impurity such as thallium, and cesium iodide doped with sodium or thallium. For these materials, the light flashes are detected by a photomultiplier (Figure 1.69). There is no dead time for a scintillation

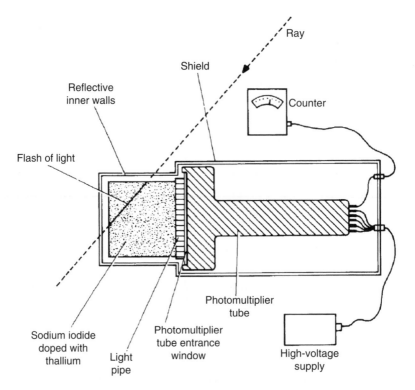

FIGURE 1.69 Schematic experimental arrangement of a scintillation counter.

detector, and the strength of the flash depends somewhat upon the original photon energy, so that some spectral resolution is possible. The noise level, however, is quite high since only about 3% of the x-ray's energy is converted into detectable radiation, with a consequent increase in the statistical fluctuations in their numbers. The spectral resolution is thus about 6% at 1 MeV. Sodium iodide or cesium iodide are useful for x-ray energies up to several hundred kilo-electron-volt, organic scintillators such as stilbene ($C_{14}H_{14}N_2$) can be used up to 10 MeV and bismuth germanate (BGO—$Bi_4Ge_3O_{12}$) for energies up to 30 or more MeV. Organically doped plastics are also useful and are planned to be used for the Telescope Array Cosmic Ray Observatory in Utah (Section 1.4). Both sodium iodide and bismuth germanate are to be used for the burst monitor on board the *Gamma-Ray Large Area Space Telescope* (*GLAST*), now scheduled for launch in 2008, to provide continuous detection from a few keV to 25 MeV. While gadolinium silicate and bismuth germinate scintillators are used on the *Suzaku* spacecraft (launched 2005) for observing over the 30–600 keV region.

Discrimination of the x-ray's arrival direction can be obtained by using sodium iodide and cesium iodide in two superimposed layers. The decay time of the pulses differs between the two compounds so that they may be separately identified, and the direction of travel of the photon inferred. This arrangement for a scintillation detector is frequently called a "phoswich" detector. Several gases such as argon, xenon, nitrogen, and their mixtures can also be used as scintillators, and combined with an optical system to produce another imaging device. The *RXTE* spacecraft has two clusters of 4 NaI/CsI phoswich detectors that cover the 15–250 keV region.

X-rays and cosmic rays (Section 1.4) may be distinguished by the differences in their resulting pulses shapes. X-rays are rapidly absorbed, and their pulses are sharp and brief. Cosmic ray particles will generally have a much longer path and so their pulses will be comparatively broader and smoother than the x-ray pulse.

1.3.2.4 Gas Scintillation Proportional Counters

A combination of the above two types of detector leads to a significant improvement in the low-energy spectral resolution. Resolutions as good as 8% at 6 keV have been achieved in practice with these devices. The x-radiation produces ion–electron pairs in an argon- or a xenon-filled chamber. The electrons are then gently accelerated until they cause scintillations of their own in the gas. These scintillations can then be observed by a conventional scintillation counter system. These have been favored detectors for launch on several of the more recent x-ray satellites because of their good spectral and positional discrimination, though in some cases, their lifetimes in orbit have been rather short.

1.3.2.5 Charge-Coupled Devices

The details of the operating principles for CCDs for optical detection are to be found in Section 1.1. They are also, however, becoming increasingly widely used as primary detectors at EUV and x-ray wavelengths. CCDs become insensitive to radiation in the blue and ultraviolet parts of the spectrum because of absorption in the electrode structure on their surfaces. They regain sensitivity at shorter wavelengths as the radiation is again able to penetrate that structure ($\lambda < 10$ nm or so, energy >120 keV). As with optical CCDs, the efficiency of the devices may be improved by using a very thin electrode structure, an electrode structure that only partially covers the surface (virtual-phase CCDs), or by illuminating the device from the back. Typically, 3.65 eV of energy is needed to produce a single electron–hole pair in silicon. So unlike optical photons, which can only produce a single such pair, the x-ray photons can each produce many pairs. The CCDs are thus used in a photon-counting mode for detecting x-rays and the resulting pulses are proportional to the x-ray's energy, giving CCDs an intrinsic spectral resolution (Section 4.1) of around 10–50. X-ray multi-mirror (XMM) Newton, for example, uses three CCDs that have an open structure and detect over the 0.15–15 keV region. While the *Chandra* spacecraft uses CCDs with 24 μm pixel size giving 0.5″ resolution in its advanced CCD imaging spectrometer (ACIS). *Suzaku* has four 1k × 1k CCDs observing from 0.2 to 12 keV, while *Swift*, launched in 2004 to find GRBs, uses a 0.6k × 0.6k CCD to cover from 0.2 to 10 keV. The reader is referred to Section 1.1.8 for further details of CCDs.

1.3.2.6 Superconducting Tunnel Junction Detectors

The details of the operating principles of STJs may be found in Section 1.1.11 They are however also sensitive to x-rays, and indeed their first applications were for x-ray detection. Since an x-ray breaks about a thousand times as many Cooper pairs as an optical photon, their spectral discrimination is about 0.1%. The need to cool them to below 1 K, and the small sizes of the arrays currently available means that they have yet to be flown on a spacecraft.

1.3.2.7 Compton Interaction Detectors

Very-high-energy photons produce electrons in a scintillator through the Compton effect, and these electrons can have sufficiently high energies to produce scintillations of their own. Two such detectors separated by a meter or so can provide directional discrimination when used in conjunction with pulse analyzers and time-of-flight measurements to eliminate other unwanted interactions.

1.3.2.8 Spark Detectors

These are discussed in more detail in Section 1.4.3.1 (Figure 1.85). They are used for the detection of γ rays of energies above 20 MeV, when the sparks follow the ionization trail left by the photon. Alternatively, a layer of tungsten or some similar material may be placed in the spark chamber. The γ ray then interacts in the tungsten to produce an electron–positron pair, and these in turn can be detected by their trails in the spark chamber. Angular resolutions of 2°, energy resolutions of 50%, and detection limits of 10^{-3} photons m^{-2} s^{-1} are currently achievable.

1.3.2.9 Čerenkov Detectors

Čerenkov radiation and the resulting detectors are referred to in detail in Section 1.4.3.3. For x-ray and γ radiation, their interest lies in the detection of particles produced by the Compton interactions of the very-high-energy photons. If these particles have a velocity higher than the velocity of light in the local medium, then Čerenkov radiation is produced. Čerenkov radiation produced in this manner high in the earth's atmosphere is detectable. Thus, the CANGAROO-II (collaboration of Australia and Nippon for a Gamma Ray Observatory in the outback) Čerenkov telescope is a 10 m optical telescope in Australia formed from one hundred and fourteen 0.8 m mirrors. It has a field of view of 2.5° and observes the outer reaches of the earth's atmosphere with 552 photomultipliers, detecting TeV γ rays from active galaxies. Similarly, the very energetic radiation imaging telescope array system (VERITAS), located in Arizona, employs four 12 m reflectors, each made up from 350 mirror segments, while the major atmospheric gamma imaging Čerenkov telescope (MAGIC) on La Palma uses a 17 m multi-segment reflector and the high-energy stereoscopic system (HESS) in Namibia has four 12 m dishes formed from 382 segments together with cameras that each have 960 photomultipliers.

Very-high-energy primary γ rays (10^{15} eV and more) may also produce cosmic ray showers and so be detectable by the methods discussed in Section 1.4.5, especially through the fluorescence of atmospheric nitrogen.

1.3.2.10 Solid-State Detectors

There are a number of different varieties of these detectors including CCDs and STJs (see Sections 1.3.2.5, 1.3.2.6, and 1.1), so that suitable ones may be found for use throughout most of the x- and γ-ray regions.

Germanium may be used as a sort of solid proportional counter. A cylinder of germanium cooled by liquid nitrogen is surrounded by a cylindrical cathode and has a central anode (Figure 1.70). A γ ray scatters off electrons in the atoms until its energy has been consumed in electron–hole pair production. The number of released electrons is proportional to the energy of the γ ray, and these are attracted to the anode where they may be detected. The spectral resolution is very high—0.2% at 1 MeV—so that detectors of this type are especially suitable for γ-ray line spectroscopy. Other materials that may replace the germanium include germanium doped

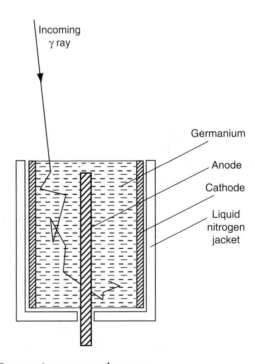

FIGURE 1.70 Germanium γ-ray detector.

with lithium, cadmium telluride, and mercury-iodine. At lower energies (0.4–4 keV), silicon-based solid-state detectors may be used similarly. Their energy resolution ranges from 4% to 30%.

Closely related detectors consist of a thick layer (up to 5 mm) of a semiconductor, such as silicon (known as silicon strip detectors or SSDs) or more recently cadmium (about 90%) plus zinc (about 10%) doped with tellurium (known as cadmium–zinc–telurium [CZT] detectors), since this latter material has a higher stopping power for the radiation. The semiconductor has a high resistivity by virtue of being back biased to a 100 V or so. The silicon surface is divided into 20–25 μm wide strips by thin layers doped with boron. Aluminum electrodes are then deposited on the surface between these layers. When an x-ray, γ ray, or charged particle passes into the material, electron–hole pairs are produced via the photoelectric effect, Compton scattering, or collisional ionization. The pulse of current is detected as the electron–hole pairs are collected at the electrodes on the top and bottom of the semiconductor slice. Processing of the pulses can then proceed by any of the usual methods, and a positional accuracy for the interaction of ±10 μm can be achieved.

Solid-state detectors have several advantages that suit them particularly for use in satellite-borne instrumentation—wide range of photon energies detected (from 1 keV to over 1 MeV), simplicity, reliability, low-power consumption, high stopping power for the radiation, room temperature operation (some varieties), no entrance window needed, high counting rates possible, etc. They also have an intrinsic spectral sensitivity since, provided that the photon is absorbed completely, the number of electron–hole pairs produced is proportional to the photon's energy. About 5 eV are need to produce an electron–hole pair in CZT, so that the spectral sensitivity to 2.5 standard deviations is potentially 18% at 1 keV and 0.5% at 1 MeV. The main disadvantages of these detectors are that their size is small compared with many other detectors, so their collecting area is also small, and that unless the photon is stopped within the detector's volume, the total energy cannot be determined. This latter disadvantage, however, also applies to most other detectors.

Solid-state detectors are increasing in popularity and recent or forthcoming examples include the International Gamma Ray Astrophysics Laboratory (INTEGRAL), launched in 2002 with a planned 10 year operating life and which carries a spectrometer using germanium detectors. The *Swift* spacecraft (see also Section 1.3.4.1) launched in 2004 carries 256 modules each with 128 CZT detectors in its burst alert telescope

(BAT), while the balloon-borne high-energy focusing telescope (HEFT) uses six 1000 pixel CZTs and the nuclear spectroscopic telescope array (NuSTAR) (due to start operating in 2011) is also planned to have two CZT detector arrays.

SSDs are to be used for the large area telescope (LAT) on board the *GLAST* spacecraft. These provide positional and directional information on the γ ray. The silicon strips are arranged orthogonally in two layers. A γ ray, or the electron–positron pair produced within a layer of lead by the γ ray, will be detected within one of the strips in each layer. The position of the ray is then localized to the crossover point of the two strips. By using a number of such pairs of layers of silicon strips piled on top of each other, the direction of the γ ray can also be determined. The LAT will use 16 such "towers," each comprising 16 layers of silicon strip pairs and lead plates, enabling it to determine γ-ray source positions to within 0.5′ to 5′.

A rather different type of solid-state detector, the micro-calorimeter, has recently been flown on the *Suzaku* spacecraft and is planned for use on the Constellation-X and the x-ray evolving universe spectrometer (XEUS) missions expected to be launched in 2010 and 2015, respectively. Micro-calorimeters have good intrinsic spectral resolution (R between 200 and 1000, see Section 4.1) and operate by detecting the change in temperature of an absorber when it captures an x-ray photon. So far their principal use has been for low- to medium-energy x-ray detection (up to 1 keV). There are three components in a micro-calorimeter: an absorber, a temperature sensor, and a thermal sink. The energy of the x-ray photon is converted into heat within the absorber and this change is picked up by the sensor. There is then a weak thermal link to the heat sink so that the temperature of the absorber gradually returns to its original value. Typical response times are a few milliseconds with up to 80 ms recovery time, so micro-calorimeters are not suitable for detecting high fluxes of x-rays. The micro-calorimeter needs to be cooled so that its stored thermal energy is small compared with that released from the x-ray photon. A wide variety of materials can be used for all three components of the device with the main requirement for the absorber being a low thermal capacity and a high stopping rate for the x-rays. Micro-calorimeters can be formed into arrays—Suzaku, for example, carried a 32-pixel micro-calorimeter comprising mercury telluride absorbers combined with silicon thermistors and cooled to 60 mK. Arrays of over a thousand pixels are currently being planned with TES devices as their temperature sensors.

Another type of detector altogether is a possibility for the future. This is a solid analogue of the nuclear physicists' cloud or bubble chamber. A superconducting solid would be held at a temperature fractionally higher than its critical temperature, or in a nonsuperconducting state and at a temperature slightly below the critical value. Passage of an ionizing ray or particle would then precipitate the changeover to the opposing state, and this in turn could be detected by the change in a magnetic field (see TES detectors, Section 1.1.13.2).

1.3.2.11   Micro-Channel Plates

For EUV and low-energy x-ray amplification and imaging, there is an ingenious variant of the photomultiplier (Section 1.1)—the MCP. The devices are also known as Multi Anode Micro channel-plate Arrays (MAMAs). A thin plate is pierced by numerous tiny holes, each perhaps only about 10 μm across or less. Its top surface is an electrode with a negative potential of some few thousand volts with respect to its base. The top is also coated with a photoelectron emitter for the x-ray energies of interest. An impinging photon releases one or more electrons that are then accelerated down the tubes. There are inevitably collisions with the walls of the tube during which further electrons are released, and these in turn are accelerated down the tube and so on (Figure 1.71). As many as 10^4 electrons can be produced for a single photon, and this may be increased to 10^6 electrons in future devices when the problems caused by ion feedback are reduced. The quantum efficiency can be up to 20%. The electrons spray out of the bottom of each tube, where they may be detected by a variety of the more conventional imaging systems, or they may be fed into a second MCP for further amplification. Early versions of these devices often used curved channels (Figure 1.71). Modern versions now employ two plates with straight channels, the second set of channels being at an angle to the direction of the first set (known as a chevron MCP). The plates are currently manufactured from a billet of glass that has an acid-resisting shell surrounding an acid soluble core. The billet is heated and drawn out until it forms a glass "wire" about 1 mm across. The wire is cut into short sections and the sections stacked to form a new billet. That billet in turn is heated and drawn out, and so on. The holes in the final plate are produced by etching away the acid soluble portions of the glass. Holes down to 6 μm in diameter may be produced in this fashion and the fraction of the plate occupied by the holes can reach 65%. Plates can be up to 0.1 m^2. For example, the high-resolution x-ray camera on board the *Chandra*

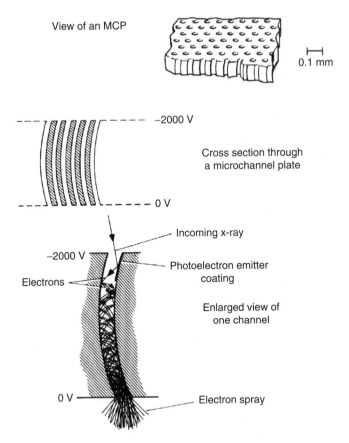

FIGURE 1.71 Schematic view of the operation of an MCP.

spacecraft uses a 93 mm^2 chevron MCP detector, with 69 million 10 μm holes, and can provide a resolution of 0.5″. A single x-ray results in about 30 million electrons that are collected by a grid of wires at the exit holes from the second stage plate. Other spacecraft that have carried MCPs include *Einstein* and *ROSAT*. MCPs can also be used in optical and near ultraviolet; however, they are then used simply to amplify the photoelectrons produced from a photocathode deposited onto the entrance window of the detector, and are thus similar to image intensifiers (Section 2.3).

The MCPs have much in common with micropore optics, which are a type of lobster-eye focusing collimator and which are discussed below.

1.3.2.12 Future Possibilities

The Quatratran detector, discussed as an infrared detector in Section 1.1.13 may also be useable for x-ray detection. Additionally, thick (20 μm) films of conjugated polymers have recently been shown to develop photocurrents when illuminated by x-rays. These are Schottky-type devices that seem likely to find medical applications, but it is not yet clear if they will be of much use for x-ray astronomy.

1.3.3 Shielding

Very few of the detectors that we have just reviewed are used in isolation. Usually several will be used together in modes that allow the rejection of information on unwanted interactions. This is known as active or anti-coincidence shielding of the detector. The range of possible configurations is very wide, and a few examples will suffice to give an indication of those in current use.

1. Germanium solid-state detectors must intercept all the energy of an incoming photon if they are to provide a reliable estimate of its magnitude. If the photon escapes, then the measured energy will be too low. A thick layer of a scintillation crystal therefore surrounds the germanium. A detection in the germanium simultaneously with two in the scintillation crystal is rejected as an escapee.

2. Solid angle viewed by a germanium solid-state detector can be limited by placing the germanium at the bottom of a hole in a scintillation crystal. Only those detections not occurring simultaneously in the germanium and the crystal are used and these are the photons that have entered down the hole. Any required degree of angular resolution can be obtained by sufficiently reducing the size of the entrance aperture.

3. Spark counters are surrounded by scintillation counters and Čerenkov detectors to discriminate between γ-ray events and those due to cosmic rays.

4. Sodium iodide scintillation counter may be surrounded, except for an entrance aperture, by one formed from cesium iodide to eliminate photons from the unwanted directions.

5. *GLAST* spacecraft will use segmented plastic scintillator tiles for its anti-coincidence shielding. These will be read out by photomultipliers.

Recently, nuclear power sources on board other spacecraft have caused interference with γ-ray observations. These pulses have to be separated from the naturally occurring ones by their spectral signatures.

Passive shields are also used, and these are just layers of an absorbing material that screen out unwanted rays. They are especially necessary for the higher energy photon detectors to reduce the much greater flux of lower energy photons. The mass involved in an adequate passive shield, however, is often a major problem for spacecraft-based experiments with their very tight mass budgets arising from their launcher capabilities.

1.3.4 Imaging

Imaging of high-energy photons is a difficult task because of their extremely penetrating nature. Normal designs of telescope are impossible since in a reflector the photon would just pass straight through the mirror, while in a refractor it would be scattered or unaffected rather than refracted by the lenses. At energies under a few kilo-electron-volts, forms of reflecting telescope can be made which work adequately, but at higher energies, the only options are occultation, collimation, and coincidence detection.

1.3.4.1 Collimation

A collimator is simply any device that physically restricts the field of view of the detector without contributing any further to the formation of an image. The image is obtained by scanning the system across the object. Almost all detectors are collimated to some degree, if only by their shielding. The simplest arrangement is a series of baffles (Figure 1.72) that may be formed into a variety of configurations, but all of which follow the same basic principle. Their general appearance leads to their name, honeycomb collimator, even though the cells are usually square rather than hexagonal. They can restrict angles of view down to a few minutes of arc, but they are limited at low energies by reflection off their walls and at high energies by penetration of the radiation through the walls. At high energies, the baffles may be formed from a crystal scintillator and pulses from there used to reject detections of radiation from high inclinations (cf. active shielding).

At the low energies, the glancing reflection of the radiation can be used to advantage, and a more truly imaging collimator produced. This is called a "lobster-eye" focusing collimator (also called micropore optics), and is essentially a honeycomb collimator curved into a portion of a sphere (Figure 1.73), with a position-sensitive detector at its focal surface. The imaging is not of very high quality, and there is a high background from the unreflected rays and the doubly reflected rays (the latter not shown in

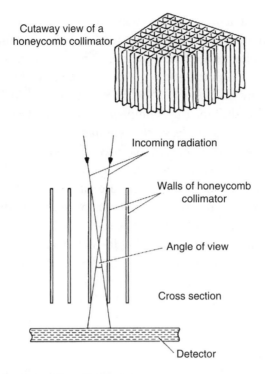

FIGURE 1.72 Honeycomb collimator.

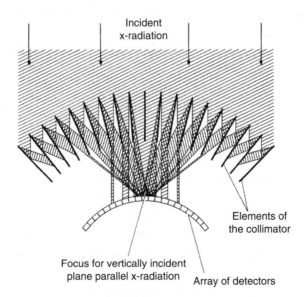

FIGURE 1.73 Cross section through a "lobster-eye" focusing wide-angle x-ray collimator.

Figure 1.73). But it is a cheap system to construct compared with the others that are discussed later in this section, and it has the potential for covering very wide fields of view (tens of degrees) at very high resolutions (seconds of arc). In practice, the device is constructed by fusing together thousands of tiny glass rods in a matrix of a different glass. The composite is then heated and shaped by slumping onto a spherical mould and the glass rods etched away to leave the required hollow tubes in the matrix glass. This is the same process as that used to produce MCPs (see above), and the only difference is that for a lobster-eye collimator, the channels are of square cross section, whereas they are circular in MCPs. Multi-layer coatings (see below) may be applied to enhance reflectivity and increase the waveband covered. Although lobster-eye collimators have yet to be used in space, an instrument (LOBSTER—not an acronym) that will have six x-ray telescopes each with 60 individual lobster-eye collimators is currently being designed at the University of Leicester, possibly for use on the International space station (ISS). While the same university is also proposing a lunar-based x-ray telescope, to be called MagEX (magnetosheath explorer in x-rays) for observing the earth's magnetosheath, also using a lobster-eye imager.

Another system that is known as a modulation collimator or Fourier transform telescope uses two or more parallel gratings that are separated by a short distance (Figure 1.74). Since the bars of the gratings alternately obscure the radiation and allow it to pass through, the output as the system scans a point source is a sine wave (Figure 1.74). The resolution is given by the angle α

$$\alpha = \frac{d}{s} \tag{1.88}$$

To obtain unambiguous positions for the sources, or for the study of multiple or extended sources, several such gratings of different resolutions are combined. The image may then be retrieved from the Fourier components of the output (cf. Section 2.5.5). Two such grating systems at right angles can give a two-dimensional image. With these systems, resolutions of a few tens of seconds of arc can be realized even for the high-energy photons.

A third type of collimating imaging system is a simple pinhole camera. A position-sensitive detector such as a resistive anode proportional counter or an MCP is placed behind a small aperture. The quality of the image

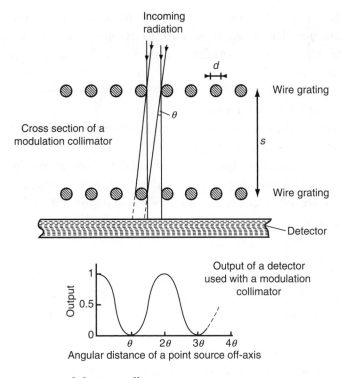

FIGURE 1.74 A modulation collimator.

is then just a function of the size of the hole and its distance in front of the detector. Unfortunately, the small size of the hole also gives a low flux of radiation so that such systems are limited to the brightest objects. A better system replaces the pinhole with a mask formed from clear and opaque regions. The pattern of the mask is known, so that when sources cast shadows of it onto the detector, their position and structure can be reconstituted in a similar manner to that used for the modulation colli-mator. The technique is known as coded mask imaging, and resolutions of 10′ or better can be achieved. Since only half the aperture is obscured, the technique obviously uses the incoming radiation much more efficiently than the pinhole camera. A somewhat related technique known as Hadamard mask imaging is discussed in Section 2.4. The disadvantage of coded mask imaging is that the detector must be nearly the same size as the mask. The *Swift* spacecraft, for example, uses a coded mask with an area of 3.2 m^2, allied with nearly 33,000 CZT detectors (see Section 1.3.2.10) covering an area of 0.52 m^2 in its BAT The BAT detects γ-ray bursts and it has a field of view of 2 sr, thus it is able to pinpoint the position of a γ-ray burst to within 4′ within 15 s of its occurrence. The image from a coded aperture telescope is extracted from the data by a cross correlation between the detected pattern and the mask pattern.

1.3.4.2 Coincidence Detectors

A telescope, in the sense of a device that has directional sensitivity, may be constructed for use at any energy, and with any resolution, by using two or more detectors in a line, and by rejecting all detections except those which occur in both detectors and separated by the correct flight time. Two separated arrays of detectors can similarly provide a two-dimensional imaging system.

1.3.4.3 Occultation

Although it is not a technique that can be used at will on any source, the occultation of a source by the moon or other object can be used to give very precise positional and structural information. The use of occultations in this manner is discussed in detail in Section 2.7. The technique is less important now than in the past because other imaging systems are available, but it was used in 1964, for example, to provide the first indication that the x-ray source associated with the Crab Nebula was an extended source.

1.3.4.4 Reflecting Telescopes

At energies below about 100 keV, photons may be reflected with up to 50% efficiency off metal surfaces, when their angle of incidence approaches 90°. Mirrors in which the radiation just grazes the surface (grazing incidence optics) may therefore be built and configured to form reflecting telescopes. The angle between the mirror surface and the radiation has to be very small though: less than 2° for 1 keV x-rays, less than 0.6° for 10 keV x-rays, and less than 0.1° for 100 keV x-rays. Several systems have been devised, but the one which has achieved most practical use is formed from a combination of annular sections of very deep paraboloidal and hyperboloidal surfaces (Figure 1.75), and known as a Wolter type I telescope after its inventor. Other designs that were also invented by Wolter are shown in Figure 1.76, and have also been used in practice, although less often than the first design. A simple paraboloid can also be used, though the resulting telescope then tends to be excessively long. The aperture of such telescopes is a thin ring, since only the

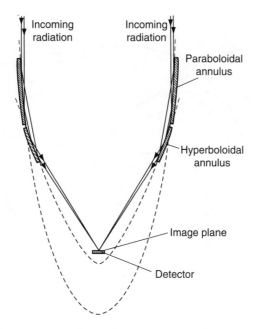

FIGURE 1.75 Cross section through a grazing incidence x-ray telescope.

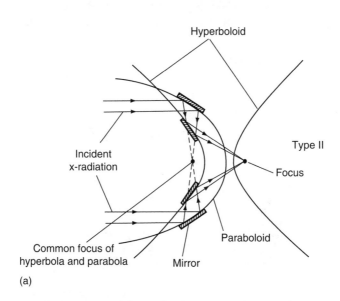

(a)

FIGURE 1.76 Cross sections through alternative designs for grazing incidence x-ray telescopes.

(*continued*)

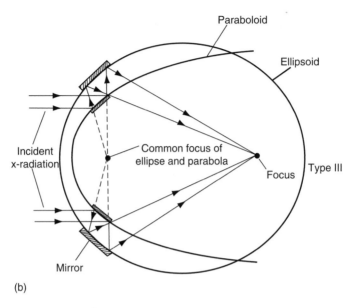

(b)

FIGURE 1.76 (continued)

radiation incident onto the paraboloidal annulus is brought to the focus. To increase the effective aperture, and hence the sensitivity of the system, several such confocal systems of differing radii may be nested inside each other (Figure 1.77). For the *XMM–Newton* spacecraft, a total of 58 such

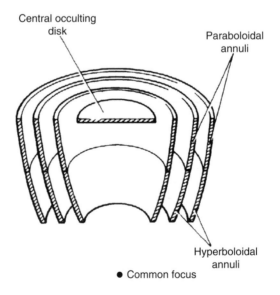

FIGURE 1.77 Section through a nested grazing incidence x-ray telescope.

nested telescope shells gave a total collecting area of about half a square meter. *Chandra* has 4 and *Swift* has 12 nested shells. A position-sensitive detector at the focal plane can then produce images with resolutions as high as a second of arc. The limit of resolution is due to surface irregularities in the mirrors, rather than the diffraction limit of the system. The irregularities are about 0.3 nm in size for the very best of the current production techniques, that is, comparable with the wavelength of photons of about 1 keV. The mirrors are produced by electro-deposition of nickel onto a mandrel of the inverse shape to that required for the mirror, the mirror shell then being separated from the mandrel by cooling. A coating of a dense metal such as gold, iridium, or platinum may be applied to improve the x-ray reflectivity. At energies of a few kilo-electron-volts, glancing incidence telescopes with low but usable resolutions can be made using foil mirrors. The incident angle is less than 1° so a hundred or more mirrors are needed. They may be formed from thin aluminum foil with a lacquer coating to provide the reflecting surface. The angular resolution of around 1′ means that the mirror shapes can be simple cones, rather than the paraboloids and hyperboloids of a true Wolter telescope and so fabrication costs are much reduced.

An alternative glancing incidence system, known as the Kirkpatrick–Baez design, that has fewer production difficulties is based upon cylindrical mirrors. Its collecting efficiency is higher but its angular resolution poorer than the three-dimensional systems discussed above. Two mirrors are used that are orthogonal to each other (Figure 1.78). The surfaces can

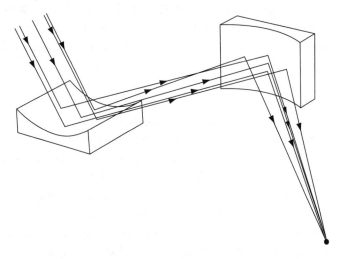

FIGURE 1.78 X-ray imaging by a pair of orthogonal cylindrical mirrors.

have a variety of shapes, but two paraboloids are probably the commonest arrangement. The mirrors can again be stacked in multiples of the basic configuration to increase the collecting area.

At lower energies, in the EUV and soft x-ray regions, near normal incidence reflection with efficiencies up to 20% is possible using multi-layer coatings. These are formed from tens, hundreds, or even thousands of alternate layers of, for example, tungsten and carbon, aluminum and gold, or magnesium and gold, each about 1 nm thick. The reflection is essentially monochromatic, with the wavelength depending upon the orientation of the crystalline structure of the layers, and upon their thickness. Reflection of several wavelengths may be accomplished by changing the thickness of the layers through the stack. The thickest layers are at the top and reflect the longest, least penetrating wavelengths, while further down, narrower layers reflect shorter wavelengths. Alternatively, several small telescopes may be used each with a different wavelength response, or as in the *TRACE* (*Transition Region and Corona Explorer*) spacecraft, different multi-layer coatings be applied in each of four quadrants of the optics. Telescopes of relatively conventional design are used with these mirrors and direct images of the Sun at wavelengths down to 4 nm can be obtained.

Multi-layer coatings and grazing incidence optics can be combined to improve reflectivity and extend the usage to higher energies—perhaps to 300 keV in the future. The coatings are alternate layers of high and low atomic number materials such as molybdenum and silicon or tungsten and boron carbide. The layers vary in thickness so that a wide range of x-ray wavelengths are reflected. HEFT uses 72 nested glass multi-layer coated mirror shells in each of its three telescopes and similar telescopes are planned for the NuSTAR mission. The balloon-borne InFOCμS (international focussing optics collaboration for μ-Crab sensitivity) instrument uses grazing incidence optics with multi-layer coatings of carbon and platinum.

At energies of tens or hundreds of kilo-electron-volts, Bragg reflection (as in the multi-layer coatings) and Laue diffraction can be used to concentrate the radiation and to provide some limited imaging. For example, the Laue diffraction pattern of a crystal comprises a number of spots into which the incoming radiation has been concentrated (Figure 1.79). With some crystals, such as germanium and copper, only a few spots are produced. A number of crystals can be mutually aligned so that one of the spots from each crystal is directed toward the same point (Figure 1.80), resulting in a crude type of lens. With careful design,

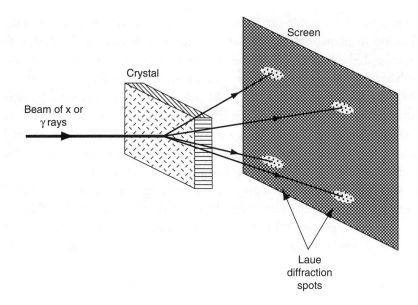

FIGURE 1.79 Laue diffraction of x and γ rays.

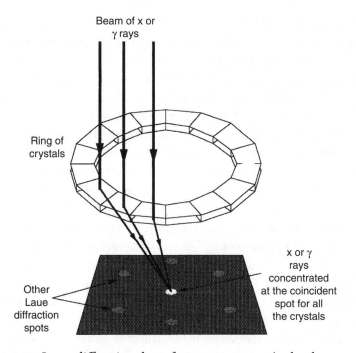

FIGURE 1.80 Laue diffraction lens for x or γ rays (only three crystals shown illuminated for clarity).

efficiencies of 25%–30% can be achieved at some wavelengths. As with the glancing incidence telescopes, several such ring lenses can be nested to increase the effective area of the telescope. A position-sensitive detector such as an array of germanium crystals (Figure 1.70) can then provide direct imaging at wavelengths down to 0.001 nm (1 MeV). A study for a possible future space mission using a Laue lens suggests that the focal length of the lens would be some 500 m. Thus, two spacecraft—one to carry the lens and one to carry the detectors—flying in tandem would be needed. X-rays with energies as high as 1 MeV might be imaged by such a system.

A completely different approach to imaging γ-ray sources is based upon cloud or bubble chambers. In these devices, the tracks of the γ rays are seen directly and can be followed back to their point of origin in the sky. Liquid xenon seems likely to be particularly useful at million electron volts energies since it has up to 50% detection efficiency and can provide information upon the spectral distribution and the polarization properties of the photons.

1.3.5 Resolution and Image Identification

At the low-energy end of the x-ray spectrum, resolutions and positional accuracies comparable with those of earth-based optical systems are possible, as we have seen. There is therefore usually little difficulty in identifying the optical counterpart of an x-ray source, should it be bright enough to be visible. The resolution rapidly worsens, however, as higher energies are approached. The position of a source that has been detected can therefore only be given to within quite broad limits. It is customary to present this uncertainty in the position of a source by specifying an error box. This is an area of the sky, usually rectangular in shape, within which there is a certain specified probability of finding the source. For there to be a reasonable certainty of being able to find the optical counterpart of a source unambiguously, the error box must be smaller than about 1 square minute of arc. Since high-energy photon imaging systems have resolutions measured in tens of minutes of arc or larger, such identification is not usually possible. The exception to this occurs when an unusual optical object lies within the error box. This object may then, rather riskily, be presumed also to be the x-ray source. Examples of this are the Crab Nebula and Vela supernova remnant which lie in the same direction as two strong sources at 100 MeV and are therefore assumed to be those sources, even though the uncertainties are measured in many degrees.

1.3.6 Spectroscopy

Many of the detectors that were discussed earlier are intrinsically capable of separating photons of differing energies. The STJ detector (Section 1.1.11 and above) has an intrinsic spectral resolution in the x-ray region that potentially could reach 10,000 (i.e., 0.0001 nm at 1 nm wavelength). However, the need to operate the devices at lower than 1 K seems likely to prevent their use on spacecraft for some time to come. Micro-calorimeters though, which also need to operate at low temperatures, have been flown on the *Suzaku* mission and are planned to be used on other spacecraft (see above). They have intrinsic spectral resolutions of up to 1000. At photon energies above about 10 keV, it is only this inherent spectral resolution that can provide information on the energy spectrum, although some additional resolution may be gained by using several similar detectors with varying degrees of passive shielding—the lower energy photons will not penetrate through to the most highly shielded detectors. At low and medium energies, the gaps between the absorption edges of the materials used for the windows of the detectors can form wideband filters. Devices akin to the more conventional idea of a spectroscope, however, can only be used at the low end of the energy spectrum and these are discussed below.

1.3.6.1 Grating Spectrometers

Gratings may be either transmission or grazing incidence reflection. The former may be ruled gratings in a thin metal film, which has been deposited onto a substrate transparent to the x-rays, or they may be formed on a pre-ruled substrate by vacuum deposition of a metal from a low angle. The shadow regions where no metal is deposited then form the transmission slits of the grating. The theoretical background for x-ray gratings is identical with that for optical gratings and is discussed in detail in Section 4.1. Typical transmission gratings have around 1000 lines mm^{-1}. The theoretical spectral resolution (Section 4.1) is between 10^3 and 10^4, but is generally limited in practice to 50–100 by other aberrations.

Reflection gratings are also similar in design to their optical counterparts. Their dispersion differs, however, because of the grazing incidence of the radiation. If the separation of the rulings is d, then from Figure 1.81, we may easily see that the path difference, ΔP, of two rays that are incident onto adjacent rulings is

$$\Delta P = d[\cos \theta - \cos (\theta + \phi)] \qquad (1.89)$$

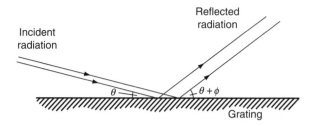

FIGURE 1.81 Optical paths in a grazing incidence reflection grating.

We may expand this via the Taylor series and neglect powers of θ and ϕ higher than two since they are small angles, to obtain

$$\Delta P = \tfrac{1}{2}d(\phi^2 - 2\theta\phi) \tag{1.90}$$

In the mth order spectrum, constructive interference occurs for radiation of wavelength λ if

$$m\lambda = \Delta P \tag{1.91}$$

so that

$$\phi = \left(\frac{2m\lambda}{d} + \theta^2\right)^{1/2} - \theta \tag{1.92}$$

and

$$\frac{d\phi}{d\lambda} = \left(\frac{m}{2d\lambda}\right)^{1/2} \tag{1.93}$$

where we have neglected θ^2, since θ is small. The dispersion for a glancing incidence reflection grating is therefore inversely proportional to the square root of the wavelength, unlike the case for near normal incidence, when the dispersion is independent of wavelength. The gratings may be plane or curved, and may be incorporated into many different designs of spectrometer (Section 4.2). Resolutions of up to 10^3 are possible for soft x-rays, but again this tends to be reduced by the effects of other aberrations. Detection may be by scanning the spectrum over a detector (or vice versa), or by placing a position-sensitive detector in the image plane so that the whole spectrum is detected in one go. The *XMM–Newton* spacecraft, for example, uses two reflection grating arrays. Each array has

182 individual gratings and they are incorporated into a Rowland-circle (Figure 4.6) spectroscope. CCDs are used as the detectors and spectral resolutions up to 800 are achieved over the 0.33–2.5 keV region.

1.3.6.2 Bragg Spectrometers

Background Distances ranging from 0.1 to 10 nm separate the planes of atoms in a crystal. This is comparable with the wavelengths of x-rays, and so a beam of x-rays interacts with a crystal in a complex manner. The details of the interaction were first explained by the Braggs (father and son). Typical radiation paths are shown in Figure 1.82. The path differences for rays such as *a*, *b*, and *c*, or *d* and *e*, are given by multiples of the path difference, ΔP, for two adjacent layers

$$\Delta P = 2d \sin \theta \qquad (1.94)$$

There will be constructive interference for path differences that are whole numbers of wavelengths. So that the reflected beam will consist of just those wavelengths, λ, for which this is true

$$M\lambda = 2d \sin \theta \qquad (1.95)$$

Spectrometer The Bragg spectrometer uses a crystal to produce monochromatic radiation of known wavelength. If a crystal is

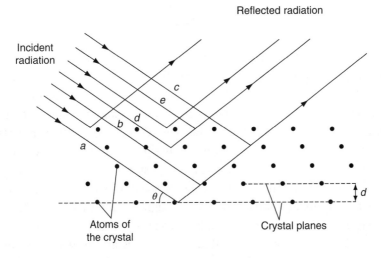

FIGURE 1.82 Bragg reflection.

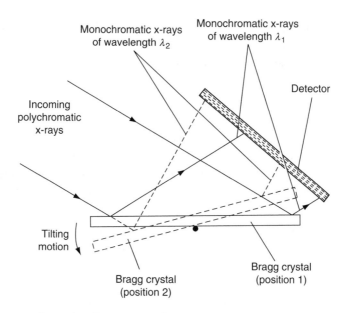

FIGURE 1.83 Scanning Bragg crystal x-ray spectrometer.

illuminated by a beam of x-rays of mixed wavelengths, at an approach angle, θ, then only those x-rays whose wavelength is given by Equation 1.95 will be reflected. The first-order reflection ($m = 1$) is by far the strongest and so the reflected beam will be effectively monochromatic with a wavelength of

$$\lambda_\theta = 2d \sin \theta \qquad (1.96)$$

The intensity of the radiation at that wavelength may then be detected with, say, a proportional counter, and the whole spectrum scanned by tilting the crystal to alter θ (Figure 1.83). An improved version of the instrument uses a bent crystal and a collimated beam of x-rays so that the approach angle varies over the crystal. The reflected beam then consists of a spectrum of all the wavelengths (Figure 1.84), and this may then be detected by a single observation using a position-sensitive detector. The latter system has the major advantage for satellite-borne instrumentation of having no moving parts and so has a significantly higher reliability. It also has good time resolution. High spectral resolutions are possible—up to 10^3 at 1 keV—but large crystal areas are necessary for good sensitivity, and this may present practical difficulties on a satellite. Many crystals may

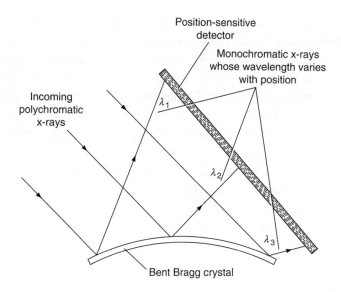

FIGURE 1.84 Bent Bragg crystal x-ray spectrometer.

be used. Among the commonest are lithium fluoride, lithium hydride, tungsten disulphide, graphite, and potassium acid phthalate (KAP). A Bragg spectrometer using six different types of crystals was flown on the *Einstein* spacecraft in 1978 to cover the range from 0.42 to 2.6 keV. Other spacecraft using Bragg spectrometers have included the *Solar Maximum Mission, Hinotori, Yohkoh*, and *CORONAS-F*.*

Many variants upon the basic spectrometer can be devised. It may be used as a monochromator and combined with a scanning telescope to produce spectroheliograms, for faint sources it may be adapted for use at the focus of a telescope and so on. Designs are rapidly changing and the reader desiring complete up-to-date information must consult the current literature.

1.3.7 Polarimetry

Bragg reflection of x-rays is polarization dependent. For an angle of incidence of 45°, photons that are polarized perpendicularly to the plane containing the incident and reflected rays will be reflected, while those polarized in this plane will not. Thus, a crystal and detector at 45° to the incoming radiation and which may be rotated around the optical axis will

---

* Russian acronym for complex orbital observatory for a study of the active Sun.

function as a polarimeter. The efficiency of such a system would be very low, however, due to the narrow energy bandwidth of Bragg reflection. This is overcome by using many randomly orientated small crystals. The crystal size is too small to absorb radiation significantly. If it should be aligned at the Bragg angle for the particular wavelength concerned, however, then that radiation will be reflected. The random orientation of the crystals ensures that overall, many wavelengths are reflected, and we have a polarimeter with a broad bandwidth.

A second type of polarimeter looks at the scattered radiation due to Thomson scattering in blocks of lithium or beryllium. If the beam is polarized, then the scattered radiation is asymmetrical, and this may be measured by surrounding the block with several pairs of detectors.

1.3.8 Observing Platforms

The Earth's atmosphere completely absorbs x- and γ-ray radiation, so that all the detectors and other apparatus discussed above, with the exception of the Čerenkov detectors, have to be lifted above at least 99% of the atmosphere. The three systems for so lifting equipment are balloons, rockets, and spacecraft. Spacecraft give the best results in terms of their height, stability, and the duration of the mission. Their cost is very high, however, and there are weight and space restrictions to be complied with; nonetheless, there have been quite a number of satellites, many of which have been mentioned above, launched for exclusive observation of the region of x- and γ-rays, and appropriate detectors have been included on a great many other missions as secondary instrumentation. Balloons can carry much heavier equipment far more cheaply than satellites, but only to a height of about 40 km, and this is too low for many of the observational needs of this spectral region. Their mission duration is also comparatively short being a few days to a week for even the most sophisticated of the self-balancing versions. Complex arrangements have to be made for communication and for retrieval of the payload because of the unpredictable drift of the balloon during its mission. The platform upon which the instruments are mounted has to be actively stabilized to provide sufficient pointing accuracy for the telescopes, etc. Thus, there are many drawbacks to set against the lower cost of a balloon. Sounding rockets are even cheaper, and they can reach heights of several hundred kilometers without difficulty. But their flight duration is measured only in minutes and the weight and size restrictions are far tighter even than for spacecraft. Rockets still, however, find uses as rapid response systems for events such as solar flares. A balloon- or spacecraft-borne detector may not be available when the event occurs, or may take some time to bear onto the target, whereas a rocket may be held on stand-by on Earth, during which time the cost commitment is minimal, and then launched when required at only a few minutes notice.

1.4 COSMIC RAY DETECTORS
1.4.1 Background
Cosmic rays comprise two quite separate populations of particles: primary cosmic rays and secondary cosmic rays. The former are the true cosmic rays, and consist mainly of atomic nuclei in the proportions 84% hydrogen, 14% helium, 1% other nuclei, and 1% electrons and positrons, moving at velocities from a few percent of the speed of light to 0.999, 999, 999, 999, 999, 999, 999, 9c. That is, the energies per particle range from 10^{-13} to 50 J. As previously mentioned (Section 1.3), a convenient unit for measuring the energy of subnuclear particles is the electron volt. This is the energy gained by an electron in falling through a potential difference of 1 V, and is about 1.602×10^{-19} J. Thus, the primary cosmic rays have energies in the range 10^6 to a few times 10^{20} eV, with the bulk of the particles having energies near 10^9 eV, or a velocity of 0.9c. These are obviously relativistic velocities for most of the particles, for which the relationship between kinetic energy and velocity is

$$v = \frac{(E^2 + 2Emc^2)^{1/2}c}{E + mc^2} \tag{1.97}$$

where
E is the kinetic energy of the particle
m is the rest mass of the particle

The flux in space of the primary cosmic rays of all energies is about 10^4 particles m^{-2} s^{-1}. A limit, the Greisen Zatsepin Kuzmin (GZK),* in the cosmic ray energy spectrum is expected above about 4×10^{19} eV (6 J). At this energy, the microwave background photons are blue shifted to γ rays of sufficient energy to interact with the cosmic ray particle to produce pions. This will slow the particle to energies lower than GZK limit in

---

* Named for Ken Greisen, Timofeevich Zatsepin, and Vadim Kuzmin.

around 100 million years. However, surprisingly, higher energy cosmic rays are observed, implying that there must be one or more sources of ultra-high-energy cosmic rays within 100 Mly (30 Mpc) of the earth. The Pierre Auger Cosmic Ray Observatory (see below) confirmed in 2007 that the most likely sources of cosmic ray particles with energies above the GZK limit were relatively nearby active galactic nuclei, although the mechanism for accelerating the particles to such enormous energies remains a mystery.

Secondary cosmic rays are produced by the interaction of the primary cosmic rays with nuclei in the earth's atmosphere. A primary cosmic ray has to travel through about 800 kg m^{-2} of matter on average before it collides with a nucleus. A column of the earth's atmosphere 1 m^2 in cross section contains about 10^4 kg, so the primary cosmic ray usually collides with an atmospheric nucleus at a height of 30–60 km. The interaction results in numerous fragments: nucleons, pions, muons, electrons, positrons, neutrinos, γ rays, etc. and these in turn may still have sufficient energy to cause further interactions, producing more fragments, and so on. For high-energy primaries ($>10^{11}$ eV), some of the secondary particles will survive down to sea level and be observed as secondary cosmic rays. At higher energies ($>10^{13}$ eV), large numbers of the secondary particles ($>10^9$) survive to sea level, and a cosmic ray shower or extensive air shower is produced. At altitudes higher than sea level, the secondaries from the lower energy primary particles may also be found. Interactions at all energies produce electron and muon neutrinos as a component of the shower. Almost all of these survive to the surface, even those passing through the whole of the earth, though some convert to tau neutrinos during that passage. The detection of these particles is considered in Section 1.5.

1.4.2 Detectors

The methods of detecting cosmic rays may be divided into

1. Real-time methods. These observe the particles instantaneously and produce information on their direction as well as their energy and composition. They are the standard detectors of nuclear physicists, and there is considerable overlap between these detectors and those discussed in Section 1.3.2. They are generally used in conjunction with passive or active shielding so that directional and spectral information can also be obtained.

2. Residual track methods. The path of a particle through a material may be found some time (hours to millions of years) after its passage.

3. Indirect methods. Information on the flux of cosmic rays at large distances from the earth, or at considerable times in the past may be obtained by studying the consequent effects of their presence.

1.4.3   Real-Time Methods

1.4.3.1   Spark Detectors

A closely related device to the Geiger counter in terms of its operating principle is the spark chamber. This comprises a series of plates separated by small gap. Alternate plates are connected together and one set charged to a high potential while the other set is earthed. The passage of an ionizing particle through the stack results in a series of sparks that follow the path of the particle (Figure 1.85). The system is also used for the detection of γ rays (Section 1.3). Then a plate of a material of a high atomic weight such as tantalum may be interposed, and the ray detected by the resulting electron–positron pairs rather than by its own ionizations. The device requires another detector to act as a trigger for the high voltage to be applied, since spontaneous discharges prevent the use of a permanent high voltage. For the energetic gamma ray experiment telescope

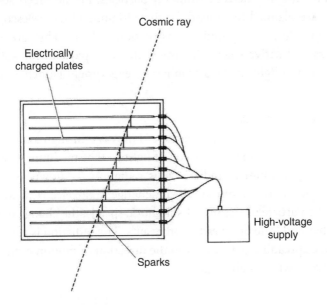

FIGURE 1.85 Spark chamber.

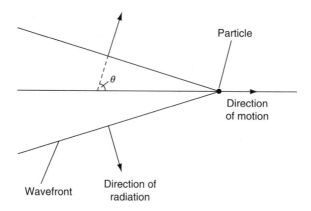

FIGURE 1.86 Čerenkov radiation.

(EGRET) detector on the Compton γ-Ray Observatory, the trigger was two sheets of plastic scintillator. The trigger is also needed to actuate cameras to image the spark tracks.

1.4.3.2 Scintillation Detectors

See Section 1.3.2.3 for a discussion of these devices. There is little change required for them to detect cosmic ray particles. Plastic sheets scintillator detectors are planned to be used for the 576 units of the Telescope Array Cosmic Ray Observatory under construction in Utah. They are also used as triggers and active shields for the PAMELA (payload for anti-matter exploration and light nuclei astrophysics) experiment launched in 2006.

1.4.3.3 Čerenkov Detectors

Background   When a charged particle is moving through a medium with a speed greater than the local speed of light in that medium, it causes the atoms of the medium to radiate. This radiation is known as Čerenkov radiation, and it arises from the abrupt change in the electric field near the atom as the particle passes by. At subphotic speeds, the change in the field is smoother and little or no radiation results. The radiation is concentrated into a cone spreading outward from the direction of motion of the particle (Figure 1.86), whose half angle, θ, is

$$\theta = \tan^{-1}\left(\mu_\nu^2 \frac{v^2}{c^2} - 1\right)^{1/2} \tag{1.98}$$

where

μ_ν is the refractive index of the material at frequency ν

v is the particle's velocity $(v > c/\mu_\nu)$

Its spectrum is given by

$$I_\nu = \frac{e^2 \nu}{2\varepsilon_0 c^2} \left(1 - \frac{c^2}{\mu_\nu^2 v^2} \right) \qquad (1.99)$$

where I_ν is the energy radiated at frequency ν per unit frequency interval, per unit distance traveled by the particle through the medium. The peak emission depends upon the form of the variation of refractive index with frequency. For a proton in air, peak emission occurs in the visible when its energy is about 2×10^{14} eV.

Detectors   Čerenkov detectors are very similar to scintillation detectors and are sometimes called that although the flashes of visible radiation are produced by different physical mechanisms in the two detectors (see above). A commonly used system employs a tank of very pure water surrounded by photomultipliers for the detection of the heavier cosmic ray particles, while high-pressure carbon dioxide is used for the electrons and positrons. With adequate observation and the use of two or more different media, the direction, energy, and possibly the type of the particle may be deduced. The Pierre Auger cosmic ray array in Argentina uses both water-based Čerenkov detectors and atmospheric fluorescence detectors (see below). There are 1600 Čerenkov detectors each containing 12 t of water and the flashes are detected by photomultipliers. The detectors, which are solar powered, are spread out in a grid over a 3000 km^2 area of grazing land.

As mentioned previously (Section 1.3.2.9), the flashes produced in the atmosphere by the primary cosmic rays can be detected. A large light "bucket" and a photomultiplier are needed, preferably with two or more similar systems observing the same part of the sky so that non-Čerenkov events can be eliminated by anti-coincidence discrimination. Thus, the 10 m Whipple telescope on Mount Hopkins is formed from around a hundred small hexagonal mirrors feeding a common focus where there is 150 pixel detector giving a 3.5° field of view. The stellar intensity interferometer (Section 2.4) was also an example of such a system, although there the Čerenkov events were a component of its noise spectrum, and not usually detected for their own interest.

An intriguing aside to Čerenkov detectors arises from the occasional flashes seen by astronauts when in space. These are thought to be Čerenkov radiation from primary cosmic rays passing through the material of the eyeball. Thus, cosmic ray physicists could observe their subjects directly. It is also just possible that they could listen to them as well. A large secondary cosmic ray shower hitting a water surface will produce a "click" sound that is probably detectable by the best of the current hydrophones. Unfortunately, a great many other events produce similar sounds so that the cosmic ray clicks are likely to be well buried in the noise. Nonetheless, the very highest energy cosmic ray showers might even be audible directly to a skin diver. Recently, several attempts have been made to detect these acoustic pulses and those from ultra-high-energy neutrinos, but so far without success. Hydrophones have been included or are to be added to the photomultipliers in several water/ice neutrino detectors including NT200 + and ANTARES (Section 1.5.3).

1.4.3.4 Solid-State Detectors

These have been discussed in Section 1.3.2.10. No change to their operation is needed for cosmic ray detection. Their main disadvantages are that their size is small compared with many other detectors, so that their collecting area is also small, and that unless the particle is stopped within the detector's volume, only the rate of energy loss may be found and not the total energy of the particle. This latter disadvantage, however, also applies to most other detectors. The PAMELA experiment launched in 2006 uses a number of different types of detectors as detectors, trackers, and active shields to find primary cosmic ray anti-protons and positrons with energies up to 270 GeV. Among these are silicon detectors, plastic sheet scintillator detectors, and tungsten–silicon micro-calorimeters (Section 1.3.2.10).

1.4.3.5 Nucleon Detectors

Generally only the products of the high-energy primaries survive to be observed at the surface of the earth. A proton will lose about 2×10^9 eV by ionization of atoms in the atmosphere even if it travels all the way to the surface without interacting with a nucleus, and there is only about 1 chance in 10,000 of it doing this. Thus, the lower energy primaries will not produce secondaries that can be detected at ground level, except if a neutron is one of its products. There is then a similar chance of noninteraction with atmospheric nuclei, but since the neutron will not ionize the atoms, even the low-energy ones will still be observable at sea level if they do miss the nuclei. Thus, neutrons and their reaction products from cosmic ray primaries of any energy can be detected using nucleon detectors.

These instruments detect low-energy neutrons by, for example, their interaction with boron in boron fluoride (BF_3):

$$^{10}_{5}B + n \rightarrow {}^{7}_{3}Li + {}^{4}_{2}He + \gamma \qquad (1.100)$$

The reaction products are detected by their ionization of the boron fluoride when this is used as a proportional counter (Section 1.3). The slow neutrons are produced by the interaction of the cosmic ray neutrons and the occasional proton with lead nuclei, followed by deceleration in a hydrogen-rich moderator such as polyethylene (Figure 1.87).

1.4.4 Residual Track Detectors

1.4.4.1 Photographic Emulsions

In a sense, the use of photographic emulsion for the detection of ionizing radiation is the second oldest technique available to the cosmic ray

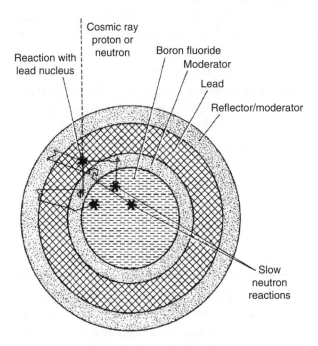

FIGURE 1.87 Cross section through a nucleon detector.

physicist after the electroscope, since Henri Becquerel discovered the existence of ionizing particles in 1896 by their effect upon a photographic plate. However, the early emulsions only hinted at the existence of the individual tracks of the particles, and were far too crude to be of any real use. For cosmic ray detection, blocks of photographic emulsion with high silver bromide content were exposed to cosmic rays at the tops of mountains, flown on balloons, or sent into space on some early spacecraft. The cosmic rays could be picked up and their nature deduced to some extent by the tracks left in the emulsion after it had been developed. This type of detector is little used now.

1.4.4.2 Ionization Damage Detectors

These provide a selective detector for nuclei with masses above about 150 amu. Their principle of operation is allied to that of the nuclear emulsions in that it is based upon the ionization produced by the particle along its track through the material. The material is usually a plastic with relatively complex molecules. As an ionizing particle passes through it, the large complex molecules are disrupted, leaving behind short, chemically reactive segments, radicals, etc. Etching the plastic reveals the higher chemical reactivity along the track of a particle and a conical pit develops along the line of the track. By stacking many thin layers of plastic, the track may be followed to its conclusion. The degree of damage to the molecules, and hence the characteristics of the pit which is produced, is a function of the particle's mass, charge, and velocity. A particular plastic may be calibrated in the laboratory so that these quantities may be inferred from the pattern of the sizes and shapes of the pits along the particle's track. Cellulose nitrate and polycarbonate plastics are the currently favored materials. The low weight of the plastic and the ease with which large-area detectors can be formed make this a particularly suitable method for use in space when there is an opportunity for returning the plastic to Earth for processing, as, for example, when the flight is a manned one.

Similar tracks may be etched into polished crystals of minerals such as feldspar and rendered visible by infilling with silver. Meteorites and lunar samples can thus be studied and provide data on cosmic rays which extend back into the past for many millions of years. The majority of such tracks appear to be attributable to iron group nuclei, but the calibration is very uncertain. Because of the uncertainties involved, the evidence has so far been of more use to the meteoriticist in dating the meteorite, than to the cosmic ray astronomer.

1.4.5 Indirect Detectors

1.4.5.1 100 MeV γ Rays

Primary cosmic rays occasionally collide with nuclei in the interstellar medium. Even though the chance of this occurring is only about 0.1% if the particle were to cross the galaxy in a straight line, it happens sufficiently often to produce detectable results. In such collisions π° mesons will frequently be produced, and these will decay rapidly into two γ rays, each with an energy of about 100 MeV. π° mesons may also be produced by the interaction of the cosmic ray particles and the 3 K microwave background radiation. This radiation when "seen" by a 10^{20} eV proton is Doppler shifted to a γ ray of 100 MeV energy, and neutral pions result from the reaction

$$p^+ + \gamma \rightarrow p^+ + \pi^\circ \tag{1.101}$$

Inverse Compton scattering of starlight or the microwave background by cosmic ray particles can produce an underlying continuum around the line emission produced by the pion decay.

γ Rays with energies as high as these are little affected by the interstellar medium so that they may be observed by spacecraft (Section 1.3) wherever they may have originated within the galaxy. The 100 MeV γ-ray flux thus gives an indication of the cosmic ray flux throughout the galaxy and beyond.

1.4.5.2 Radio Emission

Cosmic ray electrons are only a small proportion of the total flux, and the reason for this is that they lose significant amounts of energy by synchrotron emission as they interact with the galactic magnetic field. This emission lies principally between 1 MHz and 1 GHz and is observable as diffuse radio emission from the galaxy. However, the interpretation of the observations into electron energy spectra, etc. is not straightforward, and is further complicated by the lack of a proper understanding of the galactic magnetic field.

Primary cosmic rays have also recently been detected from their low-frequency emissions as they collide with the earth's atmosphere.

The LOFAR array (Sections 1.2.3 and 2.5) has detected bright flashes from cosmic rays that occur about once a day and last for a few tens of nanoseconds.

1.4.5.3 Fluorescence

The very highest energy extensive air showers are detectable via weak fluorescent light from atmospheric nitrogen. This is produced through the excitation of the nitrogen by the electron–photon component of the cascade of secondary cosmic rays. The equipment required is a light bucket and a detector (cf. Čerenkov detectors above and in Section 1.3.2.9), and detection rates of a few tens of events per year for particles in the 10^{19} to 10^{20} eV range are achieved by several automatic arrays. Fluorescence detectors require very dark sites and moonless and cloudless nights and so are not in continuous operation.

The Fly's Eye detector in Utah used sixty-seven 1.5 m mirrors feeding photomultipliers on two sites 4 km apart to monitor the whole sky. It operated from 1982 to 1992 and in 1991 it recorded the highest energy cosmic ray yet found: 3×10^{20} eV. This is well above the GZK limit (see above). Fly's Eye was upgraded to HiRes which had a baseline of 12.5 km, with 64 quadruple mirror units feeding 256 photomultipliers. HiRes, in turn, is now being superseded by the telescope array which will have three fluorescence detectors sited at the tops of hills some 25 km apart. The Pierre Auger Observatory has four fluorescence detectors separated by about 50 km, each using six 4 m mirrors and feeding cameras containing 440 photomultipliers.

1.4.5.4 Solar Cosmic Rays

Very high fluxes of low-energy cosmic rays can follow the eruption of a large solar flare. The fluxes can be intense enough to lower the earth's ionosphere and to increase its electron density. This in turn can be detected by direct radar observations, or through long-wave radio communication fade-outs, or through decreased cosmic radio background intensity as the absorption of the ionosphere increases.

1.4.5.5 Carbon-14

The radioactive isotope $^{14}_{6}C$ is produced from atmospheric $^{14}_{7}N$ by neutrons from cosmic ray showers

$$^{14}_{7}N + n \rightarrow {}^{14}_{6}C + p^{+} \tag{1.102}$$

The isotope has a half-life of 5730 years and has been studied intensively as a means of dating archaeological remains. Its existence in ancient organic remains shows that cosmic rays have been present in the earth's vicinity for at least 20,000 years. The flux seems, however, to have varied markedly at times from its present-day value, particularly between about 4000 and 1000 years BC. But this is probably attributable to increased shielding of the earth from the low-energy cosmic rays at times of high solar activity, rather than to a true variation in the number of primary cosmic rays.

1.4.6 Arrays

Primary cosmic rays may be studied by single examples of the detectors that we have considered above. The majority of the work on cosmic rays, however, is on the secondary cosmic rays, and for these a single detector is not very informative. The reason is that the secondary particles from a single high-energy primary particle have spread over an area of 10 km^2 or more by the time they have reached ground level from their point of production some 50 km up in the atmosphere. To deduce anything about the primary particle that is meaningful, the secondary shower must be sampled over a significant fraction of its area. Thus, arrays of detectors are used rather than single ones (see Section 1.5 for the detection of cosmic-ray produced neutrinos). Plastic or liquid scintillators (Section 1.3), Čerenkov and fluorescence detectors are frequently chosen as the detectors but any of the real-time instruments can be used and these are typically spread out over an area of hundreds to thousands of square kilometers. The resulting several hundred to a thousand individual detectors are all then linked to a central computer for data analysis.

The Akeno giant air shower array (AGASA) in Japan, which has now ceased operations, for example, used 111 scintillation detectors spread over 100 km^2, and detected the second highest energy cosmic ray particle known at 2×10^{20} eV. The Pierre Auger Observatory is a 3000 km^2 array in Argentina with 1600 water-based Čerenkov detectors and twenty-four 12 m^2 optical telescopes to detect nitrogen fluorescence at distances of up to 30 km. A second similar array is planned for construction in Colorado to provide coverage of the Northern Hemisphere. The telescope array in Utah uses three fluorescence detectors and five hundred and sixty-four 2×3 m plastic scintillators spread over a 1000 km^2 area.

The analysis of the data from such arrays is difficult. Only a very small sample, typically less than 0.01%, of the total number of particles in the shower is normally caught. The properties of the original particle have then to be inferred from this small sample. However, the nature of the shower varies with the height of the original interaction, and there are numerous corrections to be applied as discussed below. Thus, normally the observations are fitted to a grid of computer-simulated showers. The original particle's energy can usually be obtained within fairly broad limits by this process, but its further development is limited by the availability of computer time, and by our lack of understanding of the precise nature of these extraordinarily high-energy interactions.

1.4.7 Correction Factors

1.4.7.1 Atmospheric Effects

The secondary cosmic rays are produced within the earth's atmosphere, and so its changes may affect the observations. The two most important variations are caused by air mass and temperature.

The air mass depends upon two factors—the zenith angle of the axis of the shower, and the barometric pressure. The various components of the shower are affected in different ways by changes in the air mass. The muons are more penetrating than the nucleons and so the effect upon them is comparatively small. The electrons and positrons come largely from the decay of muons, and so their variation tends to follow that of the muons. The corrections to an observed intensity are given by

$$I(P_o) = I(P)e^{K(P-P_o)/P_o} \tag{1.103}$$

$$I(0) = I(\theta)e^{K(\sec\theta-1)} \tag{1.104}$$

where

P_o is the standard pressure

P is the instantaneous barometric pressure at the time of the shower

$I(P_o)$ and $I(P)$ are the shower intensities at pressures P_o and P, respectively

$I(0)$ and $I(\theta)$ are the shower intensities at zenith angles of zero and θ, respectively

K is the correction constant for each shower component

K has a value of 2.7 for the muons, electrons, and positrons and 7.6 for the nucleons. In practice, a given detector will also have differing sensitivities for different components of the shower, and so more precise correction factors must be determined empirically.

The atmospheric temperature changes primarily affect the muon and electron components. The scale height of the atmosphere, which is the height over which the pressure changes by a factor of e^{-1}, is given by

$$H = \frac{kR^2T}{GMm} \qquad (1.105)$$

where
R is the distance from the center of the earth
T the atmospheric temperature
M the mass of the earth
m the mean particle mass for the atmosphere

Thus, the scale height increases with temperature and so a given pressure will be reached at a greater altitude if the atmospheric temperature increases. The muons, however, are unstable particles and will have a longer time in which to decay if they are produced at greater heights. Thus, the muon and hence the electron and positron intensity decreases as the atmospheric temperature increases. The relationship is given by

$$I(T_o) = I(T)e^{0.8(T-T_o)/T} \qquad (1.106)$$

where
T is the atmospheric temperature
T_o the standard temperature
$I(T_o)$ and $I(T)$ are the muon (or electron) intensities at temperatures T_o
    and T, respectively

Since the temperature of the atmosphere varies with height, Equation 1.106 must be integrated up to the height of the muon formation to provide a reliable correction. The temperature profile will, however, be only poorly known, so it is not an easy task to produce accurate results.

1.4.7.2 Solar Effects
The Sun affects cosmic rays in two main ways. Firstly, it is itself a source of low-energy cosmic rays whose intensity varies widely. Secondly, the extended solar magnetic field tends to shield the earth from the lower energy primaries. Both these effects vary with the sunspot cycle and also on other timescales, and are not easily predictable.

1.4.7.3 Terrestrial Magnetic Field

The earth's magnetic field is essentially dipolar in nature. Lower energy charged particles may be reflected by it, and so never reach the earth at all, or particles that are incident near the equator may be channeled toward the poles. There is thus a dependence of cosmic ray intensity on latitude (Figure 1.88). Furthermore, the primary cosmic rays are almost all positively charged and they are deflected so that there is a slightly greater intensity from the west. The vertical concentration (Equation 1.104 and Figure 1.89) of cosmic rays near sea level makes this latter effect only of importance for high altitude balloon or satellite observations.

Exercise 1.9

Show that the true counting rate, C_t, of a Geiger counter whose dead time is of length Δt, is related to its observed counting rate, C_o, by

$$C_t = \frac{C_o}{1 - \Delta t\, C_o}$$

(Section 1.3 is also relevant to this problem).

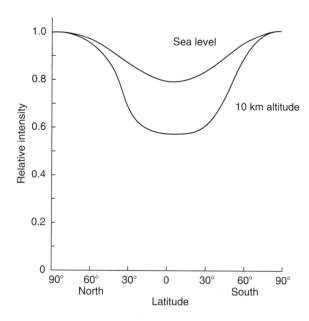

FIGURE 1.88 Latitude effect on cosmic rays.

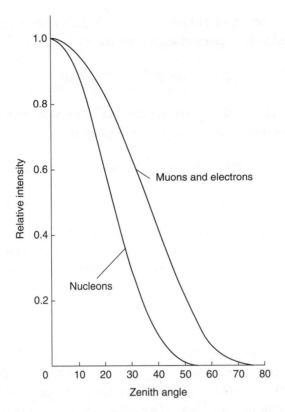

FIGURE 1.89 Zenithal concentration of secondary cosmic ray shower components as given by Equation 1.104.

If the effective range of a Geiger counter is limited to $C_t \leq 2C_o$, calculate the maximum useful volume of a Geiger counter used to detect secondary cosmic rays at sea level if its dead time is 250 μs.

Exercise 1.10

The minimum particle energy required for a primary cosmic ray to produce a shower observable at ground level when it is incident vertically onto the atmosphere is about 10^{14} eV. Show that the minimum energy required to produce a shower when the primary particle is incident at a zenith angle θ, is given by

$$E_{min}(\theta) = 6.7 \times 10^{12} e^{2.7 \sec \theta} \text{ eV}$$

(Hint: use Equation 1.104 for muons and assume that the number of particles in the shower at its maximum is proportional to the energy of the primary particle.)

The total number of primary particles, $N(E)$, whose energy is greater than or equal to E, is given at high energies by

$$N(E) \approx 10^{22}E^{-1.85} \text{ m}^{-2} \text{ s}^{-1} \text{ str}^{-1}$$

for E in eV. Hence, show that the number of showers, $N(\theta)$, observable from the ground at a zenith angle of θ is given by

$$N(\theta) \approx 0.019e^{-5.0\sec\theta} \text{ m}^{-2} \text{ s}^{-1} \text{ str}^{-1}$$

Exercise 1.11

By numerical integration, or otherwise, of the formula derived in Exercise 1.10, calculate the total flux of showers of all magnitudes onto a detector array covering 1 km^2. (Assume that the primary particle flux is isotropic.)

1.5 NEUTRINO DETECTORS

1.5.1 Background

Neutrino astronomy is in the doldrums at the moment. Only two astronomical sources of neutrinos have ever been observed—the Sun and supernova 1987A in the large Magellanic Cloud. Since the solar neutrino problem was solved a decade ago (see below) there is little new to be learnt about the Sun from neutrinos. Nonetheless, neutrino detectors are still thriving. This is partly in order to study neutrinos in their own right and partly to study cosmic rays via the neutrinos that they produce during their interactions with the earth's atmosphere. However, as detector sensitivities improve, other types of astronomical object may become detectable. Thus, ICECUBE (currently under construction, see below) is expected to be able to pick up neutrinos from AGNs and our own galaxy and may be able to detect GRBs—and, of course, another nearby supernova could happen at any moment. Thus, there is still an astronomical interest in current and forthcoming neutrino detectors.

Wolfgang Pauli postulated the neutrino in 1930 to retain the principle of conservation of mass and energy in nuclear reactions. It was necessary to provide a mechanism for the removal of residual energy in some β-decay reactions. From other conservation laws, the neutrino's properties could be defined quite well; zero charge, zero or very small rest mass, zero electric moment, zero magnetic moment, and a spin of one-half. Over a quarter of a century passed before the existence of this hypothetical particle was confirmed experimentally (in 1956). The reason for the long delay in its confirmation lay in the very low probability of the interaction of neutrinos with other matter (neutrinos interact via the weak nuclear force only). A neutrino originating at the center of the Sun would have only one chance in 10^{10} of interacting with any other particle during the whole of its 700,000 km journey to the surface of the Sun. The interaction probability for particles is measured by their cross-sectional area for absorption, σ, given by

$$\sigma = \frac{1}{\lambda N} \tag{1.107}$$

where
N is the number density of target nuclei
λ is the mean free path of the particle

Even for high-energy neutrinos, the cross section for the reaction

$$\nu + {}^{37}_{17}\text{Cl} \rightarrow {}^{37}_{18}\text{Ar} + e^- \tag{1.108}$$

that was originally used for their detection (see later in this section) is only 10^{-46} m^2 (Figure 1.90), so that such a neutrino would have a mean free path of over 1 pc even in pure liquid ${}^{37}_{17}\text{Cl}$.

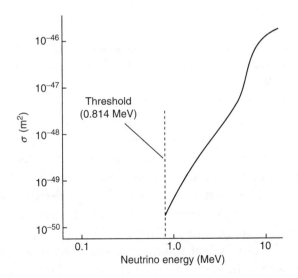

FIGURE 1.90 Neutrino absorption cross sections for the ${}^{37}_{17}\text{Cl} \rightarrow {}^{37}_{18}\text{Ar}$ reaction.

Three varieties (or flavors) of neutrino are known, plus their anti-particles. The electron neutrino, ν_e, is the type originally postulated to save the β reactions, and which is therefore involved in the archetypal decay: that of a neutron

$$n \rightarrow p^+ + e^- + \tilde{\nu}_e \qquad (1.109)$$

where $\tilde{\nu}_e$ is an anti-electron neutrino. The electron neutrino is also the type commonly produced in nuclear fusion reactions and hence to be expected from the Sun and stars. For example, the first stage of the proton–proton cycle is

$$p^+ + p^+ \rightarrow {}_1^2\text{H} + e^+ + \nu_e \qquad (1.110)$$

and the second stage of the carbon cycle

$${}_7^{13}\text{N} \rightarrow {}_6^{13}\text{C} + e^+ + \nu_e \qquad (1.111)$$

and so on.

The other two types of neutrino are the muon neutrino, ν_μ, and the tau neutrino, ν_τ. These are associated with reactions involving the heavy electrons called muons and tauons. For example, the decay of a muon

$$\mu^+ \rightarrow e^+ + \nu_e + \tilde{\nu}_\mu \qquad (1.112)$$

involves an anti-muon neutrino among other particles. The muon neutrino was detected experimentally in 1962, but the tau neutrino was not found until 2000.

The first neutrino detector started operating in 1968 (see chlorine-37 detectors below), and it quickly determined that the observed flux of solar neutrinos was too low compared with the theoretical predictions by a factor of 3.3. This deficit was confirmed later by other types of detectors such as Soviet–American gallium experiment (SAGE), GALLEX (gallium experiment), and Kamiokande (Kamioka neutrino detector) and became known as the solar neutrino problem. However, these detectors were only sensitive to electron neutrinos. Then in 1998, the Super Kamiokande detector found a deficit in the number of muon neutrinos produced by cosmic rays within the earth's atmosphere. Some of the muon neutrinos were converting to tau neutrinos during their flight time to the detector. The Super Kamiokande and SNO detectors have since confirmed that all

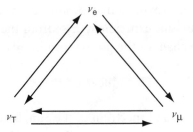

FIGURE 1.91 Neutrino oscillation.

three neutrinos can exchange their identities (Figure 1.91). The solar neutrino problem has thus disappeared, since two-thirds of the electron neutrinos produced by the Sun have changed to muon and tau neutrinos by the time that they reach the earth, and so were not detected by the early experiments. The flux of solar neutrinos (of all types) at the earth is now measured to be about 6×10^{14} m^{-2} s^{-1}, in good agreement with the theoretical predictions. The experiments also show that neutrinos have mass, although only upper limits have been determined so far. These are currently <2.2 eV (less than 0.0006% of the mass of the electron) for the electron neutrino and <0.17 MeV and <15.5 MeV for the muon and tau neutrinos, respectively. It has been suggested that there might be a fourth type of neutrino, either extremely lightweight or super massive. The lightweight neutrinos would interact with ordinary matter only through gravity, and so be even less easy to detect than "normal" neutrinos, they have thus been termed "sterile" neutrinos. The super massive ($>10^{21}$ eV) neutrino would oscillate into being from one of the other types for an imperceptibly brief instant of time only. However recent experiments using beams of anti-electron neutrinos from nuclear reactors observed by the KamLAND detector, have almost certainly ruled out the existence of a fourth neutrino.

1.5.2 Chlorine-37 Detectors

The problems involved in any attempt to detect neutrinos arise mainly from the extreme rarity of the interactions that are involved. This not only dictates the use of very large detectors and very long integration times but also means that other commoner radioactive processes may swamp the neutrino-induced reactions. The first neutrino "telescope" to be built was designed to detect electron neutrinos through the reaction given in Equation 1.108. The threshold energy of the neutrino for this reaction is

0.814 MeV (see Figure 1.90), so that 80% of the neutrinos from the Sun that are detectable by this experiment arise from the decay of boron in a low probability side chain of the proton–proton reaction

$$\prescript{8}{5}{B} \rightarrow \prescript{8}{4}{Be} + e^+ + \nu_e \tag{1.113}$$

The full predicted neutrino spectrum of the Sun, based upon the conventional astrophysical models is shown in Figure 1.92.

The chlorine-based detector operated from 1968 to 1998 under the auspices of Ray Davis Jr. and his group and so is now largely of historical interest, although many of the operating principles and the precautions that are needed apply to other types of neutrino detector. It consisted of a large tank of over 600 t of tetrachloroethene (C_2Cl_4) located 1.5 km down in the Homestake gold mine in South Dakota. About one chlorine atom in four is the $\prescript{37}{17}{Cl}$ isotope; so on average each of the molecules contains one

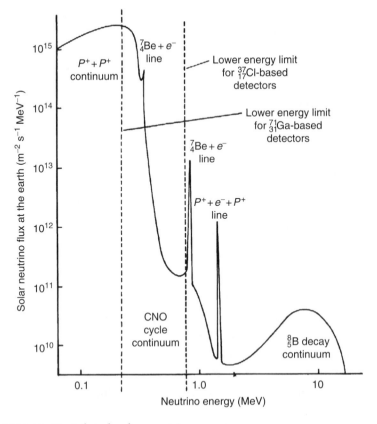

FIGURE 1.92 Postulated solar neutrino spectrum.

of the required atoms, giving a total of about 2×10^{30} $^{37}_{17}Cl$ atoms in the tank. Tetrachloroethene was chosen rather than liquid chlorine for its comparative ease of handling, and because of its cheapness: it is a common industrial solvent. The interaction of a neutrino (Equation 1.108) produces a radioactive isotope of argon. The half-life of the argon is 35 days, and it decays back to $^{37}_{17}Cl$ by capturing one of its own inner orbital electrons, at the same time ejecting a 2.8 keV electron. The experimental procedure was to allow the argon atoms to accumulate for some time, then to bubble helium through the tank. The argon was swept up by the helium and carried out of the tank. It was separated from the helium by passing the gas stream through a charcoal cold trap. The number of $^{37}_{18}Ar$ atoms was then counted by detecting the 2.8 keV electrons as they were emitted.

The experimental procedure had to be carried out with extraordinary care if the aim of finding a few tens of atoms in 10^{31} was to succeed (similar precautions have to be taken for all neutrino detectors). Preeminent among the precautions were

1. Siting the tank 1.5 km below the ground to shield it against cosmic rays and natural and artificial radiation sources. The neutrinos of course pass almost without effect through this depth of material; indeed they can pass right through the earth without any measurable drop in their intensity.

2. Surrounding the tank with a thick water jacket to shield against neutrons.

3. Monitoring the efficiency of the argon extraction by introducing a known quantity of $^{36}_{18}Ar$ into the tank before it is swept with helium, and determining the percentage of argon that was recovered.

4. Shielding the sample of $^{37}_{18}Ar$ during counting of its radioactive decays and using anti-coincidence techniques to eliminate extraneous pulses.

5. Use of long integration times (50 to 100 days) to accumulate the argon in the tank.

6. Measurement of and correction for the remaining noise sources.

The Homestake detector typically caught one neutrino every other day. This corresponds to a detection rate of 2.23 ± 0.22 SNU. (A solar neutrino unit, SNU, is 10^{-36} captures per second per target atom.) The expected rate from the Sun is 7.3 ± 1.5 SNU, and the apparent disparity between the observed and predicted rates before the confirmation of the oscillation of neutrino types gave rise to the previously mentioned solar neutrino problem. In 2002, Ray Davis shared the Nobel physics prize for his work on neutrinos.

1.5.3 Water-Based Detectors

The next generation of working neutrino telescopes did not appear until nearly two decades after the chlorine-based detector. Neither of these instruments, the Kamiokande buried 1 km below Mount Ikenoyama in Japan nor the Irvine–Michigan–Brookhaven (IMB) detector, 600 m down a salt mine in Ohio, were built to be neutrino detectors. Both were originally intended to look for proton decay, and were only later converted for use with neutrinos. The design of both detectors was similar (Figure 1.93), they differed primarily only in size (Kamiokande: 3000 t, IMB: 8000 t).

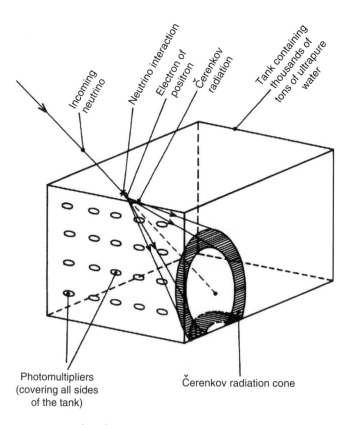

FIGURE 1.93 Principle of neutrino detection by water-based detectors.

The principle of their operation was the detection of Čerenkov radiation from the products of neutrino interactions. These may take two forms, electron scattering and inverse β-decay. In the former process, a collision between a high-energy neutrino and an electron sends the latter off at a speed in excess of the speed of light in water (225,000 km s^{-1}) and in roughly the same direction as the neutrino. All three types of neutrinos scatter electrons, but the electron neutrinos are 6.5 times more efficient at the process. In inverse β-decay, an energetic positron is produced via the reaction

$$\tilde{\nu}_e + p \rightarrow n + e^+ \tag{1.114}$$

Inverse β-decays are about 100 times more probable than the scattering events, but the positron can be emitted in any direction, and thus gives no clue to the original direction of the incoming neutrino. Both positron and electron, traveling at superphotic speeds in the water, emit Čerenkov radiation (Section 1.4) in a cone around their direction of motion. That radiation is then picked up by arrays of photomultipliers (Section 1.1) that surround the water tank on all sides. The pattern of the detected radiation can be used to infer the energy of the original particle, and in the case of a scattering event, also to give some indication of the arrival direction. The minimum detectable energy is around 5 MeV due to background noise.

Both detectors were fortunately in operation in February 1987 and detected the burst of neutrinos from the supernova in the large Magellanic Cloud (SN1987A). This was the first (and so far only) detection of neutrinos from an astronomical source other than the Sun, and it went far toward confirming the theoretical models of supernova explosions.

Both IMB and Kamiokande have now ceased operations,* but a number of detectors operating on the same principle are functioning or are under construction. Foremost among these is Super Kamiokande that contains 50,000 t of pure water and uses 13,000 photomultiplier tubes in a tank buried 1 km below Mount Ikenoyama in Japan. Super Kamiokande is able to detect muon neutrinos as well as electron neutrinos. The muon neutrino interacts with a proton to produce a relativistic muon. The Čerenkov radiation from the muon produces a well-defined ring of light. The electron resulting from an electron neutrino scattering event by contrast generates a much fuzzier ring of light. This is because the primary electron

---

* The IMB photomultipliers were re-used as an active shield for Super Kamiokande.

produces γ rays that in turn produce electron–positron pairs. The electron–positron pairs then create additional Čerenkov radiation cones, thus blurring the ring resulting from the primary electron.

This ability to distinguish between electron and muon neutrinos enabled Super Kamiokande to provide the first evidence of neutrino oscillations (see Section 1.5.1). Muon neutrinos produced in the earth's atmosphere by high-energy cosmic rays should be twice as abundant as electron neutrinos.* But in 1998 Super Kamiokande found that although this was the case for neutrinos coming from above (a distance of about 60 km), there were roughly equal numbers of muon neutrinos and electron neutrinos coming from below (i.e., having traveled 12,000 km across the earth). Since there were no extra electron neutrinos, some of the muon neutrinos must be oscillating to tau neutrinos during the flight time across the earth to the detector. This result was confirmed in 2000 by a 30% shortfall in the number of muon neutrinos observed from an artificial source at the KEK laboratory at Tsukuba some 250 km away from the detector. Further confirmation of the oscillation of all three types of neutrino has recently come from comparison of the Super Kamiokande and SNO results and the KamLAND experiment (see below). The designer of the Kamiokande and Super Kamiokande detectors, Matatoshi Koshiba, shared the Nobel physics prize in 2002 for his work.

Man-made detectors larger than Super Kamiokande are probably impractical. However, much greater quantities of water (or ice, which is equally good) may be monitored using parts of lakes, the sea, or the Antarctic ice cap. Of the water detectors, NT-200 + is currently operating, while the Astronomy with a Neutrino Telescope and Abyssal Environmental Research (ANTARES) and the Neutrino Extended Submarine Telescope with Oceanographic Research (NESTOR) instruments are under construction. The Antarctic Muon and Neutrino Detector Array 2 (AMANDA2) utilizes the Antarctic ice cap and is currently being upgraded to an array called ICECUBE.

The NT-200 detector comprises 192 sets of photomultipliers suspended on eight strings forming an array some 45 m across and 100 m long. It is

---

* The cosmic ray interaction first produces (amongst other particles) pions. A positively charged pion usually then decays in 2.6×10^{-8} s to a positive muon and a muon neutrino, and the muon decays in 2.2×10^{-6} s to a positron, a muon anti-neutrino and an electron neutrino. A negative pion likewise decays to an electron, a muon neutrino, a muon anti-neutrino and an electron anti-neutrino. There should thus be two muon neutrinos to each electron neutrino in a secondary cosmic ray shower (Section 1.4).

immersed at a depth of 1.1 km in Lake Baikal and monitors 200,000 m^3 of water for multi-TeV muons coming from neutrino interactions within the rock below it. The neutrinos themselves originate from high-energy cosmic ray impacts near the top of the earth's atmosphere. The higher energy particles can be detected coming from interactions above the detector, since the background noise at that energy is low, but at lower energies, the detector observes upwardly moving particles originating in the atmosphere on the far side of the earth. In 2005, it was up-graded to NT200 + by the addition of three more strings each with 12 photomultipliers and now monitors a volume of some 1×10^7 m^3. A possible further extension to provide coverage of 1000 million cubic meters (1 km^3) is under consideration.

ANTARES and NESTOR are being constructed in the Mediterranean at depths of 2500 and 4000 m off Marseilles and the Peloponnese, respectively. The detectors are similar to NT200+, using photomultipliers on long strings to monitor a volume of seawater for Čerenkov radiation, again arising from high-energy muons. ANTARES detected its first neutrinos in 2007. When completed in 2008 it will be 150 m across by 300 m high and will use 900 photomultipliers to observe a volume of 1×10^7 m^3. NESTOR started small-scale operations in 2003 and eventually will be 32 m across by 240 m high and will monitor 200,000 cubic meters of water using 168 photomultipliers. In 2006, a design study was commissioned for KM3NET, a project similar to ANTARES and NESTOR but enlarged to monitor a cubic kilometer of water. This project could be operational by 2013.

AMANDA2 operated from 2000 to 2005, and succeeded a smaller trial version of the same type of detector called AMANDA. The ice at a depth of one or more kilometers in the Antarctic ice cap is highly transparent, since there are few impurities and the pressure of the overlying layers has squeezed out any air bubbles. So the Čerenkov radiation can easily be seen. Like the seawater detectors, AMANDA2 used 677 photomultipliers suspended on 19 strings. There was an outer array 200 m across extending from a depth of 1150–2350 m in the ice, surrounding a smaller array 120 m across that extends from 1500 m down to 2000 m, resulting in a total volume of ice of 40 million cubic meters being monitored. Unlike the seawater detectors though, the strings of photomultipliers are frozen in place, having been lowered down holes drilled through the ice by a hot water drill. AMANDA2 is currently being upgraded to 4800 photomultipliers on 70 strings to monitor 1000 million cubic meters (1 km^3) of ice. The upgraded detector is to be called ICECUBE.

Although the peak Čerenkov emission produced by multi-TeV particles is in the optical region, radio waves are produced as well (Equation 1.99). At the temperature of the ice cap (about $-50°C$), the ice is almost completely transparent to meter-wavelength radio waves, and so these can be detected by receivers at the surface. A radio receiver borne by a high altitude balloon would enable the entire Antarctic ice cap to be watched: a volume of around a million cubic kilometers. In 2006, just such a receiver was flown. Called ANITA (Antarctic impulsive transient antenna), during a 35-day flight at a height of 35 km, it picked up 8 million events. Most of these events though were background non-neutrino pulses. A second flight for the instruments is planned for 2008.

A neutrino detector based upon 1000 t of heavy water (D_2O)* operated at Sudbury in Ontario from 1999 to 2007. Known as SNO (Sudbury Neutrino Observatory), the heavy water was contained in a 12 m diameter acrylic sphere immersed in an excavated cavity filled with 7000 t of highly purified normal water. The normal water shielded the heavy water from radioactivity in the surrounding rock. It was located 2 km down within the Creighton copper and nickel mine and used 9600 photomultipliers to detect Čerenkov emissions within the heavy water. Neutrinos may be detected through electron scattering as with other water-based detectors, but the deuterium in the heavy water provides two other mechanisms whereby neutrinos may be found. The first of these senses just electron neutrinos. An electron neutrino interacts with a deuterium nucleus to produce two protons and a relativistic electron

$$\nu_e + {}_1^2\text{H} \rightarrow p^+ + p^+ + e^- \tag{1.115}$$

and the electron is observed via its Čerenkov radiation. This mechanism is called the charged current reaction since the charged W Boson mediates it. The second mechanism is mediated by the neutral Z boson, and is thus called the neutral current reaction. In it, a neutrino of any type simply splits the deuterium nucleus into its constituent proton and neutron. The neutron is thermalized in the heavy water, and eventually combines with another nucleus with the emission of γ rays. The γ rays in turn produce relativistic electrons via Compton scattering and these are finally detected from their Čerenkov radiation. Although the neutron can combine with a

---

* On loan from the Canadian government. It has a value of around $300 million.

deuterium nucleus, the capture efficiency is low so that 75% of the neutrons will escape from the detector. The SNO detector could therefore add two t of salt to the heavy water to enable the neutron to be captured more easily by a $^{35}_{17}Cl$ nucleus (converting it to $^{36}_{17}Cl$) and reducing to 17% the number of escaping neutrons. The thermalized neutrons can also be detected using $^{3}_{2}He$ proportional counters (Section 1.3.2.2). The neutron combines with $^{3}_{2}He$ to produce a proton and tritium, and these then ionize some of the remaining gas to give an output pulse.

Since SNO could detect the number of electron neutrinos separately from the total for all types of neutrino, it provided the definitive data for the solar neutrino problem (see above). The standard solar model predicts that SNO should detect about 30 charged current or neutral current reactions per day and about 3 electron-scattering events. However, the reality of the oscillation of neutrinos between their different type, and hence the solution to the solar neutrino problem was demonstrated by the very first results from SNO. It was initially operated in the charged current mode (i.e., without salt being added to the heavy water) to detect just the solar electron neutrinos. This gave a flux that was lower than that measured by Super Kamiokande, since the latter also detects a proportion of the muon and tau neutrinos as well as the electron neutrinos. Some solar electron neutrinos must therefore have oscillated to the other types during their journey from the center of the Sun. SNO then made measurements of the total solar neutrino flux using the neutral current mode (i.e., with salt added to the heavy water), and in April 2002 finally demonstrated that the total neutrino flux is as predicted by the standard solar model. SNO ceased operations in May 2007 and the heavy water was drained from its tank.

1.5.4 Gallium-Based Detectors

There are two neutrino detectors based upon gallium. The SAGE detector (1990–2006) used 60 t of liquid metallic gallium buried beneath the Caucasus Mountains, and detected neutrinos via the reaction

$$^{71}_{31}Ga + n_e \rightarrow {}^{71}_{32}Ge + e^- \tag{1.116}$$

The germanium product of the reaction is radioactive, with a half-life of 11.4 days. It is separated out by chemical processes on lines analogous to those of the chlorine-based detector. The detection threshold is only 0.236 MeV so that the p–p neutrinos (Figure 1.92) are detected directly.

The other gallium-based detector, gallium neutrino observatory (GNO), which is still in operation, is based upon the same detection mechanism as SAGE, but uses 30 t of gallium in the form of gallium trichloride ($GaCl_3$). It replaces the Gallium experiment (GALLEX) and is located within the Gran Sasso Tunnel under the Apennines, 150 km east of Rome. Both detectors have measured a solar neutrino flux around 50% of that predicted by the standard solar model.

1.5.5 Scintillator-Based Detectors

The high-energy photons and particles resulting from neutrino interactions may be observed using scintillation detectors (Section 1.3). The Soudan II detector, located 700 m below the surface in an old iron mine at Soudan in Minnesota uses 5000 t of alternating sheets of steel and a plastic scintillator. It was originally built to try and detect the decay of protons, but can also detect the muons resulting from muon neutrino interactions within the steel plates. It is currently used to detect neutrinos arising from cosmic ray interactions. Also in the Soudan mine, a new detector, the main injector neutrino oscillation search (MINOS), operating on the same lines as Soudan II is currently being built. It will contain 5400 t of steel plates and scintillator sheets. Together with an almost identical detector at Fermilab, it is intended to refine the measurements of the parameters of neutrino oscillations by observing a beam of muon neutrinos produced at Fermilab in Illinois, some 730 km away.

Also measuring the parameters of neutrino oscillations is the KamLAND experiment in Japan. This uses two nested plastic spheres, 13 and 18 m in diameter surrounded by 2000 photomultipliers. The spheres are filled with iso-paraffin mineral oil and the inner one is also doped with a liquid scintillator. KamLAND detects anti-electron neutrinos from nuclear reactors some 175 km away, and has recently shown that a fourth type of neutrino (see Section 1.5.1) almost certainly does not exist.

Borexino is an experiment currently aimed at detecting the solar neutrino emission at 0.863 MeV arising from the decay of $^7_4$Be to $^7_3$Li. It detects all types of neutrinos via their scattering products within 300 t of liquid scintillator observed by 2200 photomultipliers and has a detection threshold of 250 keV. The active scintillator is surrounded by 1000 t of a liquid buffer, and that in turn by 2400 t of water. 400 photomultipliers pointing outward into the water buffer provide an active shield. It is located in the Gran Sasso tunnel in Italy and started operating in 2007.

1.5.6   Acoustic Detectors

The highest energy cosmic rays (Section 1.4) have energies measured in several joules. The Pierre Auger cosmic ray observatory has recently shown that these particles probably originate from AGNs within 30–40 Mpc or so of the earth. It is possible that such AGNs may also produce neutrinos with energies comparable to those of the cosmic ray particles (10^{20} eV). If such a neutrino were to interact with an atom on the earth, much of its energy would be converted into heating a small volume of the material around the interacting particles. The rapid heat rise would then generate a pulse of sound, which would have a bipolar profile. An array of sensitive microphones could then detect that acoustic pulse and it could be distinguished from the myriad of other pulses by its characteristic shape. Volumes of water, ice, or salt of a hundred or more cubic kilometers could thus be monitored for high-energy neutrinos (and also cosmic rays—Section 1.4). No neutrinos or cosmic rays have yet been detected by this technique but NT200+ had four hydrophones added to its photomultiplier array in 2006 in an attempt to try out the approach. It is planned to add hydrophones to the ANTARES array and a design study for a purpose-built system (ACoRvE, acoustic cosmic ray neutrino experiment) has recently been funded.

1.5.7   Other Types of Detectors

Detectors based upon $^{37}_{17}Cl$, etc. can only detect the high-energy neutrinos, while the bulk of the solar neutrinos are of comparatively low energies (Figure 1.92). Furthermore, no indication of the direction of the neutrinos is possible. Water and gallium-based detectors also look at the high-energy neutrinos. Numerous alternative detectors that are designed to overcome such restrictions are therefore proposed or imagined at the time of writing. A brief survey of the range of possibilities is given below. We may roughly divide the proposals into two types—direct interaction and geological—based upon the nature of their interactions.

1.5.7.1   Direct Interaction Type Detectors

These are mostly variants of Davis' and the gallium experiments with another element used in place of the chlorine or gallium. A hundred tons of iodine in the form of sodium iodide has been operating at the Homestake mine since 1996, detecting neutrinos via the conversion of $^{127}_{53}I$ to radioactive $^{127}_{54}Xe$. The low-energy neutrino spectroscope (LENS) is a proposal to undertake real-time neutrino spectroscopy using $^{176}_{70}$Yb. Inter-action with a neutrino converts it to $^{176}_{71}$Lu with the production of a high-energy electron and after a 50 ns delay, two γ rays at 72 and 144 keV, respectively. So there is a clear signature to a neutrino detection enabling it to be distinguished from background reactions. The electron's energy depends upon that of the neutrino enabling the latter to be determined. The ytterbium would be dissolved in a liquid scintillator, and about 10 t would be needed to give a detection every other day. It would be located in the Grand Sasso tunnel in Italy. Indium may also be used in a similar manner. The reaction in this case is

$$^{115}_{49}\text{In} + \nu_e \rightarrow {}^{115}_{50}\text{Sn} + e^- \qquad (1.117)$$

The tin nucleus is in an excited state after the interaction and emits two γ rays with energies of 116 and 498 keV, 3 μs on average, after its formation. Like ytterbium, neutrino detection therefore has a characteristic signature of an electron emission followed after a brief delay by two distinctive γ rays. The electron's energy and direction are related to those of the neutrino so allowing these to be found. The threshold is again low enough to capture some of the proton–proton neutrinos. About 10 t of indium might be needed to detect one neutrino per day.

Another possibility uses lithium and is based upon the reaction

$$^{7}_{3}\text{Li} + \nu_e \rightarrow {}^{7}_{4}\text{Be} + e^- \qquad (1.118)$$

It has the disadvantage of a high threshold (0.862 MeV), but the advantage of comparative cheapness, only about 15 t being needed for one detection per day. There is also the possibility of the detection of neutrinos via the reaction

$$^{37}_{19}\text{K} + \nu_e \rightarrow {}^{37}_{18}\text{Ar} + e^+ \qquad (1.119)$$

and a detector using 6 t of potassium hydroxide is currently operating in the Homestake mine.

Variants on the position-sensitive proportional counters (Section 1.3.2.2) are also used for neutrino and secondary cosmic ray detection, where they are generally known as drift, multi-wire or time-projection chambers (TPC). The TPC is currently widely used by particle physicists, and may find application to detecting astronomical neutrinos and cosmic rays in the future. The electron avalanche is confined to a narrow plane by two parallel negatively charged meshes. Sensing wires are spaced at distances of several centimeters. The accelerating voltage is designed so that the electrons gain energy from it at the same rate that they lose it via new ionizations, and they thus move between the meshes at a constant velocity. The time of the interaction is determined using scintillation counters around the drift chamber. The exact position of the original interaction between the sensing wires is then found very precisely by the drift time that it takes the electrons to reach the sensing wire. The imaging cosmic and rare underground signals (ICARUS T600) instrument, for example, currently uses 600 t of liquid argon and is situated in the Gran Sasso tunnel in Italy. It can detect neutrinos via high-energy electrons produced either from electron scattering events or from the conversion of argon to potassium. Rates of four detections of solar neutrinos per ton of argon per year are expected for the latter reaction and about a quarter of that for the former.

Other direct interaction detectors, which are based upon different principles have recently been proposed; for example, the detection of the change in the nuclear spin upon neutrino absorption by $^{115}_{49}$In and its conversion to $^{115}_{50}$Sn. Up to 20 solar neutrinos per day might be found by a detector based upon 10 t of superfluid helium held at a temperature below 0.1 K. Neutrinos would deposit their energy into the helium leading to the evaporation of helium atoms from the surface of the liquid. Those evaporated atoms would then be detected from the energy (heat) that they deposit into a thin layer of silicon placed above the helium.

There is also a suggestion to detect neutrinos from the Čerenkov radio radiation produced by their interaction products within natural salt domes in a similar manner to observing the Čerenkov radio emissions within ice (see above). Alternatively, electrons may be stored within a superconducting ring, and coherent scattering events detected by the change in the current. With this latter system, detection rates for solar neutrinos of one per day may be achievable with volumes for the detector as small as a few milliliters.

1.5.7.2 Geological Detectors

This class of detectors offers the intriguing possibility of looking at the neutrino flux over the past few thousand years. They rely upon determining the isotope ratios for possible neutrino interactions in elements in natural deposits. This is a function of the half-life of the product and the neutrino flux in the recent past, assuming that a state of equilibrium has been reached. Possible candidate elements are tabulated below.

Threshold Element	Reaction	Product Half-Life (Years)	Energy (MeV)
Thallium	$^{205}_{81}\text{Tl} + \nu_e \rightarrow {}^{205}_{82}\text{Pb} + e^-$	3×10^7	0.046
Molybdenum	$^{98}_{42}\text{Mo} + \nu_e \rightarrow {}^{97}_{43}\text{Tc} + n + e^-$	2.6×10^6	8.96
Bromine	$^{81}_{35}\text{Br} + \nu_e \rightarrow {}^{81}_{36}\text{Kr} + e^-$	2.1×10^5	0.490
Potassium	$^{41}_{19}\text{K} + \nu_e \rightarrow {}^{41}_{20}\text{Ca} + e^-$	8×10^4	2.36

Exercise 1.12

Show that if an element and its radioactive reaction product are in an equilibrium state with a steady flux of neutrinos, then the number of decays per second of the reaction product is given by

$$N_p T_{1/2}^{-1} \sum_{n=1}^{\infty} \frac{1}{n} \left(\frac{1}{2}\right)^n$$

when

$T_{1/2} \gg 1 \text{ s}$
N_p is the number density of the reaction product
$T_{1/2}$ is its half-life

Hence, show that the equilibrium ratio of product to original element is given by

$$\frac{N_p}{N_e} = \sigma_\nu F_\nu T_{1/2} \left\{ \left[\sum_{n=1}^{\infty} \frac{1}{n} \left(\frac{1}{2}\right)^n \right]^{-1} \right\}$$

where
N_e is the number density of the original element
σ_ν is the neutrino capture cross section for the reaction
F_ν is the neutrino flux

Exercise 1.13

Calculate the equilibrium ratio of $^{41}_{20}\text{Ca}$ to $^{41}_{19}\text{K}$ for $^{8}_{5}\text{B}$ solar neutrinos ($T_{1/2} = 80{,}000$ years, $\sigma_\nu = 1.45 \times 10^{-46} \text{ m}^{-2}$, $F_\nu = 3 \times 10^{10} \text{ m}^{-2} \text{ s}^{-1}$).

Exercise 1.14

If Davis' chlorine-37 neutrino detector were allowed to reach equilibrium, what would be the total number of $^{37}_{18}\text{Ar}$ atoms to be expected? (Hint: see Exercise 1.12, and note that $\sum_{n=1}^{\infty} \frac{1}{n} \left(\frac{1}{2}\right)^n = 0.693$).

1.6 GRAVITATIONAL RADIATION

1.6.1 Introduction

When the first edition of this book was written 25 years ago, this section began "This section differs from the previous ones because none of the techniques that are discussed have yet indisputably detected gravitational radiation." That is still the case. It is possible that the upgrades to some of the current interferometric gravity wave detectors (see Section 1.6.2) will be sensitive enough to make the first detections by 2010 to 2012. It has also been stated that a third-generation interferometric gravity wave detector (tentatively called the Einstein telescope) would have a "guaranteed" discovery rate of thousands of events per year. Thus, we may hope that perhaps a sixth, seventh, or eighth edition may finally be able to change the opening statement of this section.

One change to the introduction to this section from those in previous editions is possible though. No gravitational waves may have been detected, but nonetheless LIGO (Laser Interferometer Gravitational-wave Observatory) (see Section 1.6.2.2) has produced a result. On the February 1, 2007 a short GRB occurred whose position aligned with one of the Andromeda galaxy's (M31) spiral arms. No gravitational wave was picked up from this event, but if the GRB resulted from the merger of two compact objects (neutron stars and/or black holes) as many astronomers surmise, then it should easily have been detected by LIGO. A negative result therefore implies either that the GRB originated via some other process, if it was indeed located within M31, such as being a soft γ-ray repeater, or the alignment with M31 was just due to chance and the GRB was actually at the usual distance of hundreds to thousands of megaparsecs away from us.

The basic concept of a gravity wave* is simple; if an object with mass changes its position then, in general, its gravitational effect upon another

---

* The term gravity wave is also used to describe oscillations in the earth's atmosphere arising from quite different processes. There is not usually much risk of confusion.

object will change, and the information on that changing gravitational field propagates outward through the space–time continuum at the speed of light. The propagation of the changing field obeys equations analogous to those for electromagnetic radiation provided that a suitable adaptation is made for the lack of anything in gravitation that is equivalent to positive and negative electric charges. Hence, we speak of gravitational radiation and gravitational waves. Their frequencies for astronomical sources are anticipated to run from a few kilohertz for collapsing or exploding objects to a few microhertz for binary star systems. A pair of binary stars coalescing into a single object will emit a burst of waves whose frequency rises rapidly with time: a characteristic "chirp" if we could hear it.

Theoretical difficulties with gravitational radiation arise from the multitudinous metric, curved space–time theories of gravity that are currently extant. The best known of these are due to Einstein (general relativity), Brans and Dicke (scalar–tensor theory), and Hoyle and Narlikar (C-field theory), but there are many others. Einstein's theory forbids dipole radiation, but this is allowed by most of the other theories. Quadrupole radiation is thus the first allowed mode of gravitational radiation in general relativity, and will be about two orders of magnitude weaker than the dipole radiation from a binary system that is predicted by the other theories. Furthermore, there are only two polarization states for the radiation as predicted by general relativity, compared with six for most of the other theories. The possibility of decisive tests for general relativity is thus a strong motive for aspiring gravity wave observers, in addition to the information that may be contained in the waves on such matters as collapsing and colliding stars, supernovae, close binaries, pulsars, early stages of the Big Bang, etc.

The detection problem for gravity waves is best appreciated with the aid of a few order-of-magnitude calculations on their expected intensities. The quadrupole gravitational radiation from a binary system of moderate eccentricity ($e \leq 0.5$), is given in general relativity by

$$L_G \approx \frac{2 \times 10^{-63} M_1^2 M_2^2 (1 + 30e^3)}{(M_1 + M_2)^{2/3} P^{10/3}} \quad \text{W} \tag{1.120}$$

where
M_1 and M_2 are the masses of the components of the binary system
e is the orbital eccentricity
P is the orbital period

Thus, for a typical dwarf nova system with

$$M_1 = M_2 = 1.5 \times 10^{30} \text{ kg} \qquad (1.121)$$

$$e = 0 \qquad (1.122)$$

$$P = 10^4 \text{ s} \qquad (1.123)$$

we have

$$L_G = 2 \times 10^{24} \text{ W} \qquad (1.124)$$

But for an equally typical distance of 250 pc, the flux at the earth is only

$$F_G = 3 \times 10^{-15} \text{ W m}^{-2} \qquad (1.125)$$

This energy will be radiated predominantly at twice the fundamental frequency of the binary with higher harmonics becoming important as the orbital eccentricity increases. Even for a nearby close binary such as ι Boo (distance 23 pc, $M_1 = 2.7 \times 10^{30}$ kg, $M_2 = 1.4 \times 10^{30}$ kg, $e = 0$, $P = 2.3 \times 10^4$ s), the flux only rises to

$$F_G = 5 \times 10^{-14} \text{ W m}^{-2} \qquad (1.126)$$

Rotating elliptical objects radiate quadrupole radiation with an intensity given approximately by

$$L_G \approx \frac{GM^2\omega^6 r^4(A+1)^6(A-1)^2}{64c^5} \text{ W} \qquad (1.127)$$

where
M is the mass
ω the angular velocity
r the minor axis radius
A the ratio of the major and minor axis radii

So that for a pulsar with

$$\omega = 100 \text{ rad s}^{-1} \qquad (1.128)$$

$$r = 15 \text{ km} \qquad (1.129)$$

$$A = 0.99998 \qquad (1.130)$$

we obtain

$$L_G = 1.5 \times 10^{26} \text{ W} \tag{1.131}$$

which for a distance of 1000 pc leads to an estimate of the flux at the earth of

$$F_G = 10^{-14} \text{ W m}^{-2} \tag{1.132}$$

Objects collapsing to form black holes within the galaxy or nearby globular clusters or coalescing binary systems may perhaps produce transient fluxes up to three orders of magnitude higher than these continuous fluxes.

Now these fluxes are relatively large compared with, say, those of interest to radio astronomers, whose faintest sources may have an intensity of 10^{-29} W m^{-2} Hz^{-1}. But the gravitational detectors built to date and those planned for the future have all relied upon detecting the strain ($\delta x/x$) produced in a test object by the tides of the gravitational wave rather than the absolute gravitational wave flux, and this in practice means that measurements of changes in the length of a 1 m long test object of only 5×10^{-21} m, or about 10^{-12} times the diameter of the hydrogen atom, must be obtained even for the detection of the radiation from ι Boo.

In spite of the difficulties that the detection of such small changes must obviously pose, a detector was built by Joseph Weber that appeared to have detected gravity waves from the center of the galaxy early in 1969. Unfortunately, other workers did not confirm this, and the source of Weber's pulses remains a mystery, although they are not now generally attributed to gravity waves. The pressure to confirm Weber's results, however, has led to a wide variety of gravity wave detectors being proposed and built. The detectors so far built, under construction or proposed are of two types: direct detectors in which there is an attempt to detect the radiation experimentally, and indirect detectors in which the existence of the radiation is inferred from the changes that it incidentally produces in other observable properties of the object. The former group may be further subdivided into resonant, or narrow bandwidth detectors, and nonresonant or wideband detectors.

1.6.2 Detectors

1.6.2.1 Direct Resonant Detectors

The vast majority of the early working gravity wave telescopes fall into this category. They are similar to, or improvements on, Weber's original system. This used a massive (>1 t) aluminum cylinder, which was isolated by all possible means from any external disturbance, and whose shape was monitored by piezo-electric crystals attached around its equator. Two such cylinders separated by up to 1000 km were utilized, and only coincident events were regarded as significant to eliminate any remaining external interference. With this system, Weber could detect a strain of 10^{-16} (cf. 5×10^{-21} for the detection of ι Boo). Since his cylinders had a natural vibration frequency of 1.6 kHz, this was also the frequency of the gravitational radiation that they would detect most efficiently. Weber detected about three pulses per day at this frequency with an apparent sidereal correlation and a direction corresponding to the galactic center or a point 180° away from the center. If originating in the galactic center, some 500 solar masses would have to be totally converted into gravitational radiation each year to provide the energy for the pulses. Unfortunately (or perhaps fortunately from the point of view of the continuing existence of the galaxy), the results were not confirmed even by workers using identical equipment to Weber's, and it is now generally agreed that there have been no definite detections of gravitational waves to date.

Adaptations of Weber's system have been made by Aplin, who used two cylinders with the piezo-electric crystals sandwiched in between them, and by Papini, who used large crystals with very low damping constants so that the vibrations would continue for days following a disturbance. The Explorer detector at CERN near Geneva, and the NAUTILUS detector near Rome use 2300 kg bars of aluminum cooled to 2 and 0.1 K, respectively. They have resonant frequencies around 900 Hz. Other systems based upon niobium bars have also been used. Ultimately, bar detectors might achieve strain sensitivities of 10^{-21} or 10^{-22}. At Leiden, a detector based upon a spherical mass of copper–aluminum alloy, known as MiniGRAIL is currently working at a temperature of 5 K and a strain sensitivity of 1.5×10^{-20} at a resonant frequency of 3 kHz. It is expected to reach a strain sensitivity of 4×10^{-21} when operated at 0.05 K, and has the advantage of being equally sensitive to gravity waves from any direction.

1.6.2.2 Direct, Nonresonant Detectors

The direct, nonresonant detectors are of two types and have the potential advantage of being capable of detecting gravity waves with a wide range of frequencies. The first uses a Michelson type interferometer (Section 2.5) to detect the relative changes in the positions of two or more test masses. A possible layout for the system is shown in Figure 1.94, usually, however,

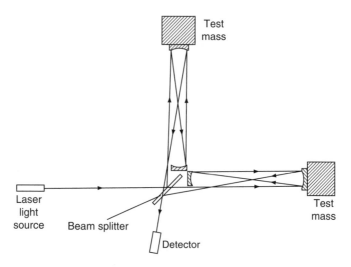

FIGURE 1.94 Possible layout for an interferometric gravity wave detector.

the path length will be amplified by up to a factor of 300 by multiple reflections from the mirrors though these are not shown in the figure to preserve simplicity. The mirror surfaces have to be flat to 0.5% of the operating wavelength. The light sources are high-stability neodymium–yttrium–garnet lasers with outputs of 100 W or more. The mirrors are mounted on test masses on pendulum suspensions, and the whole system, including the light paths, operates in a vacuum. Path length changes are detected by looking at the interference pattern between the two orthogonal beams. Ultimate strain sensitivities of 10^{-22} over a bandwidth of 1 kHz are predicted for these systems, enabling detections of collapsing neutron stars, supernovae, and coalescing binaries to be made out to distances of 10^7 to 10^8 pc. Terrestrial detectors are limited to frequencies above about 10 Hz because of noise. A proposed space-based system, LISA, which will be able to detect much lower frequencies, is discussed below.

Two Michelson-interferometer type gravity wave detectors, known as LIGO, have recently been completed in the United States. Their interferometer arms are 4 km long and they are sited 3000 km apart in Washington state, and in Louisiana, so that gravity waves may be distinguished from other disturbances through coincidence techniques. The Washington machine has a second interferometer with 2 km arms that will also help to separate out gravity wave disturbances from other effects. This is because the gravity wave should induce changes in the 2 km machine that have half the amplitude of those in the 4 km machine, whereas other disturbances are likely to have comparable effects on the two machines. The light beams make 75 round trips along the arms before recombination so that the arms are in effect 600 km long. When no gravitational disturbance is present there is destructive interference between the two beams at the detector (i.e., no signal) and the light can be returned to the arms of the interferometer, thus increasing the power available—a technique called power or signal recycling. It is marginal whether LIGO will initially be sensitive enough to detect gravity waves and as of 2007 had not done so, although it achieved its designed strain sensitivity of 10^{-21} in 2005. Upgrades of LIGO are planned—Enhanced LIGO in 2008 which among other improvements will use more powerful lasers and improve the strain sensitivity by a factor of two or three, and Advanced LIGO which will start construction in 2008 and should achieve a strain sensitivity of 10^{-22}. Anyone with a computer and an Internet link can help with the data processing part of LIGO and GEO600 through the Einstein@-home project. This currently has some 180,000 volunteers whose computers process data for the project when they would otherwise be idle. An Internet search for einstein@home will take anyone interested in joining the project straight to its home page.

Also currently starting to operate or nearing completion are the Italian–French VIRGO with 20 multiple reflections along its 3 km arms. It is sited near Pisa and had its first science run in 2007. The German–British GEO600 at Hanover with 600 m arms started to operate jointly with LIGO in 2005. Like LIGO it also uses power recycling to enhance its sensitivity. The Japanese TAMA near Tokyo has arms 300 m long and plans are being considered for the large-scale cryogenic gravity wave telescope (LCGT) that will have two sets of 3 km long arms, be situated in the Kamioka mine, and perhaps be operational by 2009. The distribution of these detectors over the earth will not only provide confirmation of detections but also enable the arrival directions of the waves to be pinpointed to a few minutes of arc. Consideration is just starting to be given to the third generation of interferometric gravity wave detectors with the recent start of the European Einstein telescope feasibility study. However, at the time of writing, little progress has been made beyond identifying a name for the project.

There is also a proposal for a space-based interferometer system called LISA. This would employ three drag-free spacecraft (Figure 1.96) at the corners of an equilateral triangle with 5,000,000 km sides. Each spacecraft would be able to operate as the vertex of the interferometer, and as the proof mass for the other spacecraft. LISA would thus have three separate interferometer systems. It would orbit at 1 AU from the Sun, but some 20° behind the earth. Its sensitivity would probably be comparable with the best of the terrestrial gravity wave detectors, but its low noise would enable it to detect the important low-frequency waves from binary stars, etc. LISA is currently planned as a joint ESA-NASA mission for 2018 with a smaller scale LISA pathfinder mission perhaps by 2011. Beyond this, there may be the Big Bang Observer which is envisaged to comprise four instruments like LISA and be able to detect gravity waves from soon after the Big Bang.

An ingenious idea underlies the second method in this class. An artificial satellite and the earth will be independently influenced by gravity waves whose wavelength is less than the physical separation of the two objects. If an accurate frequency source on the earth is broadcast to the satellite, and then returned to the earth (Figure 1.95), its frequency will be changed by the Doppler shift as the earth or satellite is moved by a gravity wave. Each gravity wave pulse will be observed three times enabling the reliability of the detection to be improved. The three detections will correspond to the effect of the wave on the transmitter, the satellite, and the receiver, although not necessarily in that order. A drag-free satellite (Figure 1.96) would be required to reduce external perturbations arising from the solar wind, radiation pressure, etc. A lengthy series of X-band observations using the *Ulysses* spacecraft aimed at detecting gravity waves by this method have been made recently, but so far without any successful detections.

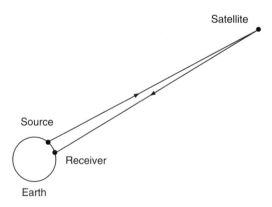

FIGURE 1.95 Arrangement for a satellite-based Doppler tracking gravity wave detector.

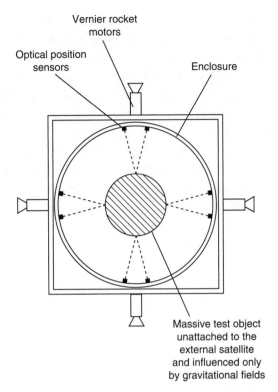

FIGURE 1.96 Schematic cross section through a drag-free satellite. The position of the massive test object is sensed optically and the external satellite driven so that it keeps the test object centered.

1.6.2.3 Indirect Detectors

Proposals for these so far only involve binary star systems. The principle of the method is to attempt to observe the effect on the period of the binary of the loss of angular momentum due to the gravitational radiation. A loss of angular momentum by some process is required to explain the present separations of dwarf novae and other close binary systems with evolved components. These separations are so small that the white dwarf would have completely absorbed its companion during its earlier giant stage, thus the separation at that time must have been greater than it is now. Gravitational radiation can provide an adequate orbital angular momentum loss to explain the observations, but it is not the only possible mechanism. Stellar winds, turbulent mass transfer, and tides may also operate to reduce the separation, so that these systems do not provide unequivocal evidence for gravitational radiation.

The prime candidate for study for evidence of gravitational radiation is the binary pulsar (PSR 1913 + 16). Its orbital period is 2.8×10^4 s, and it was found to be decreasing at a rate of 10^{-4} s year^{-1} soon after the system's discovery. This was initially attributed to gravitational radiation, and the 1993 Nobel physics prize was awarded to Russell Hulse and Joseph Taylor, for their discovery of the pulsar and their interpretation of its orbital decay. However, it now appears that a helium star may be a part of the system, and so tides or other interactions could again explain the observations. On the other hand, the effect of a rapid pulsar on the local interstellar medium might mimic the appearance of a helium star when viewed from a distance of 5000 pc. Thus, the detection of gravity waves from the binary pulsar, and indeed by all other methods remains not proven at the time of writing.

Exercise 1.15

Show that the gravitational radiation of a planet orbiting the Sun is approximately given by

$$L_G \approx 7 \times 10^{21} M^2 P^{-10/3} \text{ W}$$

where
M is the planet's mass in units of the solar mass
P the planet's orbital period in days

Exercise 1.16

Calculate the gravitational radiation for each planet in the solar system and show that every planet radiates more energy than the combined total radiated by those planets whose individual gravitational luminosities are lower than its own.

1.7 DARK MATTER AND DARK ENERGY DETECTION

1.7.1 Introduction

Main stream cosmology currently suggests that the visible portion of the universe (stars, nebulae, planets, galaxies, etc.) makes up only about 4% of the total content of the universe. Dark matter makes up about another 22% and dark energy the remaining 74%.

The existence of dark matter was first suggested by Fritz Zwicky (1898–1974) in 1933. His observations of the velocities of the galaxies within the

Coma galaxy cluster showed that they were too high for the galaxies to be retained by the cluster's gravitational field. Over a few hundred million years, the cluster would evaporate and the galaxies move away as independent entities. However, the existence of many clusters of galaxies suggests that they are stable over much longer periods of time. Zwicky therefore theorized that there must be additional material, within the cluster or galaxies or both, whose presence meant that the gravitational field of the cluster was sufficient for the cluster to be stable. Since this matter was not directly visible it became known as dark matter. Zwicky found that the amount of dark matter needed to stabilize the Coma cluster was 400 times the amount of visible matter. However, subsequent measurements by many astronomers have brought this figure down to the quantity of dark matter required generally being five to six times that of the visible matter.

Three-quarters of a century have passed since Zwicky's observations and although the existence of dark matter is confirmed by many other circumstantial observations—such as the rotation curves of galaxies—the nature of dark matter is still a mystery.* Suggestions for what it might be made up from have included mini black holes, brown dwarfs, large planets (together known as MACHOs, massive astrophysical compact halo objects), neutrinos, and WIMPs (weakly interacting massive particles). The current front runner is a variety of WIMP called a neutralino that is predicted by the particle physicists' supersymmetry theory to be produced in large numbers at the energies prevalent during the early stages of the Big Bang. Despite the plentiful secondary evidence for the existence of dark matter, three-quarters of a century is a long time for no direct evidence of its existence to have been found and some workers now question its existence at all—suggesting that some modification to the way in which the force of gravity operates could explain the observational discrepancies instead. Much effort is therefore currently being put into experiments that might detect the particles, whatever they may be, making up dark matter.

The existence of dark energy, like that of dark matter, has yet to be proven. There are two main circumstantial lines of evidence for its existence—the mean density of the universe and the brightness of Type Ia supernovae. The critical density of the universe is the density that would

---

* A similar situation though existed for the neutrino–its existence was predicted by Wolfgang Pauli in 1930, but it was not found experimentally until 1956.

enable gravity to slow down the expansion of the universe to zero after an infinite length of time. A density lower than the critical value and the universe will expand forever, a higher density and its expansion will eventually halt and it will collapse back again, perhaps ending in the "big crunch." The actual density of the matter and dark matter in the universe amounts to about 26% of the critical density. For reasons beyond the scope of this book,* if the apparent mean density is as close to the critical density as it appears to be, then the true mean density must almost certainly be exactly equal to the critical density. Thus, something other than matter must comprise 74% of the universe and that "something" is called dark energy.

The second line of evidence for dark energy emerged in the late 1990s. Observations of distant ($\geq$2000 Mpc) Type Ia supernovae seemed to suggest that after an initial period during which the expansion of the universe was decelerating, it is now accelerating. One way of causing such acceleration would be the presence within the universe of a large amount of dark energy. If the density of the universe is assumed to be the critical density, then the amount of dark energy needed to provide the observed acceleration turns out to be around 75% of the total mass/energy of the universe.

However, like the case for dark matter, an alternative explanation for dark energy can be found through a modified form of gravity (often called MOND, modified Newtonian dynamics). A second possibility is if the nature of Type Ia supernovae has changed over time so that they are seen as fainter than expected at large distances because they are fainter than modern supernovae, not because they are more distant than predicted. Thus, observational confirmation of the existence of dark energy is also eagerly being sought.

1.7.2 Dark Matter and Dark Energy Detectors

No new types of detectors are described in this section—the instruments used in the searches for dark matter and dark energy are among those that we have seen in the earlier sections of this chapter. The searches use the existing results and capabilities of those instruments to look for data with the signature of some aspects of dark matter or dark energy. For dark

---

* See *Dark Side of the Universe* by I. Nicolson (Canopus Publishing, 2007) for an excellent explanation for all of this from a layman's point of view.

matter, the search concentrates on trying to detect the neutralino (if that is what makes up dark matter) via detectors such as those used for γ rays and neutrinos, while evidence is sought for dark energy via its influence on the microwave cosmic background radiation or on galaxy formation and clustering or via improved observations of very distant supernovae.

Dark matter detectors divide into direct and indirect types. The direct types look for interactions of the dark matter particle (which we shall call the neutralino for the rest of this section even though it may be another type of particle altogether) with atoms within the test apparatus. Those interactions are primarily scintillations or ionizations arising from a high-energy normal matter particle produced by a collision with a neutralino and acoustic pulses arising from the heat deposited into the material surrounding such a collision. Discussions of these types of detectors may be found within Sections 1.3 through 1.5. The Dark matter detector (DAMA) which is located at Gran Sasso in Italy, for example, uses 250 kg of sodium iodide to try and detect scintillations arising from neutralinos, while in the United Kingdom at the Boulby salt mine another sodium iodide scintillation detector is operated alongside a gaseous drift chamber. Also housed at the Boulby mine are the various zoned proportional scintillation in liquid noble gases (ZEPLIN) detectors that are planned eventually to use a ton of xenon, again as a scintillation detector. The French Edelweiss experiment uses both ionization and heat deposition in germanium bolometers cooled to 20 mK while the cryogenic dark matter search (CDMS) uses silicon and germanium micro-calorimeters. Indirect dark matter detectors are identical to the Čerenkov γ-ray detectors discussed in Sections 1.3.2.9 and 1.4.3.3 and search for neutralino interactions high in the earth's atmosphere. Indirect detection may also be possible via the products of neutralino interactions or decays far out in space. Among these products there may be anti-protons that could be observed directly by some balloon-borne cosmic ray detectors and positrons that could be found via the 511 keV γ-ray emission line produced when positrons and electrons annihilate each other.

All dark matter detectors suffer from the problem of distinguishing real signals from background noise. Confirmation of the detection of real signals may be helped since their intensities are expected to vary with the seasons or even daily as the earth's orbital or rotation motion adds to or subtracts from the speed of the neutralinos. No neutralino detectors have yet made any definite detections, though the DAMA experiment appeared to show a seasonal variation in intensity but this has not been confirmed by other experiments.

A quite different approach to confirming the existence of dark matter would be to make some ourselves—and that is exactly what the large hadron collider (LHC) may be able to do. The LHC is a particle accelerator currently under construction at CERN on the Swiss/French border. It is scheduled to start science operation in May 2008 when two beams of protons will be collided head-on at a combined energy of 1.4×10^{13} eV (14 TeV). Among the many by-products of the collisions there may be neutralinos.

Confirmation of the existence of dark energy and of its nature (there are several possibilities such as Einstein's cosmological constant, quintessence, and phantom energy—see, for example, *Dark Side of the Universe* for further information) may come through its effect upon the large-scale structure of ordinary matter throughout the universe. Galaxies and clusters of galaxies started to form very soon after the Big Bang and their presence produces ripples in the intensity of the cosmic microwave background (CMB) radiation. Many other processes also produce ripples in the CMB and although the existence of ripples arising from dark energy has been sought in the data from the *WMAP* spacecraft, nothing definitive has yet been found. The 10 m South Pole telescope, which is being commissioned at the time of writing, may be able to improve on *WMAP*'s results, While ESA's Planck mission, due for launch in 2008, may also provide data of sufficient quality to provide better answers. Planck is planned to observe the CMB using an off-axis 1.5×1.75 m Gregorian telescope over the frequency range 27 GHz to 1 THz (11 mm to 0.3 mm) for around 21 months (the lifetime is limited by the loss of the helium coolant).

The large-scale three-dimensional structure of the universe can be studied directly by observing galaxies and clusters of galaxies. To date, surveys such as the Sloan Digital Sky Survey (SDSS) and the AAO's 2dF Galaxy Redshift Survey have suggested that dark energy may take the form of Einstein's cosmological constant. Construction of the large synoptic survey telescope (LSST) has recently begun. This will be an 8.4 m telescope designed to have a 10 square degree field of view and using a 3000 megapixel camera. It will be able to image all the available sky with 15 s exposures every three nights, covering three-quarters of the whole sky over a year. It is expected to start operating in 2012 and the resulting survey should provide a very detailed 3D map of large-scale structure of the universe. In the longer term, the SKA, if built, should provide data on at least a billion galaxies.

Studies of distant supernovae provide a history of the acceleration/deceleration of the universe over time and this in turn can put constraints on the nature of dark energy. Such data has already come from surveys like the SDSS and forms the current observational basis for the existence of dark energy. Supernova searches are continuing at several observatories with the eight 2k × 4k CCD array MOSAIC camera on the 4 m telescope at the Cerro Tololo Inter American Observatory and the CFHT's thirty-six 2k × 4k CCD MegaCAM being fairly typical. The LSST will be able to detect thousands of supernovae per year out to distances of 11,000 Mly (3300 Mpc). Additionally, NASA's design study, the Joint Dark Energy Mission is currently considering several proposals for possible launch during 2012–2015. These include the Supernova/Acceleration Probe (SNAP), which would carry a 2 m telescope and be able to measure supernova redshifts and luminosities out to 10,000 Mly (3200 Mpc) and the Joint Efficient Dark Energy Investigation (JEDI), also with a 2 m telescope but capable of picking up Type Ia supernovae out to 13,000 Mly (3900 Mpc). A similar proposal that has recently been awarded initial funding by NASA is for the advanced dark energy physics telescope (ADEPT), which would observe 100 million galaxies and 1000 supernovae using a 1.1 m NIR telescope. Neutrinos from ancient supernovae (of any type) might also provide clues to dark energy if they interact with it during their journeys through space, so that third-generation neutrino detectors (Section 1.5) may provide information about it in due course.

Clues to the large-scale structure of the universe and so to dark energy may also come from studies of the x-ray emission from clusters of galaxies, from weak gravitational lensing and from the frozen-in signatures of acoustic waves in the early universe—now about 500 Mly (160 Mpc) across.

One way or another, we may hope that the existence of dark matter and dark energy may be confirmed or refuted within the next decade and at least some clue as to their natures, if they exist, found.

CHAPTER 2

Imaging

2.1 INVERSE PROBLEM

A problem that occurs throughout much of astronomy and other remote sensing applications is how best to interpret noisy data so that the resulting deduced quantities are real and not artifacts of the noise. This problem is termed the inverse problem.

For example, stellar magnetic fields may be deduced from the polarization of the wings of spectrum lines (Section 5.2.3). The noise (errors, uncertainties) in the observations, however, will mean that a range of field strengths and orientations will fit the data equally well. The three-dimensional distribution of stars in a globular cluster must be found from observations of its two-dimensional projection onto the plane of the sky. Errors in the measurements will lead to a variety of distributions being equally good fits to the data. Similarly, the images from radio and other interferometers (Section 2.5) contain spurious features because of the side lobes of the beams. These spurious features may be removed from the image if the effects of the side lobes are known. But since there will be uncertainty in both the data and the measurements of the side lobes, there can remain the possibility that features in the final image are artifacts of the noise, or are incompletely removed side lobe effects.

The latter illustration is an instance of the general problem of instrumental degradation of data. Such degradation occurs from all measurements, since no instrument is perfect: even a faultlessly constructed telescope will spread the image of a true point source into the Airy diffraction pattern (Figure 1.29). If the effect of the instrument and other sources of blurring on a point source or its equivalent are known (the point spread function [PSF] or instrumental profile), then an attempt may be made to remove its effect from the data. The process of removing instrumental effects from data can be necessary for any type of measurement, but is perhaps best studied in relation to imaging, when the process is generally known as deconvolution.

2.1.1 Deconvolution

This form of the inverse problem is known as deconvolution, because the true image convolves with the PSF to give the observed (or dirty) image. Inversion of its effect is thus deconvolution. Convolution is most easily illustrated in the one-dimensional case (such as the image of a spectrum or the output from a Mills Cross radio array), but applies equally well to two-dimensional images.

A one-dimensional image, such as a spectrum, may be completely represented by a graph of its intensity against the distance along the image (Figure 2.1). The PSF may similarly be plotted and may be found, in the case of a spectrum, by observing the effect of the spectroscope on a monochromatic source.

If we regard the true spectrum as a collection of adjoining monochromatic intensities, then the effect of the spectroscope will be to broaden each monochromatic intensity into the PSF. At a given point (wavelength) in the observed spectrum, some of the original energy will be displaced out to nearby wavelengths, while energy will be added from the spreading out

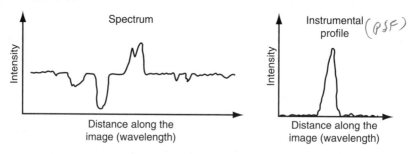

FIGURE 2.1 Representation of a one-dimensional image and the instrumental profile (PSF) plotted as intensity versus distance along image.

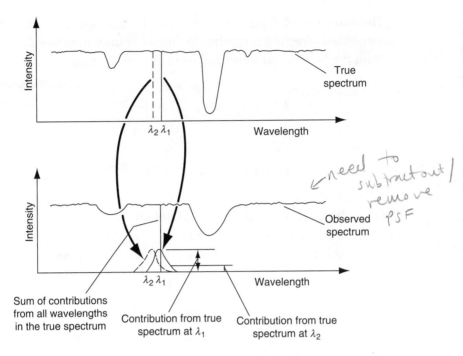

FIGURE 2.2 Convolution of the true spectrum with the PSF to produce the observed spectrum.

of nearby wavelengths (Figure 2.2). The process may be expressed mathematically by the convolution integral:

$$O(\lambda_1) = \int_0^\infty T(\lambda_2)I(\lambda_1 - \lambda_2)\,d\lambda_2 \qquad (2.1)$$

where
$O(\lambda_1)$ is the intensity in the observed spectrum at wavelength λ_1
$T(\lambda_2)$ is the intensity in the true spectrum at wavelength λ_2
$I(\lambda_1 - \lambda_2)$ is the response of the instrument (spectroscope) at a distance $(\lambda_1 - \lambda_2)$ from its center

Equation 2.1 is normally abbreviated to

$$O = T*I \qquad (2.2)$$

where * is the convolution symbol.

The inversion of Equation 2.1 to give the true spectrum cannot be accomplished directly but involves the use of Fourier transforms.

With the Fourier transform and its inverse in the form

$$F(s) = F[f(x)] = \int_{-\infty}^{\infty} f(x)e^{-2\pi ixs}\, dx \qquad (2.3)$$

$$f(x) = F^{-1}[F(s)] = \int_{-\infty}^{\infty} F(s)e^{2\pi ixs}\, ds \qquad (2.4)$$

the convolution theorem states that convolution of two functions corresponds to the multiplication of their Fourier transforms.

Thus, taking Fourier transforms of Equation 2.2, we have

$$F(O) = F(T^*I) \qquad (2.5)$$

$$= F(T) \times F(I) \qquad (2.6)$$

and so the true spectrum may be found from inverting Equation 2.6 and taking its inverse Fourier transform:

$$T = F^{-1}\left[\frac{F(O)}{F(I)}\right] \qquad (2.7)$$

In practice, obtaining the true data (or source function) via Equation 2.7 is complicated by two factors. First, data is sampled at discrete intervals and so is not the continuous function required by Equations 2.3 and 2.4, and also it is not available over the complete range from $-\infty$ to $+\infty$. Second, the presence of noise will produce ambiguities in calculating the values of T.

The first problem may be overcome by using the discrete versions of the Fourier transform and inverse transform

$$F_D(s_n) = F_D[f(x)]_n = \sum_{k=0}^{N-1} f(x_k)e^{-2\pi ikn/N}\Delta \qquad (2.8)$$

$$f(x_k) = F_D^{-1}[F_D(s_n)] = \sum_{n=0}^{N-1} F_D(s_n)e^{2\pi ikn/N}\frac{1}{N} \qquad (2.9)$$

where
$F_D(s_n)$ is the nth value of the discrete Fourier transform of $f(x)$
N is the total number of measurements

$f(x_k)$ is the kth measurement

Δ is the step length between measurements

and setting the functions to zero outside the measured range

Now a function that has a maximum frequency of f is completely determined by sampling at $2f$ (the sampling theorem). Thus, the use of the discrete Fourier transform involves no loss of information provided the sampling frequency $(1/\Delta)$ is twice the highest frequency in the source function. The highest frequency that can be determined for a given sampling interval $(1/2\Delta)$ is known as the Nyquist or critical frequency. If the source function contains frequencies higher than the Nyquist frequency, then these will not be determined by the measurements and the finer detail in the source function will be lost when it is reconstituted via Equation 2.7. Rather more seriously, however, the higher frequency components may beat with the measuring frequency to produce spurious components at frequencies lower than the Nyquist frequency. This phenomenon is known as aliasing and can give rise to major problems in finding the true source function.

The actual evaluation of the transforms and inverse transforms may nowadays be relatively easily accomplished using the fast Fourier transform algorithm on even quite small computers. The details of this algorithm are outside the scope of this book, but may be found from books on numerical computing.

The one-dimensional case just considered may be directly extended to two or more dimensions, though the number of calculations involved then increases dramatically. Thus, for example, the two-dimensional Fourier transform equations are

$$F(s_1, s_2) = F[f(x_1, x_2)] = \int\limits_{-\infty}^{\infty} \int\limits_{-\infty}^{\infty} f(x_1, x_2) e^{-2\pi i x_1 s_1} e^{-2\pi i x_2 s_2} \, dx_1 \, dx_2 \quad (2.10)$$

$$f(x_1, x_2) = F^{-1}[F(s_1, s_2)] = \int\limits_{-\infty}^{\infty} \int\limits_{-\infty}^{\infty} F(s_1, s_2) e^{2\pi i x_1 s_1} e^{2\pi i x_2 s_2} \, ds_1 \, ds_2 \quad (2.11)$$

Some reduction in the noise in the data may be achieved by operating on its Fourier transform. In particular, cutting back on or removing the corresponding frequencies in the transform domain may reduce cyclic noise such as 50 or 60 Hz mains hum or the stripes on scanned images. Random noise may be reduced by using the optimal (or Wiener) filter defined by

$$W = \frac{[F(O)]^2}{[F(O)]^2 + [F(N)]^2} \qquad (2.12)$$

where

$F(O)$ is the Fourier transform of the observations, without the effect of the random noise

$F(N)$ is the Fourier transform of the random noise

(The noise and the noise-free signal are separated by assuming the high-frequency tail of the power spectrum to be just due to noise, and then extrapolating linearly back to the lower frequencies.) Equation 2.7 then becomes

$$T = F^{-1} \left[\frac{F(O)W}{F(I)} \right] \qquad (2.13)$$

While processes, such as those outlined above, may reduce noise in the data, it can never be totally eliminated. The effect of the residual noise, as previously mentioned, is to cause uncertainties in the deduced quantities. The problem is often ill-conditioned, that is to say, the uncertainties in the deduced quantities may, proportionally, be very much greater than those in the original data.

A widely used technique especially for reducing the instrumental broadening for images obtained via radio telescope was invented by W.H. Richardson and Leon Lucy. The Richardson–Lucy (RL) algorithm is an iterative procedure in which the $n + 1$th approximation to the true image is related to the PSF and nth approximation by

$$T_{n+1} = T_n \int \frac{O}{T_n * I} I \qquad (2.14)$$

The first approximation to start the iterative process is usually taken as the observed data. The RL algorithm has the advantage, compared with some other iterative techniques, of producing a normalized approximation to the true data without any negative values, and it also usually converges quite rapidly.

Recently, several methods have been developed to aid choosing the best version of the deduced quantities from the range of possibilities. Termed "non-classical" methods they aim to stabilize the problem by introducing additional information not inherently present in the data as constraints, and thus to arrive at a unique solution. The best known of these methods is the maximum entropy method (MEM). The MEM introduces the external constraint that the intensity cannot be negative, and finds the solution that has the least structure in it that is consistent with the data. The name derives from the concept of entropy as the inverse of the structure (or information content) of a system. The maximum entropy solution is thus the one with the least structure (the least information content or the smoothest solution) that is consistent with the data. The commonest measure of the entropy is

$$s = -\sum p_i \ln p_i \qquad (2.15)$$

where

$$p_i = \frac{d_i}{\sum_j d_j} \qquad (2.16)$$

and d_i is the ith datum value, but other measures can be used. A solution obtained by an MEM has the advantage that any features in it must be real and not artifacts of the noise. However, it also has the disadvantages of perhaps throwing away information, and that the resolution in the solution is variable, being highest at the strongest values.

Other approaches can be used either separately or in conjunction with MEM to try and improve the solution. The CLEAN method, much used on interferometer images, is discussed in Section 2.5.5. The method of regularization stabilizes the solution by minimizing the norm of the second derivative of the solution as a constraint on the smoothness of the solution. The non-negative least squares approach (NNLS) solves the data algebraically, but subject to the added constraint that there are no negative elements. Myopic deconvolution attempts to correct the image when the PSF is poorly known by determining the PSF as well as the corrected image from the observed image. While blind deconvolution is used when the PSF is unknown and requires several images of the same object, preferably with dissimilar PSFs as occurs with adaptive optics images (Section 1.1).

Burger–van Cittert deconvolution first convolves the PSF with the observed image and then adds the difference between the convolved and observed images from the observed image. The process is repeated with the new image and iterations continue until the result of convolving the image with the PSF matches the observed image. The $n + 1$th approximation is thus

$$T_{n+1} = T_n + (O - T_n * I) \qquad (2.17)$$

Compared with the RL approach, the Burger–van Cittert algorithm has the disadvantage of spurious producing negative values. The recently developed technique of spectral deconvolution is based upon the relative positions of two real objects (such as a bright star with a much fainter companion) being the same at different wavelengths, while artifacts (such as internal reflections, etc.) will generally change their positions with wavelength thus enabling real objects to be separated from false ones.

2.2 PHOTOGRAPHY

2.2.1 Introduction

When the first edition of this book appeared in 1984, photography was still the main means of imaging in astronomy throughout the visual region and into the very near infrared (to about 1 μm). Charge-coupled devices (CCDs) were starting to come on the scene, but were still small and expensive. The situation is now completely reversed. Photography is hardly used at all in professional astronomy—for example, the last photograph on the Anglo-Australian telescope (AAT) was taken in 1999. CCDs now dominate imaging from the ultraviolet to the near infrared, though they are still expensive in the largest sizes and are themselves perhaps on the verge of replacement by wavelength-sensitive imaging detectors, such as superconducting tunnel junction detectors (STJs). Amongst amateur astronomers, small cooled CCDs are now quite common and can be purchased along with sophisticated software for image processing for much less than the cost of a 0.2 m Schmidt–Cassegrain telescope. Digital still and video cameras produced for the popular market that are very cheap can also be used on telescopes, although without the long exposures possible with cooled CCDs specifically manufactured for astronomical use.

However, photography is not quite dead yet! Many, perhaps most, archives are still in the form of photographs or data derived from photographs so that even recent major surveys such as the multi-mission archive for the space telescope's (MAST) digitized sky survey that produced the *Hubble Guide Star Catalog* (GSC) and the Anglo–Australian Observatory/United Kingdom Schmidt Telescope (AAO/UKST) H-α survey are based upon photographic images. Any web search for astronomy images will also show that photography is still widely used amongst amateur astronomers (and many professional astronomers are amateur astronomers in their spare time) despite the numbers who use CCDs. Thus, some coverage of photography still needs to be included, if only so that a student working on archive photographs is aware of how they were produced and what their limitations and problems may be.

Photography's preeminence as a recording device throughout most of the twentieth century arose primarily from its ability to provide a permanent record of an observation that is (largely) independent of the observer. But it also has many other major advantages including its cheapness, its ability to accumulate light over a long period of time and so to detect sources fainter than those visible to the eye, its very high information density, and its ease of storage for future reference. These undoubted advantages coupled with our familiarity with the technique have, however, tended to blind us to its many disadvantages. The most important of these are its slow speed and very low quantum efficiency. The human eye has an effective exposure time of about a tenth of a second, but a photograph would need an exposure of ten to a hundred times longer to show the same detail. The impression that photography is far more sensitive than the eye only arises because most photographs of astronomical objects have exposures ranging from minutes to hours and so show far more detail than is seen directly. The processing of a photograph is a complex procedure and there are many possibilities for the introduction of errors. Furthermore, and by no means finally in this list of problems of the photographic method, the final image is recorded as a density, and the conversion of this back to intensity is a tedious process. The whole business of photography in fact bears more resemblance to alchemy than to science, especially when it has to be used near the limit of its capabilities.

Color photography has no place in astronomy, except for taking pretty pictures of the Orion nebula, etc. The reason for this is that the color balance of both reversal (slide) and print color films is optimized for exposures of a few seconds or less. Long exposures, such as those needed in astronomy, therefore result in false colors for the object. When a color image is needed, it is normally obtained by three monochromatic (black and white) images obtained through magenta, yellow, and cyan filters that are then printed onto a single final image using the same filters.

2.2.2 Structure of the Photographic Emulsion

Many materials exhibit sensitivity to light that might potentially be used to produce images. In practice, however, a compound of silver and one or more of the halogens is used almost exclusively. This arises from the relatively high sensitivity of such compounds allied to their ability to retain the latent image for long periods of time. For most purposes, the highest sensitivity is realized using silver bromide with a small percentage of iodide ions in solid solution. The silver halogen is in the form of tiny crystals, and these are supported in a solid, transparent medium, the whole structure is called the photographic emulsion. The supporting medium is usually gelatin, which is a complex organic material manufactured from animal collagen, whose molecules may have masses up to half a million atomic mass unit. Gelatin has the advantage of adding to the stability of the latent image by absorbing the halogens that are released during its formation. It also allows easy penetration of the processing chemicals, and forms the mixing medium during the manufacture of the emulsion.

The size of the silver halide crystals, or grains as they are commonly known, is governed by two conflicting requirements. The first of these is resolution, or the ability of the film to reproduce fine details. Resolution is affected by both the grain size and by scattering of the light within the emulsion, and is measured by the maximum number of lines per millimeter that can be distinguished in images of gratings. It ranges from about 20 to 2000 lines per millimeter, with generally the smaller the grain size the higher the resolution. The modulation transfer function (MTF) specifies more precisely the resolution of an emulsion. The definition of the MTF is in terms of the recording of an image that varies sinusoidally in intensity with distance. In terms of the quantities shown in Figure 2.3, the modulation transfer, T, is given by

$$T = \frac{(A_R/I_R)}{(A_O/I_O)} \tag{2.18}$$

The MTF is then the manner in which T varies with spatial frequency. Schematic examples of the MTF for some emulsions are shown in Figure 2.4. Using an MTF curve, the resolution may be defined as the spatial frequency at which the recorded spatial variations become imperceptible. For visual observation, this usually occurs for values of T near 0.05.

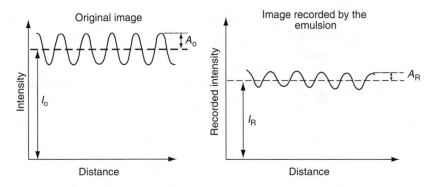

FIGURE 2.3 Schematic representation of the reproduction of an image varying spatially in a sinusoidal manner by an emulsion.

The second requirement is that of the speed of the emulsion, and this is essentially the reciprocal of a measure of the time taken to reach a given image density under some standard illumination. The two systems in common use for measuring the speed of an emulsion are ISO (international organization for standardization), based upon the exposure times for image densities of 0.1 and 0.9, and DIN (Deutsches institut für normung), based only on the exposure time to reach an image density of 0.1. ISO is an arithmetic scale, while DIN is logarithmic. Their relationship

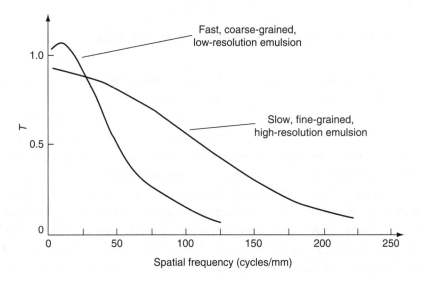

FIGURE 2.4 Schematic MTF curves for typical emulsions.

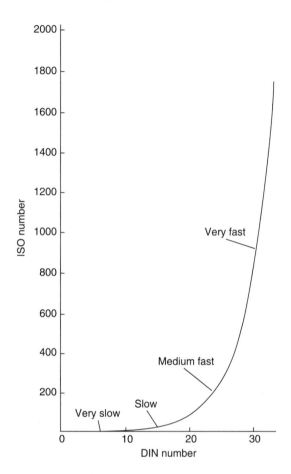

FIGURE 2.5 Emulsion speeds (approximate relationship).

is shown in Figure 2.5, together with an indication of the speeds of normal films. The two numbers are frequently combined into the exposure index number (EI) that is simply the ISO number followed by the DIN number, as in EI 100/21°. The speed of an emulsion is proportional to the volume of a grain for unsensitized emulsions, and to the surface area of a grain for dye-sensitized (see below) emulsions. Thus, in either case, higher speed requires larger grains. The conflict between these two requirements means that high-resolution films are slow and that fast films have poor resolution. Grain sizes range from 50 nm for a very high-resolution film, through 800 nm for a normal slow film to 1100 nm for a fast film. Not all the grains are the same size, except in nuclear emulsions (Section 1.4), and the standard deviation of the size distribution curve ranges from about 1% of the particle diameter for very high-resolution films to 50% or more for normal commercial films.

A normal silver halide emulsion is sensitive only to short wavelengths, that is, to the blue, violet, and near ultraviolet parts of the spectrum. Hermann Vogel made the fundamental discovery that was to render lifelike photography possible and to extend the sensitivity into the red and infrared in 1873. The discovery was that certain dyes when adsorbed onto the surfaces of the silver halide grains would absorb light at their own characteristic frequencies, and then transfer the absorbed energy to the silver halide. The latent image would then be produced in the same manner as normal. The effect of the dye can be very great; a few spectral response curves are shown in Figure 2.6 for unsensitized and sensitized emulsions. Panchromatic or isochromatic emulsions have an almost uniform sensitivity throughout the visible region, while the orthochromatic emulsions mimic the response of the human eye. Color film would of course be quite impossible without the use of dyes since it uses three emulsions with differing spectral responses. The advantages of dye sensitization are very slightly counteracted by the reduction in the sensitivity of the emulsion in its original spectral region, because of the absorption by the dye, and more seriously by the introduction of some chemical fogging.

In many applications, the response of the film can be used to highlight an item of interest, for example, comparison of photographs of the same area of sky with blue- and red-sensitive emulsions is a simple way of finding very hot and very cold stars.

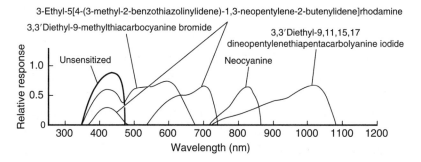

FIGURE 2.6 Effects of dyes upon the spectral sensitivity of photographic emulsion.

2.2.3 Photographic Image

When an emulsion is exposed to light, the effect is to form the latent image, that is, an image that requires the further step of development to be visible. Its production on silver halide grains is thought to arise in the following manner.

1. Absorption of a photon by an electron occurs in the valence band of the silver halide. The electron thereby acquires sufficient energy to move into the conduction band (see Section 1.1 for a discussion of solid state energy levels).

2. Removal of the electron leaves a positive hole in the valence band.

3. Both the electron and the hole are mobile. If they meet again, they will recombine and emit a photon. There will then be no permanent effect. To avoid this recombination, the electron and the hole must be separated by their involvement in alternative reactions. The electrons are immobilized at chemical or physical traps formed from impurities or crystal defects. The positive hole's motion eventually carries it to the surface of the grain. There it can be consumed directly by reaction with the gelatin, or two holes can combine releasing a halogen atom that will then in turn be absorbed by the gelatin.

4. Electron, now immobilized, neutralizes a mobile silver ion leaving a silver atom within the crystal structure.

5. Effect of the original electron trap is now enhanced by the presence of the silver atom, and it may more easily capture further electrons. In this way, a speck of pure silver containing a few to a few hundred atoms is formed within the silver halide crystal.

6. Developers are reducing solutions that convert silver halide into silver. They act only very slowly; however, on pure silver halide, the specks of silver act as catalysts, so that those grains containing silver specks are reduced rapidly to pure silver. Since a 1 μm grain will contain 10^{10} to 10^{11} silver atoms, this represents an amplification of the latent image by a factor of 10^9. Adjacent grains that do not contain an initial silver speck will be unaffected and will continue to react at the normal slow rate.

7. Thus, the latent image consists of those grains that have developable specks of silver on them. For the most sensitive emulsions, 3–6 silver atoms per grain will suffice, but higher numbers are required for less

sensitive emulsions, and specks comprising 10–12 silver atoms normally represent the minimum necessary for a stable latent image in normal emulsions. The latent image is therefore turned into a visible one by developing for a time that is long enough to reduce the grains containing the specks of silver, but that is not long enough to affect those grains without such specks. The occasional reduction of an unexposed grain occurs through other processes and is a component of chemical fog. After the development has stopped, the remaining silver halide grains are removed by a second chemical solution known as the fixer.

The final visible image thus consists of a number of silver grains dispersed in the emulsion. These absorb light, and so the image is darkest at those points where it was most brightly illuminated, that is, the image is the negative of the original. Astronomers customarily work directly with negatives since the production of a positive image normally requires the formation of a photograph of the negative with all its attendant opportunities for the introduction of distortions and errors. This practice may seem a little strange at first, but the student studying photographic images will rapidly become so familiar with negatives that the occasional positive, when encountered, requires considerable thought for its interpretation.

One of the problems encountered in using photographic images to determine the original intensities of sources is that the response of the emulsion is nonlinear. The main features of the response curve, or characteristic curve as it is more commonly known, are shown in Figure 2.7. Photographic density is plotted along the vertical axis. This is more properly called the optical density since it is definable for any partially transparent medium, and it is given by

$$D = \log_{10} \left(\frac{F_i}{F_t} \right) \qquad (2.19)$$

where
F_i is the incident flux
F_t is the transmitted flux (Figure 2.8)

The transmittance, T, of a medium is defined similarly by

$$T = \frac{F_t}{F_i} = 10^{-D} \qquad (2.20)$$

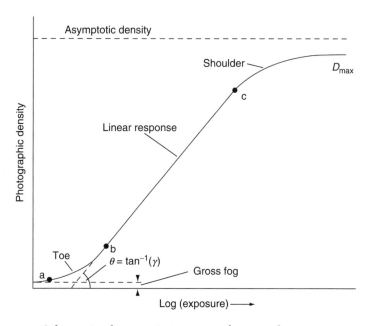

FIGURE 2.7 Schematic characteristic curve of an emulsion.

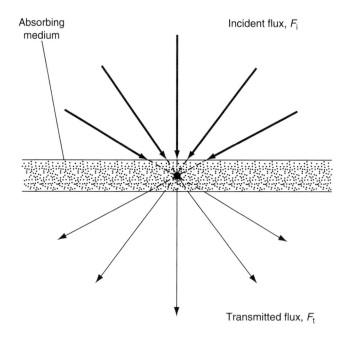

FIGURE 2.8 Quantities used to define photographic density.

and the opacity, A, by

$$A = \frac{F_i}{F_t} = T^{-1} = 10^D \tag{2.21}$$

The intensity of the exposure is plotted on the horizontal axis as a logarithm. Since density is also a logarithmic function (Equation 2.19), the linear portion of the characteristic curve (b to c in Figure 2.7) represents a genuine linear response of the emulsion to illumination. The response is nonlinear above and below the linear portion. The gross fog level is the background fog of the unexposed but developed emulsion and of its supporting material. The point marked "a" on Figure 2.7 is called the threshold and is the minimum exposure required for a detectable image. The asymptotic density is the ultimate density of which the emulsion is capable. The maximum density achieved for a particular developer, $D_{\max}$, will generally be below this value. In practice, the image is ideally formed in the linear portion of the response curve, although this may not always be possible for objects with a very wide range of intensities, or for very faint objects.

The contrast of a film is a measure of its ability to separate regions of differing intensity. It is determined by the steepness of the slope of the linear portion of the characteristic curve. A film with a high contrast has a steep linear section, and a low contrast film has a shallow linear section. Two actual measures of the contrast are used: gamma and contrast index. Gamma is simply the slope of the linear portion of the curve

$$\gamma = \tan \theta \tag{2.22}$$

The contrast index is the slope of that portion of the characteristic curve that is used to form most images. It usually includes some of the toe, but does not extend above a density of 2.0. Contrast and response of emulsions both vary markedly with the type of developer and with the conditions during development. Thus, for most astronomical applications, a calibration exposure must be obtained that is developed along with the required photograph and that enables the photographic density to be converted back into intensity.

The resolution of an emulsion is affected by the emulsion structure and by the processing stages. In addition to the grain size and light scattering in the emulsion, there may also be scattering or reflection from the

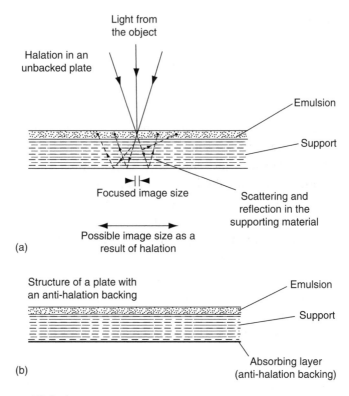

FIGURE 2.9 Halation.

supporting material (Figure 2.9). This is known as halation and it is usually reduced by the addition of an absorbing coating on the back of the film. Processing affects resolution through the change in the concentration of the processing solutions within the emulsion since these are more rapidly consumed in well-exposed regions and almost unused elsewhere. The basic effect of this on the image may best be demonstrated at a sharp boundary between a uniformly exposed region and an unexposed region. The availability of the developer from the unexposed region at the edge of the exposed region leads to its greater development and so greater density (Figure 2.10). In astronomy, this has important consequences for photometry of small images. If two small regions are given exposures of equal length and equal intensity per unit area, then the smaller image will be darker than the larger one by up to 0.3 in density units. This corresponds to an error in the finally estimated intensities by a factor of two, unless account is taken of the effect. A similar phenomenon occurs in the region between two small close images. The developer becomes exhausted

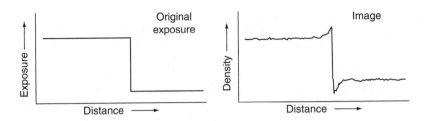

FIGURE 2.10 Edge effect in photographic emulsions.

by its work within the two images, leaving the region between them underdeveloped. Images of double stars, close spectrum emission lines, etc. may therefore appear more widely separated because of this than is the true case.

The quantum efficiency of an emulsion and its speed are two related concepts. We may measure the quantum efficiency by the number of photons required to produce a detectable image compared with an ideal detector (see also Section 1.1). Now all detectors have a certain level of background noise, and we may regard a signal as ideally detectable if it exceeds the noise level by one standard deviation of the noise. In a photographic image, the noise is due to the granularity of the image and the background fog. For a Poisson variation in granularity, the standard deviation of the noise is given by

$$\sigma = (N\omega)^{1/2} \qquad (2.23)$$

where
σ is the noise equivalent power (Section 1.1)
N is the number of grains in the exposed region
ω is the equivalent background exposure in terms of photons per grain

(All sources of fogging may be regarded as due to a uniform illumination of a fog-free emulsion. The equivalent background exposure is therefore the exposure required to produce the gross fog level in such an emulsion.) Now for our practical, and nonideal, detector let the actual number of photons per grain required for a detectable image be Ω. Then the detective quantum efficiency (Section 1.1) is given by

$$DQE = \left(\frac{\sigma}{N\Omega}\right)^2 \qquad (2.24)$$

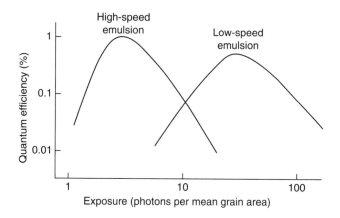

FIGURE 2.11 Quantum efficiency curves for photographic emulsions.

or

$$DQE = \left(\frac{\omega}{N\Omega^2}\right) \tag{2.25}$$

A given emulsion may further reduce its quantum efficiency by failing to intercept all the photons that fall onto it. Also, between 2 and 20 photons are required to be absorbed before any given grain reaches developability, as we have already seen. So the quantum efficiency rarely rises above 1% in practice (Figure 2.11).

The speed of an emulsion and its quantum efficiency are related since the speed is a measure of the exposure required to reach a given density in the image. Aspects of the speed have already been mentioned (Figure 2.5), and it is inherent in the position of the characteristic curve along the horizontal axis (Figure 2.7). It can also be affected by processing; the use of more active developers and/or longer developing times usually increases the speed. Hypersensitization can also improve the speed (see Section 2.2.5). Other factors that affect the speed include temperature, humidity, the spectral region being observed, the age of the emulsion, etc.

One important factor for astronomical photography that influences the speed very markedly is the length of the exposure. The change in the speed for a given level of total illumination as the exposure length changes is called reciprocity failure. If an emulsion always gave the same density of image for the same number of photons, then the exposure time would be proportional to the reciprocal of the intensity of the illumination. However,

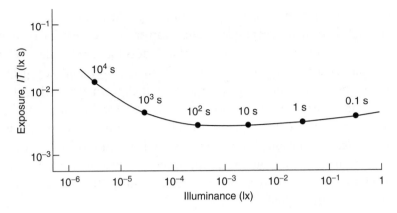

FIGURE 2.12 Typical reciprocity curve for a photographic emulsion.

for very long and very short exposures this reciprocal relationship breaks down; hence the name reciprocity failure. A typical example of the variation caused by exposure time is shown in Figure 2.12. Here the total exposure, that is, intensity × time, IT, that is required to produce a given density in the final image is plotted against the illumination. Exposure times are also indicated. If the reciprocity law held, then the graph would be a horizontal straight line. The effect of reciprocity failure is drastic; some commonly available roll films, for example, may have a speed of a few percent of their normal ratings for exposures of an hour or more.

The photographic emulsion is a weak and easily damaged material and it must therefore have a support. This is usually either a glass plate or a plastic film. Both have been used in astronomy, but plates are preferred for applications requiring high-dimensional stability and/or retention for comparative purposes for long periods of time. In both cases, great care must be taken in the storage of the material to avoid scratches, high humidity, mould, breakages, distortion, etc. that can all too easily destroy the usefulness of the information on the photograph.

2.2.4 Processing

Processing is the means whereby the latent image is converted into a visible and permanent image. For monochrome work, there are two main stages: development and fixing. Although black and white film can still be processed commercially, it is becoming increasingly hard to find laboratories that will undertake the work. Also, astronomical emulsions are often hypersensitized and/or nonstandard processing is used. It is also very irritating to be told that your magnificent star pictures have not come out because all that the processing laboratory can see is the clear background and they have not noticed the scatter of black dots that are the stars. Thus, most amateur astronomers using monochromatic film, and any professional astronomers still using it, undertake their own processing. Brief details of this are given below; for more information earlier editions of this book may be consulted, or specialist books on photography and the film suppliers' leaflets consulted. Color processing is much more complex and is rarely required for astronomical work; the interested reader is therefore referred to specialist books on the subject and to the manufacturers' literature for further information.

The action of a developer is to reduce to a grain of silver those silver halide grains that contain a latent image. Four factors principally affect the ability of a developer to carry out this function: the chemical activity of the developer, the length of the developing time, the temperature of the solution, and the degree of agitation (mixing) during development. Normally, the manufacturer's guidelines should be followed, since these are usually optimized to minimize the background fog. However, the characteristic curve may be altered to some extent by varying the developing time. In particular, the speed and contrast may be changed if so desired (Figure 2.13), though both the fog and the granularity of the image are increased by increasing the developing time; so the technique is not without its drawbacks.

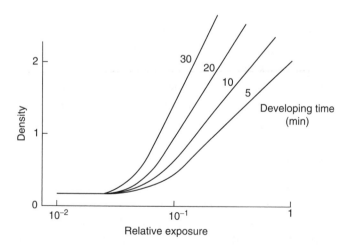

FIGURE 2.13 Effect of changing the developing time upon the characteristic curve of an emulsion.

The second main stage of processing is fixing. The purpose of this is to remove the silver halide left in the emulsion after development. Commonly, sodium thiosulphate (known as hypo) or ammonium thiosulphate are used for this purpose. Their reaction products are water soluble and they do not attack the photographic materials. However, if fixation is overextended, the fixer will start to attack the silver image, and so the manufacturers' recommended times should again be used.

If a positive image is needed, then these are obtained by re-photographing the original negative image. Usually, this is onto opaque printing paper, although film can be used when transparent images are required. The negative can simply be placed on top of the printing paper and a diffuse light shone through to produce a contact print. More normally, the negative is placed in an enlarger so that the final print is of a convenient size.

2.2.5 Hypersensitization

Hypersensitization or hypering includes any process that, when applied to an emulsion, causes it to have a faster response than normal. Its use in astronomical applications is widespread since even a 10% saving on an exposure of an hour is worthwhile, and the savings in exposure time are frequently very much higher than this. The mechanism that produces the hypersensitization is thought to be the removal of oxygen and water from the emulsion.

Many different approaches to hypersensitization have been tried in the past (for a review of these see earlier editions of this book). Today, just heating and gas immersion are employed. The shelf life of hypered plates is often severely reduced, and there is some increase in the fog level.

2.2.5.1 Heating

The commonest method of hypersensitization in use in astronomy consists simply of baking the emulsion before exposure. For blue-sensitive plates, 24–72 h at a temperature of 50°C–75°C can increase their speed by a factor of two or three. The optimum baking conditions vary between emulsions and even from one batch of a particular emulsion to the next, and so must be determined empirically in each case.

2.2.5.2 Gas Immersion Effects
1. *Nitrogen.* Immersion in an atmosphere of dry nitrogen for several days, or for shorter periods if baking is simultaneously applied, produces speed gains of up to a factor of five with little increase in the fog level.

2. *Hydrogen (Caution: possible explosion risk).* The plates are immersed in an atmosphere of nitrogen mixed with 2%–10% hydrogen (known as forming gas) at room temperature for several hours. This is frequently combined with nitrogen baking for some of the commonly used astronomical emulsions, with a resulting gain in sensitivity by about a factor of 25.

3. *Vacuum.* Lodging the plates in a soft or hard vacuum for up to 24 h, possibly combined with baking, can give gains of up to a factor of four.

2.2.6 Techniques of Astronomical Photography

The prime difference between astronomical photography and more normal types of photography lies in the exposure length. Because of this, not only must the telescope be driven to follow the motion of the stars, but also it must be continually guided to correct any errors in the drive. Usually, guiding is by means of a second and smaller telescope attached to the main instrument. The astronomer guides the telescope by keeping cross wires in the eyepiece of the secondary telescope centered on the object, or by using automatic systems (Section 1.1). For faint and/or diffuse objects, offset guiding may be necessary. That is, the guide telescope is moved from its alignment with the main telescope until it is viewing a nearby bright star while the main telescope continues to point at the required object. Guiding then continues using this bright star. Sometimes the main telescope can act as its own guide by using a guider that inserts a small mirror into the light beam to divert light not producing the main image into the eyepiece.

The determination of the required length of the exposure may be difficult; exposures may be interrupted by cloud, or the star may be a variable whose brightness is not known accurately. Previous experience may provide a guide and records of all exposures, whether successful or not, should always be kept for this reason. Alternatively, several photographs may be taken with differing exposures. The linear part of the characteristic curve is sufficiently long that the exposures can differ by a factor of two or even four if this approach is tried.

Amongst amateur astronomers, a commonly used technique is to take photographs by eyepiece projection, instead of at prime focus. A camera without its normal lens is positioned a little distance behind the eyepiece, and the focus is adjusted to produce a sharp image on the film. This provides much higher plate scales so that even small objects such as the planets can be resolved with a small telescope. A guide to the extra magnification of eyepiece projection over prime focus photography, and to the change in the exposure that may be required are given by the effective focal ratio

$$\text{EFR} = \frac{fd}{e} \qquad (2.26)$$

where
f is the focal ratio of the objective
d is the projection distance (i.e., the distance from the optical center of the eyepiece to the film plane)
e is the focal length of the eyepiece

If T is the time required for an exposure on an extended object at prime focus, then T', the exposure time for the projected image is given by

$$T' = T\left(\frac{\text{EFR}}{f}\right)^2 \qquad (2.27)$$

It is also possible just to point a camera through the eyepiece of a telescope to obtain images. Both the camera and the telescope need to be focused on infinity, and a single-lens reflex (SLR) or a digital camera with a real-time display is essential so that the image can be focussed and positioned correctly. This technique is sometimes called magnified imaging, and is a very poor second to either direct or eyepiece projection imaging.

The observational limit to the length of exposure times is usually imposed by the background sky brightness. This produces additional fog on the emulsion, and no improvement in the signal-to-noise ratio will occur once the total fog has reached the linear portion of the characteristic curve. Since the brightness of star images is proportional to the square of the objective's diameter, while that of an extended object, including the sky background, is inversely proportional to the square of the focal ratio (Section 1.1), it is usually possible to obtain adequate photographs of star fields by using a telescope with a large focal ratio. But for extended objects no such improvement is possible since their brightness scales in the same manner as that of the sky background. Much of the light in the sky background comes from scattered artificial light, so that observatories are normally sited in remote spots well away from built-up areas. Even so, a fast Schmidt camera may have an exposure limit of a few tens of minutes at best. Some further improvement may sometimes then be gained by the use of filters. The scattered artificial light comes largely from sodium and mercury vapor street lightings, so a filter that rejects the most important of the emission lines from these sources can improve the limiting magnitude by two or three stellar magnitudes at badly light-polluted sites. For the study of hot interstellar gas clouds, a narrow band filter centered on the Hα line will almost eliminate the sky background while still allowing about half the energy from the source to pass through it.

2.2.7 Analysis of Photographic Images

Photographs are used by the astronomer to provide two main types of information: the relative positions of objects and the relative intensities of objects. The analysis for the relative positions of objects is straightforward, although care is needed for the highest quality results. The positions of the images are measured by means of a machine that is essentially a highly accurate automated traveling microscope (Section 5.1).

To determine the relative intensities of sources from their photographs is more of a problem because of the nonlinear nature of the emulsion's characteristic curve (Figure 2.7). To convert back from the photographic density of the image to the intensity of the source, this curve must be known with some precision. It is obtained by means of a photometric calibration exposure, which is a photograph of a series of sources of differing and known intensities. The characteristic curve is then plotted from the measured densities of the images of these sources. To produce accurate results, great care must be exercised such as the following:

1. Photometric calibration film must be from the same batch of film as that used for the main exposure.

2. Treatment of the two films, including their storage, must be identical.

3. Both films must be processed, hypersensitized, etc. together.

4. Exposure lengths, intermittency effects, temperatures during exposures, etc. should be as similar as possible for the two plates.

5. If the main exposure is for a spectrum, then the calibration exposure should include a range of wavelengths as well as intensities.

6. Both plates must be measured under identical conditions.

With these precautions, a reasonably accurate characteristic curve may be plotted. It is then a simple, although tedious, procedure to convert the density of the main image back to intensity. The measurement of a plate is undertaken on a microdensitometer. This is a machine that shines a small spot of light through the image and measures the transmitted intensity. Such machines may readily be connected to a small computer, and the characteristic curve also fed into it, so that the conversion of the image to intensity may be undertaken automatically.

Photometry of the highest accuracy is no longer attempted from photographs for stars, but a rough guide to their brightnesses may be obtained from the diameters of their images, as well as by direct microphotometry. Scattering, halation, diffraction rings, etc. around star images mean that bright stars have larger images than faint stars. If a sequence of stars of known brightnesses is on the plate, then a calibration curve may be plotted, and estimates of the magnitudes of large numbers of stars made very rapidly.

2.3   ELECTRONIC IMAGING

2.3.1   Introduction

The alternatives to photography for recording images directly are almost all electronic in nature. They have two great advantages over the photographic emulsion. First, the image is usually produced as an electrical signal and can therefore be relayed or transmitted to a remote observer, which is vital for satellite-borne instrumentation and useful in many other circumstances, and also enables the data to be fed directly into a computer. Second, with many systems the quantum efficiency is up to a factor of a hundred or so higher than that of the photographic plate. Subsidiary advantages can include intrinsic amplification of the signal, linear response, and long-wavelength sensitivity.

The most basic form of electronic imaging simply consists of an array of point-source detecting elements. The method is most appropriate for the intrinsically small detectors such as photoconductive cells (Section 1.1). Hundreds or thousands or even more of the individual elements (usually termed pixels) can then be used to give high spatial resolution. The array is simply placed at the focus of the telescope or spectroscope, etc. in place of the photographic plate. Other arrays such as CCDs, infrared arrays, and STJs were reviewed in detail in Section 1.1. Millimeter-wave and radio arrays were also considered in Sections 1.1 and 1.2. Here therefore we are concerned with other electronic approaches to imaging, most of which have been superseded by array detectors, but which have historical and archival interests.

2.3.2 Television and Related Systems

Low light level television systems combined with image intensifiers and perhaps using a slower scan rate than normal have been used in the past. They were particularly found on the guide systems of large telescopes where they enabled the operator at a remote-control console to have a viewing/finding/guiding display. Also, many of the early planetary probes and some other spacecraft such as the International Ultraviolet Observer used TV cameras of various designs. These systems have now been replaced by solid state arrays.

2.3.3 Image Intensifiers

Micro-channel plates are a form of image intensifier (Section 1.3). The term is usually reserved though for the types of devices that produce an intensified image visible to the eye. Although used in the past for astronomical purposes, they are now rarely encountered and are mostly found as night sights for the military and for wildlife observers. Details may be found in earlier editions of this book if required.

2.3.4 Photon Counting Imaging Systems

A combination of a high-gain image intensifier and TV camera led in the 1970s to the development by Alec Boksenberg and others of the image photon counting system (IPCS). In the original IPCS, the image intensifier was placed at the telescope's (or spectroscope's, etc.) focal plane and produced a blip of some 10^7 photons at its output stage for each incoming photon in the original image. A relatively conventional TV camera then viewed this output and the video signal fed to a computer for storage. Again this type of device is now superseded by solid state detectors.

More recently, a number of variants on Boksenberg's original IPCS have been developed. These include micro-channel plate image intensifiers linked to CCDs, and other varieties of image intensifiers linked to CCDs or other array-type detectors. However, the principles of the devices remain unchanged.

2.4 SCANNING

This is such an obvious technique as scarcely to appear to deserve a separate section, and indeed at its simplest it hardly does so. A two-dimensional image may be built up using any point-source detector, if that detector is scanned over the image or vice versa. Many images at all wavelengths are obtained in this way. Sometimes the detector moves, sometimes the whole telescope or satellite, etc. Occasionally, it may be the movement of the secondary mirror of the telescope or of some ad hoc component in the apparatus that allows the image to be scanned. Scanning patterns are normally raster or spiral (Figure 2.14). However, Other patterns may be encountered, such as the continuously nutating roll employed by some of the early artificial satellites. The only precautions advisable are the matching of the scan rate to the response of the detector or to the integration time being used, and the matching of the separation of the scanning lines to the resolution of the system. Scanning by radio telescopes (Section 1.2) used to be by observing individual points within the scan pattern for a set interval of time. More recently, the scanning has been continuous, or on the fly, since this enables slow changes because of the atmosphere or the instrument to be eliminated. Specialized applications of scanning occur in the spectrohelioscope (Section 5.3) and the scanning spectrometer (Section 4.2). Many earth observation satellites use push-broom scanning. In this system, a linear array of detectors is aligned at right angles to the spacecraft's ground track. The image is then built up as the spacecraft's motion moves the array to look at successive slices of the swathe of ground over which the satellite is passing.

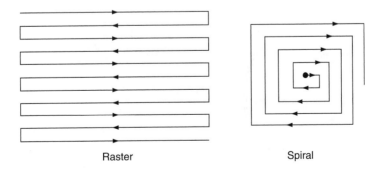

Raster   Spiral

FIGURE 2.14 Scanning patterns.

A more sophisticated approach to scanning is to modulate the output of the detector by interposing a mask of some type in the light beam. Examples of this technique are discussed in Section 1.3 and include the modulation collimator and the coded array mask used in x-ray imaging. A very great improvement over either these methods or the basic approach discussed above may be made by using a series of differing masks, or by scanning a single mask through the light beam so that differing portions of it are utilized. This improved method is known as Hadamard mask imaging. We may best illustrate its principles by considering one-dimensional images such as might be required for spectrophotometry. The optical arrangement is shown in Figure 2.15. The mask is placed in the image plane of the telescope, and the Fabry lens directs all the light that passes through the mask onto the single detector. Thus, the output from the system consists of a simple measurement of the intensity passed by the mask. If a different mask is now substituted for the first, then a new and, in

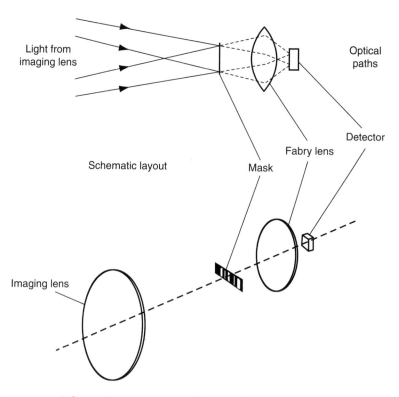

FIGURE 2.15 Schematic arrangement for a Hadamard mask imaging system.

general, different intensity reading will be obtained. If the image is to be resolved into N elements, then N such different masks must be used, and N intensity readings determined. If **D** is the vector formed from the detector output readings, **I** is the vector of the intensities of the elements of the image, and **M** is the $N \times N$ matrix whose columns each represent one of the masks, with the individual elements of the matrix corresponding to the transmissions of the individual segments of the mask, that is,

$$\mathbf{D} = [D_1, D_2, D_3, \ldots, D_N] \tag{2.28}$$

$$\mathbf{I} = [I_1, I_2, I_3, \ldots, I_N] \tag{2.29}$$

$$\mathbf{M} = \begin{bmatrix} m_{11} & m_{12} & m_{13} & \cdots & m_{1N} \\ m_{21} & & & & \\ m_{31} & & & & \\ \cdot & & & & \\ \cdot & & & & \\ \cdot & & & & \\ m_{N1} & & & & m_{NN} \end{bmatrix} \tag{2.30}$$

Then ignoring any contribution arising from noise, we have

$$\mathbf{D} = \mathbf{I}\,\mathbf{M} \tag{2.31}$$

and so

$$\mathbf{I} = \mathbf{D}\,\mathbf{M}^{-1} \tag{2.32}$$

Thus, the original image is simply obtained by inverting the matrix representing the masks. The improvement of the method over a simple scan lies in its multiplex advantage (cf. Section 4.1.4). The masks usually comprise segments that either transmit or obscure the radiation completely, that is

$$m_{ij} = 0 \text{ or } 1 \tag{2.33}$$

and on average, about half of the total image is obscured by a mask. Thus, $N/2$ image segments contribute to the intensity falling onto the detector at any one time. Hence, if a given signal-to-noise ratio is reached in a time T, when the detector observes a single image element, then the total time required to detect the whole image with a simple scan is

$$N \times T \tag{2.34}$$

and is approximately

$$\sqrt{2N} T \tag{2.35}$$

for the Hadamard masking system. Thus, the multiplex advantage is approximately a factor of

$$\sqrt{\frac{N}{2}} \tag{2.36}$$

improvement in the exposure length.

In practice, moving a larger mask across the image generates the different masks. The matrix representing the masks must then be cyclic, or in other words, each successive column is related to the previous one by being moved down a row. The use of a single mask in this way can lead to a considerable saving in the construction costs, since $2N - 1$ segments are needed for it compared with N^2 if separate masks are used. An additional constraint on the matrix $\mathbf{M}$ arises from the presence of noise in the output. The errors introduced by this are minimized when

$$\text{Tr}[\mathbf{M}^{-1}(\mathbf{M}^{-1})^T] \tag{2.37}$$

is minimized. There are many possible matrices that will satisfy these two constraints, but there is no completely general method for their generation. One method of finding suitable matrices and so of specifying the mask is based upon the group of matrices known as the Hadamard matrices (hence the name for this scanning method). These are matrices whose elements are ± 1 s and that have the property

$$\mathbf{H}\,\mathbf{H}^T = N\,\mathbf{I} \tag{2.38}$$

where
$\mathbf{H}$ is the Hadamard matrix
$\mathbf{I}$ is the identity matrix

A typical example of the mask matrix, **M**, obtained in this way might be

$$M = \begin{bmatrix} 0 & 1 & 0 & 1 & 1 & 1 & 0 \\ 0 & 0 & 1 & 0 & 1 & 1 & 1 \\ 1 & 0 & 0 & 1 & 0 & 1 & 1 \\ 1 & 1 & 0 & 0 & 1 & 0 & 1 \\ 1 & 1 & 1 & 0 & 0 & 1 & 0 \\ 0 & 1 & 1 & 1 & 0 & 0 & 1 \\ 1 & 0 & 1 & 1 & 1 & 0 & 0 \end{bmatrix} \tag{2.39}$$

for N having a value of 7.

Variations to this scheme can include making the obscuring segments out of mirrors and using the reflected signal as well as the transmitted signal, or arranging for the opaque segments to have some standard intensity rather than zero intensity. The latter device is particularly useful for infrared work. The scheme can also easily be extended to more than one dimension, although it then rapidly becomes very complex. A two-dimensional mask may be used in a straightforward extension of the one-dimensional system to provide two-dimensional imaging. A two-dimensional mask combined suitably with another one-dimensional mask can provide data on three independent variables, for example, spectrophotometry of two-dimensional images. Two two-dimensional masks could then add, for example, polarimetry to this, and so on. Except in the one-dimensional case, the use of a computer to unravel the data is obviously essential, and even in the one-dimensional case it will be helpful as soon as the number of resolution elements rises above five or six.

2.5 INTERFEROMETRY

2.5.1 Introduction

Interferometry is the technique of using constructive and destructive addition of radiation to determine information about the source of that radiation. Interferometers are also used to obtain high-precision positions for optical and radio sources, and this application is considered in Section 5.1. Two principal types of interferometer exist: the Michelson stellar interferometer, first proposed by Armand Fizeau in 1868 and the intensity interferometer proposed by Robert Hanbury-Brown in 1949. There are also a number of subsidiary techniques such as amplitude interferometry, nulling interferometry, speckle interferometry, Fourier spectroscopy, etc.

The intensity interferometer is now largely of historical interest, although a brief description is included at the end of this section for completeness (Section 2.5.6).

2.5.2 Michelson Optical Stellar Interferometer

The Michelson stellar interferometer is so called to distinguish it from the Michelson interferometer used in the Michelson–Morley experiment and in Fourier spectroscopy. Since the latter type of interferometer is discussed primarily in Section 4.1.4, we shall normally drop the "stellar" qualification for this section when referring to the first of the two main types of interferometer.

In practice, many interferometers use the outputs from numerous telescopes, however, this is just to reduce the time taken for the observations, and it is actually the outputs from pairs of telescopes that are combined to produce the interference effects. We may start therefore by considering the interference effects arising from two beams of radiation.

The complete optical system that we shall study is shown in Figure 2.16. The telescope objective, which for simplicity is shown as a lens (but exactly the same considerations apply to mirrors), is covered by an opaque screen with two small apertures, and is illuminated by two monochromatic equally bright point sources at infinity. The objective diameter is D, its focal length F, the separation of the apertures is d, the width of the apertures is Δ, the angular separation of the sources is α, and their emitted wavelength is λ.

Let us first consider the objective by itself, without the screen, and with just one of the sources. The image structure is then just the well-known diffraction pattern of a circular lens (Figure 2.17). With both sources viewed by the whole objective, two such patterns are superimposed. There are no interference effects between these two images, for their radiation is not mutually coherent. When the main maximum of one pattern is superimposed upon the first minimum of the other pattern, we have Rayleigh's criterion for the resolution of a lens (Figures 1.29 and 2.18). The Rayleigh criterion for the minimum angle between two separable sources α', as we saw in Section 1.1, is

$$\alpha' = \frac{1.22\lambda}{D} \tag{2.40}$$

Now let us consider the situation with the screen in front of the objective, and firstly consider just one aperture looking at just one of the sources.

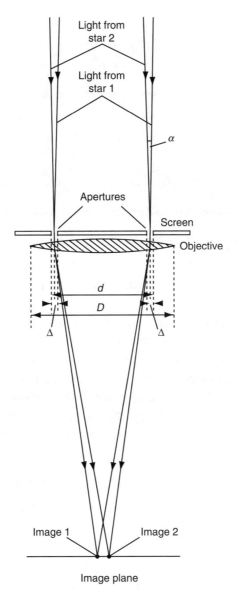

FIGURE 2.16 Optical arrangement of a Michelson interferometer.

The situation is then the same as that illustrated in Figure 2.17, but the total intensity and the resolution are both reduced because of the smaller size of the aperture compared with the objective (Figure 2.19). Although the image for the whole objective is also shown for comparison in Figure 2.19, if it were truly to scale for the situation illustrated in

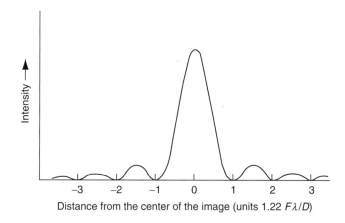

FIGURE 2.17 Image structure for one point source and the whole objective.

Figure 2.16, then it would be 1/7th of the width that is shown and 1800 times higher!

Now consider what happens when one of the sources is viewed simultaneously through both small apertures. If the apertures were infinitely small, then ignoring the inconvenient fact that no light would get through anyway, we would obtain a simple interference pattern (Figure 2.20). The effect of the finite width of the apertures is to modulate the straightforward variation of Figure 2.20 by the shape of the image for a single aperture (Figure 2.19). Since two apertures are now contributing to the

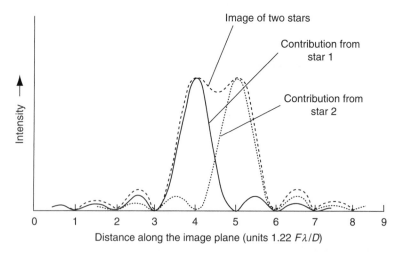

FIGURE 2.18 Image structure for two point sources and the whole objective.

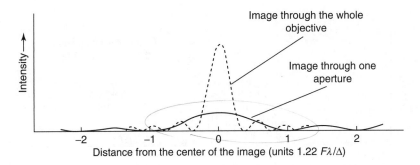

FIGURE 2.19 Image structure for one point source and one aperture.

intensity and the energy lost at the minima reappears at the maxima, the overall envelope of the image peaks at four times the intensity for a single aperture (Figure 2.21). Again, for the actual situation shown in Figure 2.16, there should be a total of 33 fringes inside the major maximum of the envelope.

Finally, let us consider the case of two equally bright point sources viewed through two apertures. Each source has an image whose structure is that shown in Figure 2.21 and these simply add together in the same manner as the case illustrated in Figure 2.18 to form the combined image. The structure of this combined image will depend upon the separation of the two sources. When the sources are almost superimposed, the image structure will be identical with that shown in Figure 2.21, except that all the intensities will have doubled. As the sources move apart, the two fringe

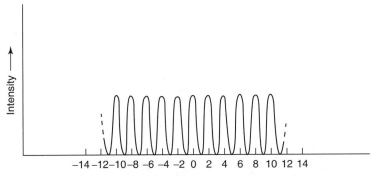

FIGURE 2.20 Image structure for a single point source viewed through two infinitely small apertures.

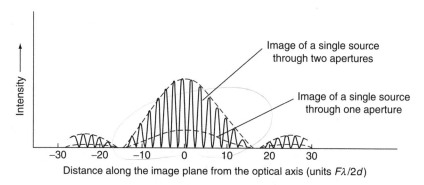

FIGURE 2.21 Image structure for one point source viewed through two apertures.

patterns will also separate, until when the sources are separated by an angle α'', given by

$$\alpha'' = \frac{\lambda}{2d} \tag{2.41}$$

the maxima of one fringe pattern will be superimposed upon the minima of the other and vice versa. The fringes will then disappear and the image will be given by their envelope (Figure 2.22). There may still be a very slight ripple on this image structure because of the incomplete filling of the minima by the maxima, but it is unlikely to be noticeable. The fringes will reappear as the sources continue to separate until the pattern is almost double that of Figure 2.21 once again. The image patterns are then

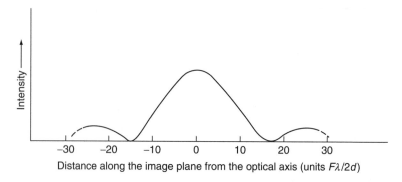

FIGURE 2.22 Image structure for two point sources separated by an angle, $\lambda/2d$, viewed through two apertures.

separated by a whole fringe width and the sources by $2\alpha''$. The fringes disappear again for a source separation of $3\alpha''$, and reach yet another maximum for a source separation of $4\alpha''$, and so on. Thus, the fringes are most clearly visible when the source's angular separation is given by $2n\alpha''$ (where n is an integer) and they disappear or reach a minimum in clarity for separations of $(2n + 1)\alpha''$. Applying the Rayleigh criterion for resolution, we see that the resolution of two apertures is given by the separation of the sources for which the two fringe patterns are mutually displaced by half a fringe width. This as we have seen is simply the angle α'', and so we have

$$\frac{\text{Resolution through two apertures}}{\text{Resolution of the objective}} = \frac{\alpha''}{\alpha'} \quad (2.42)$$

$$= \frac{2.44d}{D} \quad (2.43)$$

Imagining the two apertures placed at the edge of the objective (i.e., $d = D$), we see the quite remarkable result that the resolution of an objective may be increased by almost a factor of two and a half by screening it down to two small apertures at opposite ends of one of its diameters. The improvement in the resolution is only along the relevant axis—perpendicular to this, the resolution is just that of one of the apertures. The resolution in both axes may, however, be improved by using a central occulting disk that leaves the rim of the objective clear. This also enables us to get a feeling for the physical basis for this improvement in resolution. We may regard the objective as composed of a series of concentric narrow rings, each of which has a resolution given by Equation 2.41, with d as the diameter of the ring. Since the objective's resolution is the average of all these individual resolutions, it is naturally less than their maximum. In practice, some blurring of the image may be detected for separations of the sources that are smaller than the Rayleigh limit (Section 1.1). This blurring is easier to detect for the fringes produced by the two apertures, than it is for the images through the whole objective. The effective improvement in the resolution may therefore be even larger than that given by Equation 2.43.

Now for the situation illustrated in Figure 2.16, the path difference between the two light beams on arriving at the apertures is likely to be small, since the screen will be perpendicular to the line of sight to the objects. However, to produce fringes with the maximum clarity, that path difference must be close to zero. This is because radiation is never completely monochromatic; there is always a certain range of wavelengths present known as the bandwidth of the signal. For a zero path difference at the apertures, all wavelengths will be in phase when they are combined and will interfere constructively. However, if the path difference is not zero, some wavelengths will be in phase but others will be out of phase to a greater or lesser extent, since the path difference will equal different numbers of cycles or fractions of cycles at different wavelengths. There will thus be a mix of constructive and destructive interference, and the observed fringes will have reduced contrast. The path difference for which the contrast in the fringes reduces to zero (i.e., no fringes are seen) is called the coherence length, l, and is given by

$$l = \frac{c}{\Delta \nu} = \frac{\lambda^2}{\Delta \lambda} \tag{2.44}$$

where
$\Delta \nu$ is the frequency bandwidth of the radiation
$\Delta \lambda$ is the wavelength bandwidth of the radiation

So that for $\lambda = 500$ nm and $\Delta \lambda = 1$ nm, we have a coherence length of 0.25 mm; for white light ($\Delta \lambda \approx 300$ nm) this reduces to less than a micron; but in the radio region it can be large, for example, 30 m at $\nu = 1.5$ GHz and $\Delta \nu = 10$ MHz. However, as we shall see below, the main output from an interferometer is the fringe contrast (usually known as the fringe visibility, V, Equation 2.45), so that for this not to be degraded, the path difference to the apertures, or their equivalent, must be kept to a small fraction of the coherence length. For interferometers such as Michelson's 1921 stellar interferometer (Figure 2.23), the correction to zero path difference is small, and in that case was accomplished through the use of adjustable glass wedges. However most interferometers, whether operating in the optical or radio regions, now use separate telescopes on the ground (e.g., Figure 2.24), so the path difference to the telescopes can be over 100 m for optical interferometers, and up to thousands of kilometers for very long baseline radio interferometry. These path differences have to be corrected either during the observation by hardware, or afterwards during the data processing.

Michelson's original interferometer was essentially identical to that shown in Figure 2.16, except that the apertures were replaced by two

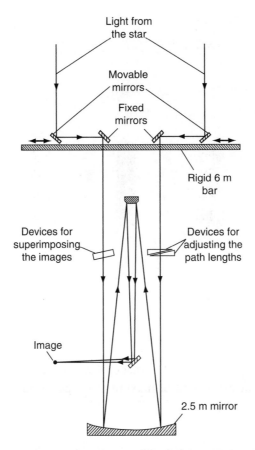

FIGURE 2.23 Schematic optical arrangement of the Michelson stellar interferometer.

movable parallel slits to increase the amount of light available, and a narrow band filter was included to give nearly monochromatic light (earlier we had assumed that the sources were monochromatic). To use the instrument to measure the separation of a double star with equally bright components, the slits are aligned perpendicularly to the line joining the stars, and moved close together, so that the image fringe patterns are undisplaced and the image structure is that of Figure 2.21. The actual appearance of the image in the eyepiece at this stage is shown in Figure 2.24. The slits are then moved apart until the fringes disappear (Figures 2.22 and 2.25). The distance between the slits is then such that the fringe pattern from one star is filling-in the fringe pattern from the other, and their separation is given by α'' (Equation 2.41, with d given by the

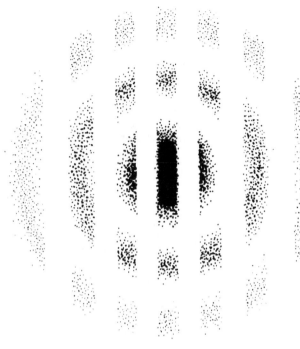

FIGURE 2.24 Appearance of the image of a source seen in a Michelson interferometer at the instant of maximum fringe visibility.

distance between the slits). If the two stars are of differing brightnesses, then the fringes will not disappear completely, but the fringe visibility, V, given by

$$V = \frac{(I_{max} - I_{min})}{(I_{max} + I_{min})} \tag{2.45}$$

where
I_{max} is the intensity of a fringe maximum
I_{min} is the intensity of a fringe minimum

will reach a minimum at the same slit distance. To measure the diameter of a symmetrical object like a star or the satellite of a planet, the same procedure is used. However, the two sources are now the two halves of the star's disk, represented by point sources at the optical centers of the two

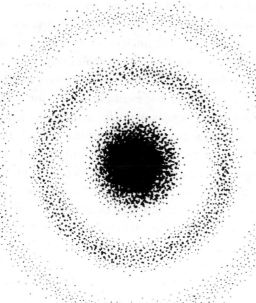

FIGURE 2.25 Appearance of the image of a source seen in a Michelson interferometer at the time of zero fringe visibility.

semicircles. The diameter of the star is then given by 2.44α'', so that in determining the diameters of objects, an interferometer has the same resolution as a conventional telescope with an objective diameter equal to the separation of the slits.

Very stringent requirements on the stability and accuracy of the apparatus are required for the success of this technique. As we have seen in Equation 2.44, the path difference to the slits must not be greater than a small fraction of the coherence length of the radiation. Furthermore that path difference must remain constant to considerably better than the wavelength of the radiation that is being used as the slits are separated. Vibrations and scintillation are additional limiting factors. In general, the paths will not be identical when the slits are close together and the fringes are at their most visible. But the path difference will then be some integral number of wavelengths and provided that it is significantly less than the coherence length the interferometer may still be used. Thus, in practice, the value that is required for d is the difference between the slit separations at minimum and maximum fringe visibilities. A system, such as has just been described, was used by Michelson in 1891 to measure the diameters of the Galilean satellites. Their diameters are in the region of 1 second of arc, so that d has a value of a few tens of millimeters.

Now the angular diameters of stellar disks are very much smaller than those of the Galilean satellites; for example, 0.047″ for α Orionis (Betelgeuse), one of the largest angular diameter stars in the sky. A separation for the apertures of several meters is thus needed if stellar diameters are to be measured. This led to the improved version of the system that is rather better known than the one used for the Galilean satellites and that is the type of interferometer that is usually intended when the Michelson stellar interferometer is mentioned. It was used by Albert Michelson and Francis Pease in 1921 to measure the diameters of half a dozen or so of the larger stars. The slits were replaced by movable mirrors on a rigid 6 m long bar mounted on top of the 2.5 m Hooker telescope (Figure 2.23). By this means, d could be increased beyond the diameter of the telescope. A system of mirrors then reflected the light into the telescope. The practical difficulties with the system were very great and although larger systems have since been attempted none has been successful, and so the design is now largely of historical interest only.

Modern optical interferometers are based upon separate telescopes rather than two apertures feeding a single instrument. They are still primarily used to measure stellar diameters, although some aperture synthesis systems (see Section 2.5.5) are now working and others are under construction. There is no difference in the principle of operation of these interferometers from that of Michelson, but there are many practical changes. The main change arises because the path difference to the two telescopes can now be many meters. A system for compensating for the path difference is thus required. In most designs, the path compensation is via a delay line; the light from one telescope is sent on a longer path than that from the other one before they are mixed together. The extra path length is arranged to be equal to the path difference between the telescopes. Since the path difference will change as the object moves across the sky, the delay must also change, and this is usually accomplished by having one or more of the mirrors reflecting the light onto the longer path mounted on a moveable carriage. Even using narrow bandwidths, the small value of the coherence length (Equation 2.44) at optical wavelengths means that the carriage must be capable of being positioned to submicron accuracy and yet be able to travel tens of meters.

The second major difference is that the telescopes are usually fixed in position or are moveable only slowly between observations. It is thus not possible to change their separation to observe maximum and minimum fringe visibilities. A change in the effective separation occurs as the angle between the baseline and the source in the sky alters, but this is generally insufficient to range from maximum to minimum fringe visibility. Thus, instead a measurement of the fringe visibility is made for just one, or a small number, of the separations. This measurement is then fitted to a theoretical plot of the variation of the fringe visibility with the telescope separation and the size of the source to determine the latter.

The third difference arises from the atmospheric turbulence. We have seen that there is an upper limit to the size of a telescope aperture, given by Fried's coherence length (Equation 1.72). In the visible region, this is only about 120 mm, though in the infrared region it can rise to 300 mm. Telescopes with apertures larger than Fried's limit will thus be receiving radiation through several, perhaps many, atmospheric cells, each of which will have different phase shifts and wavefront distortions. When the beams are combined, the desired fringes will be washed out by all these differing contributions. Furthermore, the atmosphere changes on a time scale of around 5 ms. So that even if small apertures are used, exposures have to be short enough to freeze the atmospheric motion (see Section 2.6). Thus, many separate observations have to be added together to get a value with a sufficiently high signal-to-noise ratio. To use large telescopes as a part of an interferometer, their images have to be corrected through adaptive optics (Section 1.1). Even a simple tip-tilt correction will allow the useable aperture to be increased by a factor of three. However, to use 8 and 10 m telescopes, such as European Southern Observatory's (ESO's) very large telescope (VLT) and the Keck telescopes, full atmospheric correction is needed.

When only two telescopes form the interferometer, the interruption of the phase of the signal by the atmosphere results in the loss of that information. Full reconstruction of the morphology of the original object is not then possible, only the fringe visibility can be determined (see Section 2.5.5 for a discussion of multielement interferometers, closure phase, and aperture synthesis). However, the fringe visibility still enables stellar diameters and double-star separations to be found, and so several two-element interferometers either have been until recently or are currently in operation. Thus, the infrared spatial interferometer (ISI) that has now closed down was based on Mount Wilson, California, and observed at

11 μm, initially using two 1.65 m Pfund-type telescopes with a maximum baseline of 35 m. Later a third telescope was added enabling the baseline to be increased to 70 m. The Sydney University stellar interferometer (SUSI) in Australia operates in the visible region with two 0.14 m mirrors and a maximum separation of 640 m, while grand interféromètre à 2 télescopes (GI2T) in France, now also decommissioned, operated in the visible region using two 1.5 m mirrors with a maximum baseline of 65 m. The 10 m Keck telescopes started operating as an interferometer in 2001 with an 85 m baseline. While, shortly, the large binocular telescope (LBT) on Mount Graham, Arizona, is expected to start science operations and soon after that is to be commissioned as an interferometer.

A proposal for the future that would have much in common with stellar coronagraphs (Section 5.3) is for a nulling interferometer based in space. This would use destructive interference to suppress the bright central object, thus allowing faint companions (e.g., planets) to be discerned. Earth-like planets around stars 10 or more parsecs away might be detected in observations of a few hours using a four-element interferometer with 50 m baselines. An earth-based prototype has shown that the principle works. It used two 1.8 m mirrors and operated at 10 μm wavelength. The path differences induced by the atmosphere were in this case utilized to provide the interference effects. Numerous 50 ms exposures were obtained of α Orionis (Betelgeuse), and only those with a 5 μs delay were used. The star's light was largely suppressed, but the radiation from the surrounding dust cloud underwent a different delay, and so could be observed directly. In 2003, the 6.5 m *Magellan* telescopes at Las Campanas detected a gap in the dust disk surrounding the star HD 100546 where a planet may be forming via nulling interferometry and ESA's Darwin mission plans to use the technique to detect extrasolar planets.

A variation on the system that also has some analogy with intensity interferometry (see Section 2.5.6) is known as amplitude interferometry. This has been used recently in a successful attempt to measure stellar diameters. At its heart is a device known as a Köster prism that splits and combines the two separate light beams (Figure 2.26). The interfering beams are detected in a straightforward manner using photomultipliers. Their outputs may then be compared. The fringe visibility of the Michelson interferometer appears as the anti-correlated component of the two signals, while the atmospheric scintillation, etc. affects the correlated component. Thus, the atmospheric effects can be screened out to a large extent, and the stability of the system is vastly improved.

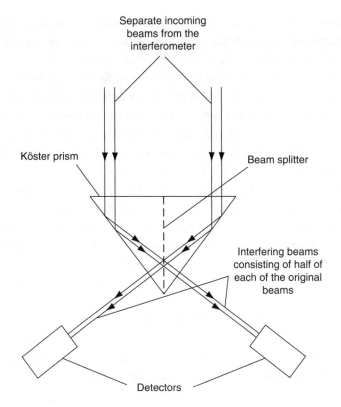

FIGURE 2.26 Köster prism.

2.5.3 Michelson Radio Interferometer

Radio aerials are generally only a small number of wavelengths in diameter (Section 1.2), and so their intrinsic resolution (Equation 1.27) is poor. The use of interferometers to obtain higher resolution therefore dates back to soon after the start of radio astronomy. There is no difference in principle between a radio and an optical interferometer when they are used to measure the separation of double sources, or the diameters of uniform objects. However, the output from a radio telescope contains both the amplitude and phase of the signal, so that complete imaging of the source is possible. Radio interferometers generally use more than two antennae, and these may be in a two-dimensional array. The basic principle, however, is the same as for just two antennae, although the calculations become considerably more involved.

The individual elements of a radio interferometer are usually fairly conventional radio telescopes (Section 1.2) and the electrical signal that they output varies in phase with the received signal. In fact, with most radio interferometers, it is the electrical signals that are mixed to produce the interference effect rather than the radio signals themselves. Their signals may then be combined in two quite different ways to provide the interferometer output. In the simplest version, the signals are simply added together before the square law detector (Figure 1.61), and the output will then vary with the path difference between the two signals. Such an arrangement, however, suffers from instability problems, particularly because of variations in the voltage gain.

A system that is often preferred to the simple adding interferometer is the correlation or multiplying interferometer. In this, as the name suggests, the intermediate frequency (IF) signals from the receivers are multiplied together. The output of an ideal correlation interferometer will contain only the signals from the source (which are correlated); the other components of the outputs of the telescopes will be zero as shown below.

If we take the output voltages from the two elements of the interferometer to be $(V_1 + V_1')$ and $(V_2 + V_2')$ where V_1 and V_2 are the correlated components (i.e., from the source in the sky) and V_1' and V_2' are the uncorrelated components, the product of the signals is then

$$V = (V_1 + V_1') \times (V_2 + V_2') \tag{2.46}$$

$$= V_1 V_2 + V_1' V_2 + V_1 V_2' + V_1' V_2' \tag{2.47}$$

If we average the output over time, then any component of Equation 2.47 containing an uncorrelated component will tend to zero. Thus,

$$\overline{V} = \overline{V_1 V_2} \tag{2.48}$$

In other words, the time-averaged output of a correlation interferometer is the product of the correlated voltages. Since most noise sources contribute primarily to the uncorrelated components, the correlation interferometer is inherently much more stable than the adding interferometer. The phase-switched interferometer (see below) is an early example of a correlation interferometer, though direct multiplying interferometers are now more common.

The systematics of radio astronomy differ from those of optical work (Section 1.2), so that some translation is required to relate optical interferometers to radio interferometers. If we take the polar diagram of a single radio aerial (e.g., Figure 1.65), then this is the radio analogue of our image structure for a single source and a single aperture (Figures 2.17 and 2.19). The detector in a radio telescope accepts energy from only a small fraction of this image at any given instant (though a few array detectors with small numbers of pixels are now coming into use; Section 1.2). Scanning the radio telescope across the sky corresponds to scanning this energy-accepting region through the optical image structure. The main lobe of the polar diagram is thus the equivalent of the central maximum of the optical image, and the side lobes are the equivalent of the diffraction fringes. Since an aerial is normally directed toward a source, the signal from it corresponds to a measurement of the central peak intensity of the optical image (Figures 2.17 and 2.19). If two stationary aerials are now considered and their outputs combined, then when the signals arrive without any path differences, the final output from the radio system as a whole corresponds to the central peak intensity of Figure 2.21. However, if a path difference does exist, then provided that it is less than the coherence length, the final output will correspond to some other point within that image. In particular, when the path difference is a whole number of wavelengths, the output will correspond to the peak intensity of one of the fringes, and when it is a whole number plus half a wavelength, it will correspond to one of the minima. Now the path differences arise in two main ways: from the angle of inclination of the source to the line joining the two antennae, and from delays in the electronics and cables between the antennae and the central processing station (Figure 2.27). The latter will normally be small and constant and may be ignored or corrected. The former will alter as the rotation of the earth changes the angle of inclination. Thus, the output of the interferometer will vary with time as the value of P changes. The output over a period of time, however, will not follow precisely the shape of Figure 2.21 because the rate of change of path difference varies throughout the day, as we may see from Equation 2.49 for P

$$P = s \cos \phi \cos (\psi - E) \qquad (2.49)$$

where μ is the altitude of the object and is given by

$$\mu = \sin^{-1} [\sin \delta \sin \phi + \cos \delta \cos \phi \cos (T - \alpha)] \qquad (2.50)$$

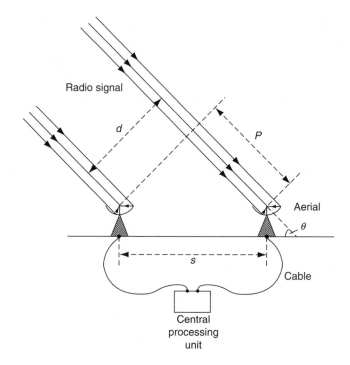

FIGURE 2.27 Schematic arrangement of a radio interferometer with fixed aerials.

ψ is the azimuth of the object and is given by

$$\psi = \cot^{-1}[\sin\phi\cot(T-\alpha) - \cos\phi\tan\delta\,\mathrm{cosec}(T-\alpha)] \qquad (2.51)$$

where
E is the azimuth of the line joining the aerials
α is the right ascension of the object
δ is the declination of the object
T is the local sidereal time at the instant of observation
ϕ is the latitude of the interferometer

The path difference also varies because the effective separation of the aerials, d, where

$$d = s[\sin^2\mu + \cos^2\mu\sin^2(\psi - E)]^{1/2} \qquad (2.52)$$

also changes with time, thus the resolution and fringe spacing (Equation 2.41) are altered. Hence, the output over a period of time from a radio

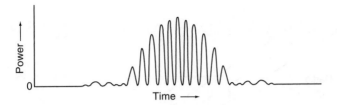

FIGURE 2.28 Output from a radio interferometer with stationary aerials, viewing a single source.

interferometer with fixed antennae is a set of fringes whose spacing varies, with maxima and minima corresponding to those of Figure 2.21, for the instantaneous values of the path difference and effective aerial separation (Figure 2.28).

An improved type of interferometer, which has increased sensitivity and stability, is the phase-switched interferometer (see earlier discussion). The phase of the signal from one aerial is periodically changed through 180° by, for example, switching an extra piece of cable half a wavelength long into or out of the circuit. This has the effect of oscillating the beam pattern of the interferometer through half a fringe width. The difference in the signal for the two positions is then recorded. The phase switching is generally undertaken in the latter stages of the receiver (Section 1.2), so that any transient effects of the switching are not amplified. The output fluctuates either side of the zero position as the object moves across the sky (Figure 2.29).

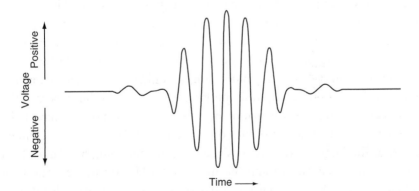

FIGURE 2.29 Output from a radio interferometer with stationary aerials and phase switching, viewing a single point source.

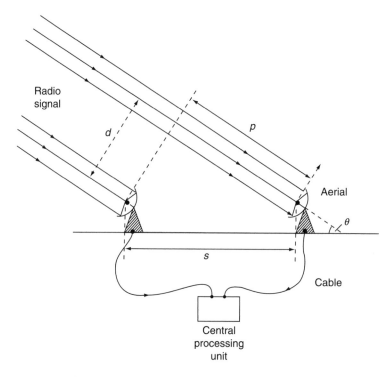

FIGURE 2.30 Schematic arrangement of a radio interferometer with tracking aerials.

When the aerials are driven so that they track the object across the sky (Figure 2.30), then the output of each aerial corresponds to the central peak intensity of each image (Figures 2.17 and 2.19). The path difference then causes a simple interference pattern (Figure 2.31 and cf. Figure 2.20) whose fringe spacing alters due to the varying rate of change of path difference and aerial effective spacing as in the previous case. The maxima are now of constant amplitude since the aerials' projected effective areas are constant. In reality, many more fringes would occur than are shown in Figure 2.31. As with the optical interferometer, the path differences at the aerials arising from the inclination of the source to the base line have to be compensated to a fraction of the coherence length. However, this is much easier at the longer wavelengths since the coherence length is large, 30 m for a 10 MHz bandwidth and 300 m for a 1 MHz bandwidth. The path difference can therefore be corrected by switching in extra lengths of cable or by shifting the recorded signals with respect to each other during data processing.

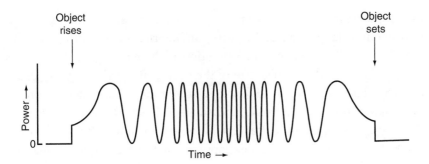

FIGURE 2.31 Output from a radio interferometer with tracking aerials viewing a single point source.

2.5.4 Reflection Interferometers

These are not an important class of interferometer, but they are included for the sake of completeness and for their historical interest. They are closely related to the phased array type of radio telescope discussed in Section 1.2. The first astronomical radio interferometer measurements were made with this type of instrument and were of the Sun. Only one aerial is required, and it receives both the direct and the reflected rays (Figure 2.32). These interfere, and the changing output as the object moves around the sky can be interpreted in terms of the structure of the object.

2.5.5 Aperture Synthesis

An interferometer works most efficiently, in the sense of returning the most information, for sources whose separation is comparable with its resolution. For objects substantially larger than the resolution, little or no

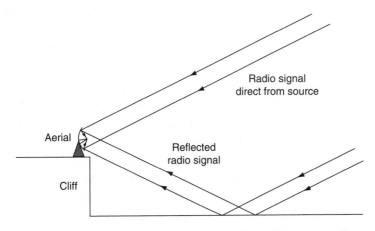

FIGURE 2.32 Schematic possible arrangement for reflection interferometry.

useful information may be obtained. We may, however, obtain information about a larger source by using an interferometer with a smaller separation of its elements. The resolution is thereby degraded until it is comparable with the angular size of the source. By combining the results of two interferometers of differing separations, one might thus obtain information on both the large- and small-scale structures of the source. Following this idea to its logical conclusion led Sir Martin Ryle in the early 1960s to the invention and development of the technique of aperture synthesis. He was awarded the Nobel physics prize in 1974 for this work.

By this technique, which also goes under the name of earth-rotation synthesis and is closely related to synthetic aperture radar (SAR) (Section 2.8), observations of a stable source by a number of interferometers are combined to give the effect of an observation using a single VLT. The simplest way to understand how aperture synthesis works is to take an alternative view of the operation of an interferometer. The output from a two-aperture interferometer viewing a monochromatic point source is shown in Figure 2.20, and is a simple sine wave. This output function is just the Fourier transform (Equation 2.3) of the source. Such a relationship between source and interferometer output is no coincidence, but is just a special case of the van Cittert–Zernicke theorem.

> The instantaneous output of a two-element interferometer is a measure of one component of the two-dimensional Fourier transform (Equation 2.10) of the objects in the field of view of the telescopes.

Thus, if a large number of two-element interferometers were available, so that all the components of the Fourier transform could be measured, then the inverse two-dimensional Fourier transform (Equation 2.11) would immediately give an image of the portion of the sky under observation.

Now the complete determination of the Fourier transform of even a small field of view would require an infinite number of interferometers. In practice, therefore, the technique of aperture synthesis is modified in several ways. The most important of these is relaxing the requirement to measure all the Fourier components at the same instant. However, once the measurements are spread over time, the sources being observed must remain unvarying over the length of time required for those measurements.

Given, then, a source that is stable at least over the measurement time, we may use one or more interferometer pairs to measure the Fourier components using different separations and angles. Of course, it is still not

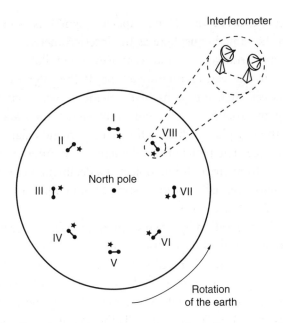

FIGURE 2.33 Changing orientation of an interferometer. The earth is viewed from above the north pole and successive positions of the interfero-meter at 3 h intervals are shown. Notice how the orientation of the starred aerial changes through 360° with respect to the other aerial during a day.

possible to make an infinite number of such measurements, but we may use the discrete versions of Equations 2.10 and 2.11 to bring the required measurements down to a finite number (cf. the one-dimensional ana-logues, Equations 2.3, 2.4, 2.8, and 2.9) though at the expense of losing the high-frequency components of the transform, and hence the finer details of the image (Section 2.1). The problem of observing with many different baselines is eased because we are observing from the rotating earth. Thus, if a single pair of telescopes observes a source over 24 h, the orientation of the baseline revolves through 360° (Figure 2.33). If the object is not at the north (or south) pole then the projected spacing of the interferometer will also vary and they will seem to trace out an ellipse. The continuous output of such an interferometer is a complex function whose amplitude is proportional to the amplitude of the Fourier transform and whose phase is the phase shift in the fringe pattern.

The requirement for 24 h of observation would limit the technique to circumpolar objects. Fortunately, however, only 12 h are actually required, the other 12 h can then be calculated by the computer from the conjugates of the first set of observations. Hence, aperture synthesis can be applied to any object in the same hemisphere as the interferometer.

Two elements arranged upon an east–west line follow a circular track perpendicular to the earth's rotational axis. If the interferometer is not aligned east–west, then the paths of its elements will occupy a volume centred upon the rotational axis. For convenience the data are usually plotted onto the u–v plane, which is the plane perpendicular to the line of sight to the source. The paths of the elements of an interferometer in the u–v plane range from circles for an object at a declination of $\pm 90°$ through increasingly narrower ellipses to a straight line for an object with a declination of $0°$ (Figure 2.34).

A single 12 h observation by a two-element interferometer thus samples all the components of the Fourier transform of the field of view covered by its track in the u–v plane. We require, however, the whole of the u–v plane to be sampled. Thus, a series of 12 h observations must be made, with the interferometer baseline changed by the diameter of one of its elements each time (Figure 2.35). The u–v plane is then sampled completely out to the maximum baseline possible for the interferometer, and the inverse Fourier transform will give an image equivalent to that from a single telescope with a diameter equal to the maximum baseline.

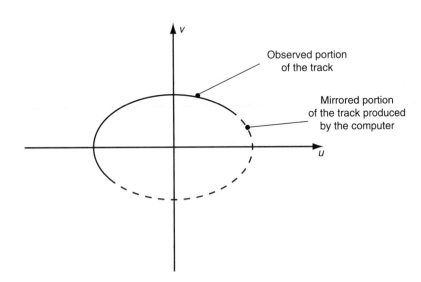

FIGURE 2.34 Track of one element of an interferometer with respect to the other in the u–v plane for an object at a declination of $\pm 35°$.

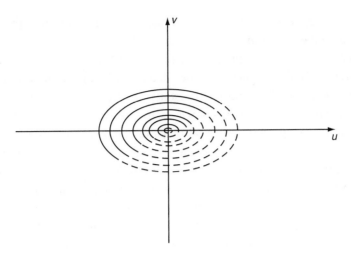

FIGURE 2.35 Successive tracks in the u–v plane of a two-element interferometer as its baseline is varied.

A two-element radio interferometer with (say) 20 m diameter aerials and a maximum baseline of 1 km would need to make fifty 12 h observations to synthesize a 1 km diameter telescope. By using more than two aerials, however, the time required for the observations can be much reduced. Thus, with six elements, there are 15 different pairings. If the spacings of these pairs of elements are all different (nonredundant spacing, something that it is not always possible to achieve) then the 50 visibility functions required by the previous example can be obtained in just four 12 h observing sessions. If the source has a reasonably smooth spectrum, the u–v plane may be sampled even more rapidly by observing simultaneously at several nearby frequencies (multifrequency synthesis). Since the path differences between the aerials are dependent upon the operating wavelength, this effectively multiplies the number of interferometer pairs with different separations by the number of frequencies being observed. The analysis of the results from such observations, however, will be considerably complicated by any variability of the source over the frequencies being used.

A radio aperture synthesis system, such as the one we have just been considering, is a filled-aperture system. That is, the observations synthesize the sensitivity and resolution of a telescope with a diameter equal to the maximum available baseline. Although many aperture synthesis systems are of this type, it becomes increasingly impractical and expensive to sample the whole u–v plane as the baseline extends beyond 10 km or so. The largest such system is the very large array (VLA) in New Mexico.

It uses twenty-seven 25 m dishes arranged in a "Y" pattern with a maximum baseline of 36 km.

Aperture synthesis systems larger than the VLA have unfilled or sparse apertures. This also applies to some smaller radio and to optical aperture synthesis systems. The Fourier transform of the source is thus not fully sampled. Special techniques, known as hybrid mapping (see below), are then required to produce the final maps of radio sources. The MERLIN* system based at Jodrell Bank in the United Kingdom is typical of these larger unfilled systems. MERLIN has seven fixed aerials with baselines ranging from 6 to 230 km, and can reach resolutions of 0.008" at a wavelength of 10 mm. While in the Netherlands, the Westerbork synthesis radio telescope (WSRT) uses ten fixed and four movable 25 m dishes over a maximum baseline of 3 km. For the future, the square kilometer array (SKA) is a proposal for some thirty 200 m diameter radio dishes that will provide a collecting area of a square kilometer with a maximum baseline of 3000 km; alternative designs for this project are for one hundred and fifty 90 m aerials or for thousands of smaller Luneburg lens-based receivers.

Beyond MERLIN and similar radio systems, we have very long baseline interferometry (VLBI). For VLBI, the elements of the interferometer may be separated by thousands of kilometers, over several continents, and provide resolutions of 0.001" or better. In VLBI, the signals from each element are separately recorded along with timing pulses from an atomic clock. The recordings are then physically brought together and processed by a computer that uses the time signals to ensure the correct registration of one radio signal with respect to another. The very long baseline array (VLBA), for example, uses ten 25 m telescopes spread out over the United States from Hawaii to the U.S. Virgin islands. Its maximum baseline is 10,000 km, and at a wavelength of 7 mm its resolution can be 0.00015". It has, for example, recently been able to measure the parallax for the Orion nebula to an accuracy of 0.0001", reducing the previously accepted distance for the nebula from 1570 ly (475 pc) to 1270 ly (385 pc). The *Japanese Halca* spacecraft that carried an 8 m radio telescope was linked with up to 40 ground-based dishes, to give baselines up to 30,000 km in length. However, the spacecraft lost attitude control in 2003 and has now ceased to operate. Recently, real-time analysis of the data from a VLBI system comprising the Arecibo radio dish and telescopes in the United Kingdom, Sweden, the

---

* Multielement radio linked interferometer—although the radio links have recently been replaced by fiber optics that have increased MERLIN's sensitivity by a factor of 30.

Netherlands, and Poland became possible when they were linked via Internet research networks—earning the system the name of e-VLBI.

At millimeter and submillimeter wavelengths, the Owens valley radio observatory in California is operating an interferometer with six 10 m telescopes and can achieve a resolution of 0.5″. While on the Plateau de Bure in France, the Institut de Radio Astronomie Millimétrique (IRAM) operates a system with six 15 m telescopes providing imaging to better than 1 second of arc. The Atacama large millimeter array (ALMA) is due to be completed by 2009. It will have fifty 12 m antennas and operate at wavelengths from 10 to 0.35 mm and have angular resolutions down to 0.01″ at its maximum aerial separation of 10 km.

Aperture synthesis at visible and near infrared wavelengths has only recently been successfully attempted, because of the very stringent requirements on the stability and accuracy of construction of the instruments imposed by the small wavelengths involved. The Cambridge optical aperture synthesis telescope (COAST) based at Cambridge, United Kingdom, produced its first images in 1995, and now uses five 0.4 m telescopes with a maximum baseline of 67 m. The Center for High Angular Resolution Astronomy (CHARA) array on Mount Wilson can reach a resolution of 0.0002″ using six 1 m telescopes with a maximum baseline of 350 m. The four 8 m telescopes of ESO's very large telescope interferometer (VLTI) can be combined with four 1.8 m auxiliary telescopes for aperture synthesis and the latter can be moved to 30 different positions offering a maximum baseline of 254 m. It is also planned that the two 10 m Keck telescopes that already can combine to form a two-element interferometer will shortly combine with four to six 1.8 m telescopes to act as an aperture synthesis system. There are several projects currently being planned for space-based interferometers. Gaia and space interferometry mission (SIM) are intended for astrometry and are discussed in Section 5.1. Terrestrial planet finder (TPF) has now been postponed indefinitely but was intended to use five spacecraft and four 3.5 m telescopes in an infrared nulling interferometer specifically designed to detect earth-like planets.

The extraction of the image of the sky from an aperture synthesis system is complicated by the presence of noise and errors. Overcoming the effects of these adds additional stages to the data reduction process. In summary it becomes

1. Data calibration

2. Inverse Fourier transform

3. Deconvolution of instrumental effects

4. Self-calibration

Data calibration is required to compensate for problems such as errors in the locations of the elements of the interferometer, and variations in the atmosphere. It is carried out by comparing the theoretical instrumental response function with the observed response to an isolated stable point source. These responses should be the same and if there is any difference then the data calibration process attempts to correct it by adjusting the amplitudes and phases of the Fourier components. Often ideal calibration sources cannot be found close to the observed field. Then, calibration can be attempted using special calibration signals, but these are insensitive to the atmospheric variations, and the result is much inferior to that obtained using a celestial source.

Applying the inverse Fourier transform to the data has already been discussed, and further details are given in Section 2.1.

After the inverse Fourier transformation has been completed, we are left with the "dirty" map of the sky. This is the map contaminated by artifacts introduced by the PSF of the interferometer. The primary components of the PSF, apart from the central response, are the side lobes. These appear on the dirty map as a series of rings that extend widely over the sky, and that are centered on the central response of the PSF. The PSF can be calculated from interferometry theory to a high degree of precision. The deconvolution can then proceed as outlined in Section 2.1.

The MEM discussed in Section 2.1 has recently become widely used for determining the best source function to fit the dirty map. However, another method of deconvolving the PSF has long been in use, and is still used by many workers, and the method is known as CLEAN.

The CLEAN algorithm was introduced by Jan Högbom in 1974. It involves the following stages:

1. Normalize the PSF (instrumental profile or dirty beam) to (gI_{MAX}), where I_{MAX} is the intensity of the point of maximum intensity in the dirty map and g is the loop gain with a value between 0 and 1.

2. Subtract the normalized PSF from the dirty map.

3. Find the point of maximum intensity in the new map—this may or may not be the same point as before—and repeat the first two steps.

4. Continue the process iteratively until I_{MAX} is comparable with the noise level.

5. Produce a final clear map by returning all the components removed in the previous stages in the form of clean beams with appropriate positions and amplitudes. The clean beams are typically chosen to be Gaussian with similar widths to the central response of the dirty beam.

CLEAN has proved to be a useful method for images made up of point sources despite its lack of a substantial theoretical basis. For extended sources, MEMs are better because CLEAN may then require thousands of iterations. As pointed out in Section 2.1 though, MEMs also suffer from problems, especially the variation of resolution over the image.

Many packages for the reduction of radio interferometric data are available within the astronomical image processing system (AIPS and AIPS++) produced by the National Radio Astronomy Observatory (NRAO). This is the most widely used set of software for radio work. It may be downloaded from the Web free of charge for use on personal computers (see http://www.aoc.nrao.edu/aips/). You will also need your computer to have a UNIX or LINUX operating system. The latter is available, also without charge for downloading onto personal computers, at http://www.linux.org/. However, you are warned that though available free of charge, these packages are not easy for an inexperienced computer user to implement successfully. However, with the increasing speed and power of computers, real-time processing of the data from radio telescopes and aperture synthesis systems is becoming more common.

The final stage of self-calibration is required for optical systems and for radio systems when the baselines become more than a few kilometers in length, since the atmospheric effects then differ from one telescope to another. Under such circumstances with three or more elements, we may use the closure phase that is independent of the atmospheric phase delays. The closure phase is defined as the sum of the observed phases for the three baselines made by three elements of the interferometer. It is independent of the atmospheric phase delays; as we may see by defining the phases for the three baselines, it in the absence of an atmosphere to be ϕ_{12}, ϕ_{23}, and ϕ_{31}, and the atmospheric phase delays at each element as a_1, a_2, and a_3. The observed phases are then

$$\phi_{12} + a_1 - a_2$$
$$\phi_{23} + a_2 - a_3$$
$$\phi_{31} + a_3 - a_1$$

The closure phase is then given by the sum of the phases around the triangle of the baselines:

$$\phi_{123} = \phi_{12} + a_1 - a_2 + \phi_{23} + a_2 - a_3 + \phi_{31} + a_3 - a_1 \qquad (2.53)$$
$$= \phi_{12} + \phi_{23} + \phi_{31} \qquad (2.54)$$

From Equation 2.54, we may see that the closure phase is independent of the atmospheric effects and is equal to the sum of the phases in the absence of an atmosphere. In a similar way, an atmosphere-independent closure amplitude can be defined for four elements:

$$G_{1234} = \frac{A_{12}A_{34}}{A_{13}A_{24}} \qquad (2.55)$$

where A_{12} is the amplitude for the baseline between elements 1 and 2, etc. Neither of these closure quantities are actually used to form the image, but they are used to reduce the number of unknowns in the procedure. At millimeter wavelengths, the principal atmospheric effects are due to water vapor. Since this absorbs the radiation, as well as leading to the phase delays, monitoring the sky brightness can provide information on the amount of water vapor along the line of sight, and so provide additional information for correcting the phase delays.

For VLBI, hybrid mapping is required since there is insufficient information in the visibility functions to produce a map directly. Hybrid mapping is an iterative technique that uses a mixture of measurements and guesswork. An initial guess is made at the form of the required map. This may well actually be a lower resolution map from a smaller interferometer. From this map, the visibility functions are predicted, and the true phases estimated. A new map is then generated by Fourier inversion. This map is then improved and used to provide a new start to the cycle. The iteration is continued until the hybrid map is in satisfactory agreement with the observations.

2.5.6 Intensity Interferometer

Until the last decade, the technical difficulties of an optical Michelson interferometer severely limited its usefulness. Most of these problems, however, may be reduced in a device that correlates intensity fluctuations. Robert Hanbury-Brown originally invented the device in 1949 as a radio interferometer, but it has found its main application in the optical region. The disadvantage of this system compared with the Michelson interferometer is that phase information is lost, and so the structure of a complex source cannot be reconstituted. The far greater ease of operation of the intensity interferometer led to it being able to measure some hundred stellar diameters between 1965 and 1972. Although it has been long closed down, the instrument is briefly discussed here for its historical interest, and because its principle of operation differs markedly from that of the Michelson interferometers.

The principle of the operation of the interferometer relies upon phase differences in the low frequency beat signals from different mutually incoherent sources at each aerial, combined with electrical filters to reject the high-frequency components of the signals. The schematic arrangement of the system is shown in Figure 2.36. We may imagine the signal from a source resolved into its Fourier components and consider the interaction of one such component from one source with another component from the other source. Let the frequencies of these two components be ν_1 and ν_2, then when they are mixed, there will be two additional frequencies—the upper and lower beat frequencies, $(\nu_1 + \nu_2)$ and $(\nu_1 - \nu_2)$ involved. For light waves, the lower beat frequency will be in the radio region, and this component may easily be filtered out within the electronics from the much higher original and upper beat frequencies. The low-frequency outputs from the two telescopes are multiplied and integrated over a short time interval and the signal bandwidth to produce the correlation function, K. Robert Hanbury-Brown and Richard Twiss were able to show that K was simply the square of the fringe visibility (Equation 2.45) for the Michelson interferometer.

The intensity interferometer was used to measure stellar diameters and $K(d)$ reaches its first zero when the angular stellar diameter, θ', is given by

$$\theta' = \frac{1.22\lambda}{d} \tag{2.56}$$

where d is the separation of the receivers. So the resolution of an intensity interferometer (and also of a Michelson interferometer) for stellar disks is the same as that of a telescope (Equation 2.40) whose diameter is equal to the separation of the receivers (Figure 2.37).

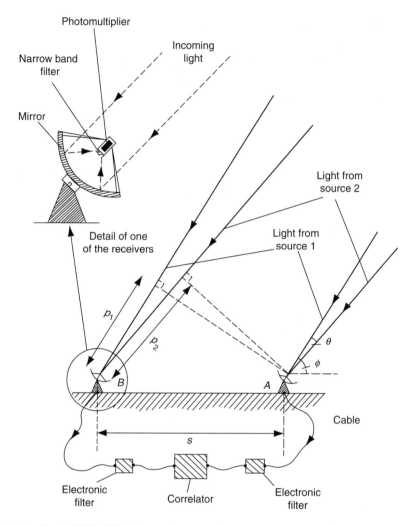

FIGURE 2.36 Schematic arrangement of an intensity interferometer.

The greater ease of construction and operation of the intensity inter-ferometer over the Michelson interferometer arises from its dependence upon the beat frequency of two light beams of similar wavelengths, rather than upon the actual frequency of the light. A typical value of the lower beat frequency is 100 MHz, which corresponds to a wavelength of 3 m. Thus, the path differences for the two receivers may vary by up to about 0.3 m during an observing sequence without ill effects. Scintillation, by the same argument, is also negligible.

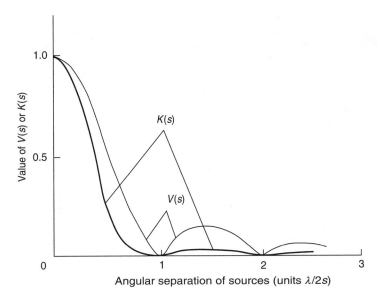

FIGURE 2.37 Comparison of the fringe visibility of a Michelson interferometer with the correlation function of an intensity interferometer.

Only one working intensity interferometer has been constructed. It was built by Robert Hanbury-Brown at Narrabri in Australia. It has now been decommissioned. It used two 6.5 m reflectors that were formed from several hundred smaller mirrors. There was no need for very high optical quality since the reflectors simply acted as light buckets and only the brightest stars could be observed. The reflectors were mounted on trolleys on a circular track 94 m in radius. The line between the reflectors could therefore always be kept perpendicular to the line of sight to the source, and their separation could be varied from 0 to 196 m (Figure 2.38). It operated at a wavelength of 433 nm, giving it a maximum resolution of about 0.0005 second of arc.

Exercise 2.1

Calculate the separation of the slits over a telescope objective in the first type of Michelson stellar interferometer, which would be required to measure the diameter of Ganymede (a) at opposition and (b) near conjunction of Jupiter. Take Ganymede's diameter to be 5000 km and the eye's sensitivity to peak at 550 nm.

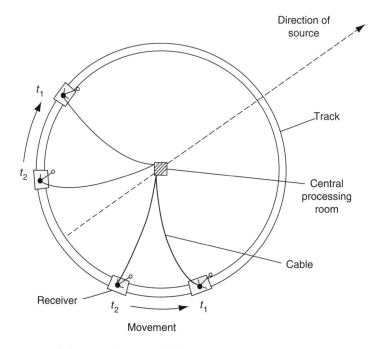

FIGURE 2.38 Schematic layout of the intensity interferometer at Narrabri, showing the positions of the receivers for observing in the same direction with different baselines. Note there were only two receivers; the diagram shows the arrangement at two separate baselines, superimposed.

Exercise 2.2

Observations of a quasar that is thought to be at a distance of 1500 Mpc just reveal structure to a VLBI working at 50 GHz. If its maximum baseline is 9000 km, what is the linear scale of this structure?

Exercise 2.3

Calculate the maximum distance at which the Narrabri intensity interferometer would be able to measure the diameter of a solar-type star.

2.6 SPECKLE INTERFEROMETRY

This technique is a kind of poor man's space telescope since it provides near diffraction-limited performance from earth-based telescopes. It works by obtaining images of the object sufficiently rapidly to freeze the blurring of the image that arises from atmospheric scintillation. The total image then consists of a large number of small dots or speckles, each of

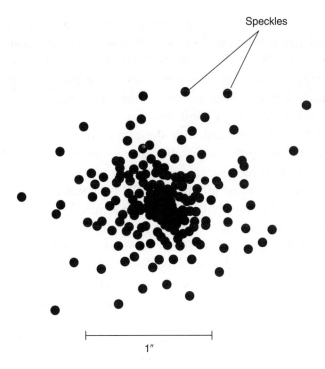

FIGURE 2.39 Schematic image of a point source obtained using a very short exposure time.

which is a diffraction-limited image for some objective diameter up to and including the diameter of the actual objective (Figure 2.39). An alternative to this technique is the adaptive optics telescope (Section 1.1), where adjusting the telescope optics to compensate for the atmospheric effects recombines the speckles. Adaptive optics, especially with artificial laser guide stars, has replaced speckle interferometry for most of the largest instruments, but the technique is still to be encountered in use on smaller telescopes—for example, the U.S. Naval Observatory's 0.66 m refractor and the 2.5 m Hooker telescope on Mount Wilson.

We may see how this speckled image structure arises by considering the effect of scintillation upon the incoming wavefront from the object. If we assume that above the atmosphere the wavefront is planar and coherent, then the main effect of scintillation is to introduce differential phase delays across it. The delays arise because the atmosphere is nonuniform with different cells within it having slightly different refractive indices. A typical cell size is 0.1 m, and the scintillation frequencies usually lie in the range 1–100 Hz. Thus, some 100 atmospheric cells will affect an average image from a 1 m telescope at any given instant. These will be rapidly changing, and over a normal exposure time, which can range from seconds to hours, they will form an integrated image that will be large and blurred compared with the diffraction-limited image. Even under the best seeing conditions, the image is rarely less than 1 second of arc across. However, an exposure of a few milliseconds is sufficiently rapid to freeze the image motion, and the observed image is then just the resultant of the contributions from the atmospheric cells across the telescope objective at that moment. Now the large number of these cells renders it highly probable that some of the phase delays will be similar to each other, and so some of the contributions to the image will be in phase with each other. These particular contributions will have been distributed over the objective in a random manner. Considering two such contributions, we have in fact a simple interferometer, and the two beams of radiation will combine in the image plane to produce results identical with those of an interferometer whose baseline is equal to the separation of the contributions on the objective. We have already seen what this image structure might be (Figure 2.21). If several collinear contributions are in phase, then the image structure will approach that shown in Figure 4.2 modulated by the intensity variation because of the aperture of a single cell. The resolution of the images is then given by the maximum separation of the cells. When the in-phase cells are distributed in two dimensions over the objective, the images have resolutions in both axes given by the maximum separations along those axes at the objective. The smallest speckles in the total image therefore have the diffraction-limited resolution of the whole objective, assuming always of course that the optical quality of the telescope is sufficient to reach this limit. Similar results will be obtained for those contributions to the image that are delayed by an integral number of wavelengths with respect to each other. Intermediate phase delays will cause destructive interference to a greater or lesser extent at the point in the image plane that is symmetrical between the two contributing beams of radiation, but will again interfere constructively at other points, to produce an interference pattern that has been shifted with respect to its normal position. Thus, all pairs of contributions to the final image interfere with each other to produce one or more speckles.

FIGURE 2.40 Schematic image of a just-resolved double source obtained with a very short exposure time.

In its most straightforward application, speckle interferometry can produce an actual image of the object. Each speckle is a photon noise limited image of the object, so that in some cases the structure can even be seen by the naked eye on the original image. A schematic instantaneous image of a double star that is just at the diffraction limit of the objective is shown in Figure 2.40. Normally however, the details of the image are completely lost in the noise. Then, many images, perhaps several thousand, must be averaged before any improvement in the resolution is gained.

More commonly, the data are Fourier analyzed, and the power spectrum obtained. This is the square of the modulus of the Fourier transform of the image intensity (Equation 2.3). The Fourier transform can be obtained directly using a large computer, or optical means can be used. In the latter case, the image is illuminated by collimated coherent light (Figure 2.41). It is placed at one focus of an objective and its Fourier transform is imaged at the back focus of the objective. A spatial filter at this point can be used to remove unwanted frequencies if required, and then the Fourier transform reimaged. However it may be obtained, the power spectrum can then be inverted to give centrisymmetric information such as diameters, limb darkening, oblateness, and binarity. Non-centrisymmetric structure can only be obtained if there is a point source close enough for its image to be obtained simultaneously with that of the object and for it to

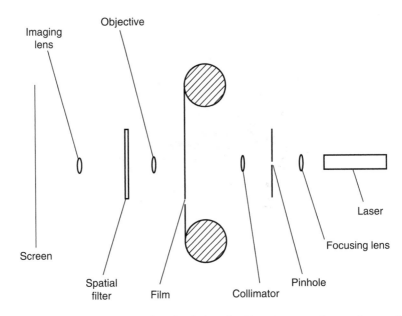

FIGURE 2.41 Arrangement for obtaining the Fourier transform of a speckle photograph by optical means.

be affected by the atmosphere in the same manner as the image of the object (i.e., the point source is within the same isoplanatic patch of sky as the object of interest, Figure 1.53). Then deconvolution (Section 2.1.1) can be used to retrieve the image structure.

The practical application of the technique requires the use of large telescopes and high-gain image intensifiers. Not only are very short exposures required (0.001–0.1 s), but very large plate scales (0.1–1 second of arc per millimeter) are also needed to separate the individual speckles. Furthermore, the wavelength range must be restricted by a narrow band filter to 20–30 nm. Even with such a restricted wavelength range, it may still be necessary to correct any remaining atmospheric dispersion using a low-power direct vision spectroscope. Thus, the limiting magnitude of the technique is currently about $+18^{m}$, and it has found its major applications in the study of red supergiants, the orbits of close binary stars, Seyfert galaxies, asteroids, and fine details of solar structure. Speckle interferometry, however, is a relatively recent invention and it is still developing; with CCD detectors now being used in place of photographic emulsions.

2.7 OCCULTATIONS

2.7.1 Background

An occultation is the more general version of an eclipse. It is the obscuration of a distant astronomical object by the passage in front of it of a closer object. Normally, the term implies occultation by the Moon, or more rarely by a planet, since other types of occultation are almost unknown. It is one of the most ancient forms of astronomical observation, with records of lunar occultations stretching back some two and a half millennia. More recently, interest in the events has been revived for their ability to give precise positions for objects observed at low angular resolution in, say, the x-ray region, and for their ability to give structural information about objects at better than the normal diffraction-limited resolution of a telescope. To see how the latter effect is possible, we must consider what happens during an occultation.

First, let us consider Fresnel diffraction at a knife-edge (Figure 2.42) for radiation from a monochromatic point source. The phase difference, δ, at a point P between the direct and the diffracted rays is then

$$\delta = \frac{2\pi}{\lambda}\left\{d_1 + \left[d_2^2 + (d_2\tan\theta)^2\right]^{1/2} - \left[(d_1 + d_2)^2 + (d_2\tan\theta)^2\right]^{1/2}\right\} \quad (2.57)$$

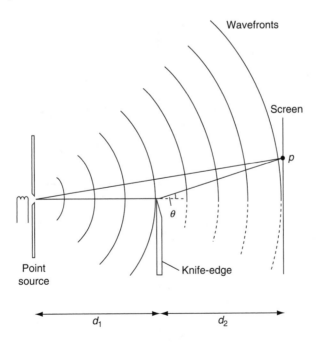

FIGURE 2.42 Fresnel diffraction at a knife-edge.

which, since θ is very small, simplifies to

$$\delta = \frac{\pi d_1 d_2}{\lambda(d_1 + d_2)} \theta^2 \tag{2.58}$$

The intensity at a point in a diffraction pattern is obtainable from the Cornu spiral (Figure 2.43), by the square of the length of the vector, **A**. P' is the point whose distance along the curve from the origin, l, is given by

$$l = \left[\frac{2d_1 d_2}{\lambda(d_1 + d_2)} \right]^{1/2} \theta \tag{2.59}$$

and the phase difference at P' is the angle that the tangent to the curve at that point makes with the x-axis, or from Equation 2.58

$$\delta = \frac{1}{2} \pi l^2 \tag{2.60}$$

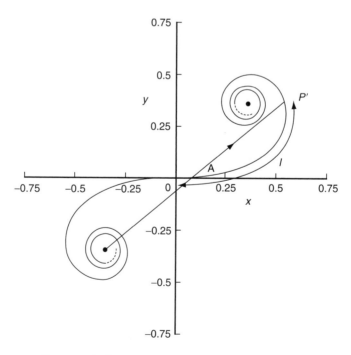

FIGURE 2.43 Cornu spiral.

The coordinates of P', which is the point on the Cornu spiral giving the intensity at P, x, and y, are obtainable from the Fresnel integrals

$$x = \frac{1}{\sqrt{2}} \int_0^l \cos\left(\frac{1}{2}\pi l^2\right) dl \tag{2.61}$$

$$y = \frac{1}{\sqrt{2}} \int_0^l \sin\left(\frac{1}{2}\pi l^2\right) dl \tag{2.62}$$

whose pattern of behavior is shown in Figure 2.44.

If we now consider a star occulted by the Moon, then we have

$$d_1 \gg d_2 \tag{2.63}$$

so that

$$l \approx \left(\frac{2d_2}{\lambda}\right)\theta \tag{2.64}$$

but otherwise the situation is unchanged from the one we have just discussed. The edge of the Moon of course is not a sharp knife-edge, but

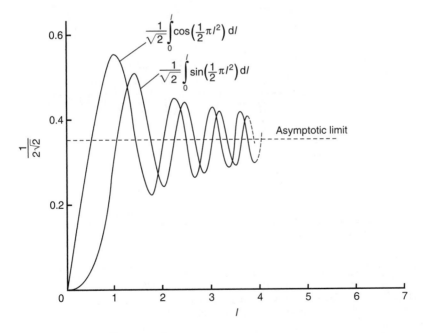

FIGURE 2.44 Fresnel integrals.

since even a sharp knife-edge is many wavelengths thick, the two situations are not in practice any different. The shadow of the Moon cast by the star onto the earth therefore has a standard set of diffraction fringes around its edge. The intensities of the fringes are obtainable from the Cornu spiral and Equations 2.61 and 2.62, and are shown in Figure 2.45. The first minimum occurs for

$$l = 1.22 \tag{2.65}$$

so that for the mean Earth–Moon distance

$$d_2 = 3.84 \times 10^8 \, \text{m} \tag{2.66}$$

we obtain

$$\theta = 0.0064'' \tag{2.67}$$

at a wavelength of 500 nm. Therefore, the fringes have a linear width of about 12 m, when the shadow is projected onto a part of the earth's surface

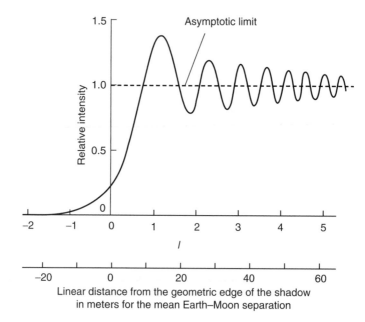

FIGURE 2.45 Fringes at the edge of the lunar shadow.

that is perpendicular to the line of sight. The mean rate of angular motion of the Moon, $\dot{\theta}$, is

$$\dot{\theta} = 0.55'' \, s^{-1} \tag{2.68}$$

so that at a given spot on the earth, the fringes will be observed as intensity variations of the star as it is occulted, with a basic frequency of up to 85 Hz. Usually, the basic frequency is lower than this since the earth's rotation can partially offset the Moon's motion, and because the section of the lunar limb that occults the star will generally be inclined to the direction of the motion.

If the star is not a point source, then the fringe pattern shown in Figure 2.45 becomes modified. We may see how this happens by imagining two point sources separated by an angle of about $0.0037''$ in a direction perpendicular to the limb of the Moon. At a wavelength of 500 nm, the first maximum of one star is then superimposed upon the first minimum of the other star, and the amplitude of the resultant fringes is much reduced compared with the single point-source case. The separation of the sources parallel to the lunar limb is, within reason, unimportant in terms of its effect on the fringes. Thus, an extended source can be divided into strips parallel to the lunar limb (Figure 2.46), and each strip then behaves during the occultation as though it was a centered point source of the relevant intensity. The fringe patterns from all these point sources are then superimposed in the final tracing of the intensity variations. The precise nature of the alteration from the point-source case will depend upon the size of the star and upon its surface intensity variations through such factors as limb darkening and gravity darkening, etc. as discussed later in this section. A rough guide, however, is that changes from the point-source pattern are just detectable for stellar diameters of $0.002''$, while the fringes disappear completely for object diameters of about $0.2''$. In the latter situation, the diameter may be recoverable from the total length of time that is required for the object to fade from sight. Double stars may also be distinguished from point sources when their separation exceeds about $0.002''$.

2.7.2 Techniques

In complete contrast to almost all other areas of astronomy, the detection of occultations in the optical region is best undertaken using comparatively small telescopes. This arises from the linear size of the fringes that

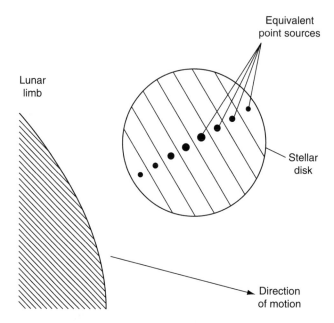

FIGURE 2.46 Schematic decomposition of a nonpoint source into equivalent point sources for occultation purposes.

we have seen is about 12 m. Telescopes larger than about 1 m will therefore simultaneously sample widely differing parts of the fringe so that the detected luminosity variations become smeared. Since the signal-to-noise ratio decreases with decreasing size of telescope, the optimum size for observing occultations usually lies between 0.5 and 2 m.

Most of the photometers described in Section 3.2 can be used to detect the star. However, we have seen that the observed intensity variations can have basic frequencies of 85 Hz with harmonics of several hundred hertz. The photometer and its associated electronics must therefore be capable of responding sufficiently rapidly to pick up these frequencies. The photometers used to detect occultations thus require response times of one to a few milliseconds.

Fringes are also blurred by the waveband over which the photometer is operating. We have seen from Equation 2.64 that the monochromatic fringe pattern is wavelength dependent. A bichromatic source with wavelengths differing by a factor of 2.37 would have the first maximum for one wavelength superimposed upon the first minimum for the other, so that the fringes would almost disappear. Smaller bandwidths will still degrade the visibility of the fringes although to a lesser extent. Using the standard

UBV filters (Section 3.1) only five or six fringes will be detectable even in the absence of all other noise sources. A bandwidth of about 20 nm must be used if the fringes found using a medium-sized telescope are not to deteriorate through this effect.

CCDs can also be used to determine the diffraction pattern of an occulted star. The diffraction pattern moves over the CCD at a calculable velocity, and the charges in the pixels are moved through the device at the same rate (cf. image tracking for liquid mirrors in Section 1.1.17.2). By this process of time delay and integration (TDI), high signal-to-noise ratios can be reached because each portion of the diffraction pattern is observed for a longer time than when a single element detector is used.

Since the observations are obviously always carried out within a few minutes of arc of the brightly illuminated portion of the Moon, the scattered background light is a serious problem. It can be minimized by using clean, dust free optics, precisely made light baffles, and a very small entrance aperture for the photometer, but it can never be completely eliminated. The total intensity of the background can easily be many times that of the star even when all these precautions have been taken. Thus, electronic methods must be used to compensate for its effects. The simplest system is to use a differential amplifier that subtracts a preset voltage from the photometer's signal before amplification. It must be set manually a few seconds before the occultation since the background is continually changing. A more sophisticated approach uses a feedback to the differential amplifier that continually adjusts the preset voltage so that a zero output results. The time constant of this adjustment, however, is arranged to be long, so that while the slowly changing background is counteracted, the rapid changes during the occultation are unaffected. The latter system is particularly important for observing disappearances, since it does not require last minute movements of the telescope onto and off the star.

An occultation can be observed at the leading edge of the Moon (a disappearance event) or at the trailing edge (when it is a reappearance event), provided that the limb near the star is not illuminated. Disappearances are by far the easiest events to observe since the star is visible and can be accurately tracked right up to the instant of the occultation. Reappearances can be observed only if a very precise offset is available for the photometer or telescope, or if the telescope can be set onto the star before its disappearance and then kept tracking precisely on its position for tens of minutes until it reappears. The same information content is available, however, from either event. The Moon often has quite a large motion in declination, so that it is occasionally possible for both a disappearance and the following reappearance to occur at a dark limb. This is a particularly important situation since the limb angles will be different between the two events, and a complete two-dimensional map of the object can be found.

Scintillation is another major problem in observing an occultation. The frequencies involved in scintillation are similar to those of the occultation so that they cannot be filtered out. The noise level is therefore largely because of the effects of scintillation, and it is rare for more than four fringes to be detectable in practice. Little can be done to reduce the problems caused by scintillation since the usual precautions of observing only near the zenith and on the steadiest nights cannot be followed; occultations have to be observed when and where they occur.

Occultations can be used to provide information other than whether the star is a double or not, or to determine its diameter. Precise timing of the event can give the lunar position to 0.05″ or better, and can be used to calibrate ephemeris time. Until recently, when navigation satellites became available, lunar occultations were used by surveyors to determine the positions on the earth of isolated remote sites, such as oceanic islands. Occultations of stars by planets are quite rare but are of considerable interest. Their main use is as a probe for the upper atmosphere of the planet, since it may be observed through the upper layers for some time before it is completely obscured. There has also been the serendipitous discovery of the rings of Uranus in 1977 through their occultation of a star. In spectral regions other than the visible region, lunar occultations have in the past found application in the determination of precise positions of objects.

2.7.3 Analysis

The analysis of the observations to determine the diameters and/or duplicity of the sources is simple to describe, but very time consuming to carry out. It involves the comparison of the observed curve with synthesized curves for various models. The synthetic curves are obtained in the manner indicated in Figure 2.46, taking account also of the bandwidth and the size of the telescope. One additional factor to be taken into account is the inclination of the lunar limb to the direction of motion of the Moon. If the Moon was a smooth sphere this would be a simple calculation, but the smooth limb is distorted by lunar surface features. Sometimes the actual inclination at the point of contact may be determinable from artificial satellite photographs of the Moon. On other occasions, the lunar slope must appear as an additional unknown to be determined from the analysis. Very rarely, a steep slope may be encountered and the star may reappear briefly before its final extinction. More often, the surface may be too rough on a scale of tens of meters to provide a good straightedge. Both these cases can usually be recognized from the data, and have to be discarded, at least for the purposes of determination of diameters or duplicity.

The analysis for positions of objects simply comprises determining the precise position of the edge of the Moon at the instant of occultation. One observation limits the position of the source to a semicircle corresponding to the leading (or trailing) edge of the Moon. Two observations from separate sites or from the same site at different lunations fix the object's position to the one or two intersections of two such semicircles. This will usually be sufficient to allow an unambiguous optical identification for the object, but if not, further observations can continue to be added to reduce the ambiguity in its position and to increase the accuracy, for as long as the occultations continue. Generally, a source that is occulted one month as observed from the earth or from a close earth-orbiting satellite will continue to be occulted for several succeeding lunations. Eventually, however, the rotation of the lunar orbit in the Saros cycle of 18 2/3 years will move the Moon away from the object's position, and no further occultations will occur for several years.

2.8 RADAR

2.8.1 Introduction

Radar astronomy and radio astronomy are very closely linked because the same equipment is often used for both purposes. However, radar astronomy is less familiar to the astrophysicist, for only one star, the Sun, has ever been studied with it, and this is likely to remain the case until other stars are visited by spacecraft, which will be a while yet. Other aspects of astronomy benefit from the use of radar to a much greater extent, so the technique is included, despite being almost outside the limits of this book, for completeness, and because its results may find applications in some areas of astrophysics.

Radar, as used in astronomy, can be fairly conventional equipment for use on board spacecraft, or for studying meteor trails, or it can be of highly specialized design and construction for use in detecting the Moon, planets,

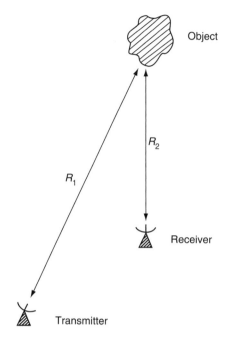

FIGURE 2.47 Schematic radar system.

Sun, etc. from the earth. Its results potentially contain information on three aspects of the object being observed: distance, surface fine scale structure, and relative velocity.

2.8.2 Theoretical Principles

Consider a radar system as shown in Figure 2.47. The transmitter has a power, P, and is an isotropic emitter. The radar cross section of the object is α, defined as the cross-sectional area of a perfectly isotropically scattering sphere, which would return the same amount of energy to the receiver as the object. Then the flux at the receiver, f, is given by

$$f = \frac{P\alpha}{4\pi R_1^2 4\pi R_2^2} \tag{2.69}$$

Normally, we have the transmitter and receiver close together, if they are not actually the same antenna, so that

$$R_1 = R_2 = R \tag{2.70}$$

Furthermore, the transmitter would not in practice be isotropic, but would have a gain, g (see Section 1.2.3). If the receiver has an effective collecting area of A_e then the received signal, F, is given by

$$F = \frac{A_e \alpha P g}{16 \pi^2 R^4} \tag{2.71}$$

For an antenna, we have the gain from Equation 1.82

$$g = \frac{4 \pi \nu^2 A'_e}{c^2} \tag{2.72}$$

where A'_e is the effective area of the transmitting antenna. Much of the time in radar astronomy, the transmitting and receiving dishes are the same, so that

$$A'_e = A_e \tag{2.73}$$

and so

$$F = \frac{P \alpha A_e^2 \nu^2}{4 \pi c^2 R^4} \tag{2.74}$$

Equation 2.74 is valid for objects that are not angularly resolved by the radar. For targets that are comparable with the beam width or larger, the radar cross section, α, must be replaced by an appropriate integral. Thus, for a spherical target which has the radar beam directed toward its center, we have

$$F \approx \frac{P A_e^2 \nu^2}{4 \pi c^2 R^4} \int\limits_0^{\pi/2} 2 \pi r \alpha(\phi) \sin \phi \left[s \left(\frac{r \sin \phi}{R} \right) \right]^2 d\phi \tag{2.75}$$

where
r is the radius of the target and is assumed to be small compared with the distance, R
ϕ is the angle at the center of the target to a point on its surface illuminated by the radar beam
$\alpha(\phi)$ is the radar cross section for the surface of the target when the incident and returned beams make an angle ϕ to the normal to the surface. The function is normalized so that the integral tends to α as the beam width increases
$s(\theta)$ is the sensitivity of the transmitter/receiver at an angle θ to its optical axis (cf. Equation 1.83)

The amount of flux that is received is not the only criterion for the detection of objects as we saw in Section 1.2.2 (Equation 1.76). The flux must also be sufficiently stronger than the noise level of the whole system if the returned pulse is to be distinguishable. Now we have seen in Section 1.2.2 that the receiver noise may be characterized by comparing it with the noise generated in a resistor at some temperature. We may similarly characterize all the other noise sources, and obtain a temperature, T_s, for the whole system, which includes the effects of the target and of the transmission paths as well as the transmitter and receiver. Then from Equation 1.77, we have the noise in power terms, N

$$N = 4kT_s \Delta\nu \tag{2.76}$$

where $\Delta\nu$ is the bandwidth of the receiver in frequency terms. The signal-to-noise ratio is therefore

$$\frac{F}{N} = \frac{P\alpha A_e^2 \nu^2}{16\pi kc^2 R^4 T_s \Delta\nu} \tag{2.77}$$

and must be unity or larger if a single pulse is to be detected. Since it is the selected target that fixes α and R, the signal-to-noise ratio only may be increased by increasing the power of the transmitter, the effective area of the antenna, or the frequency, or by decreasing the system temperature or the bandwidth. Only over the last of these items does the experimenter have any real control, all the others will be fixed by the choice of antenna. However, the bandwidth is related to the length of the radar pulse. Even if the transmitter emits monochromatic radiation, the pulse will have a spread of frequencies given by the Fourier transform of the pulse shape. To a first approximation

$$\Delta\nu \approx \frac{1}{\tau} \text{ Hz} \tag{2.78}$$

where τ is the length of the transmitted pulse in seconds. Thus, increasing the pulse length can increase the signal-to-noise ratio. Unfortunately for accurate ranging, the pulse needs to be as short as possible, and so an optimum value that minimizes the conflict between these two requirements must be sought, and this will vary from radar system to radar system and from target to target.

When ranging is not required, so that information is only being sought on the surface structure and the velocity, the pulse length may be increased very considerably. Such a system is then known as a continuous wave (CW) radar, and its useful pulse length is only limited by the stability of the transmitter frequency, and by the spread in frequencies introduced into the returned pulse through Doppler shifts arising from the movement and rotation of the target. In practice, CW radars work continuously even though the signal-to-noise ratio remains the same as that for the optimum pulse. The alternative to CW radar is pulsed radar. Astronomical pulsed radar systems have a pulse length typically between 10 μs and 10 ms, with peak powers of several tens of megawatts. For both CW and pulsed systems, the receiver passband must be matched to the returned pulse in both frequency and bandwidth. The receiver must therefore be tunable over a short range to allow for the Doppler shifting of the emitted frequency.

The signal-to-noise ratio may be improved by integration. With CW radar, samples are taken at intervals given by the optimum pulse length. For pulsed radar, the pulses are simply combined. In either case, the noise is reduced by a factor of $N^{1/2}$, where N is the number of samples or pulses that are combined. If the radiation in separate pulses from a pulsed radar system is coherent, then the signal-to-noise ratio improves directly with N. However, coherence may only be retained over the time interval given by the optimum pulse length for the system operating in the CW mode. Thus, if integration is continued over several times the coherence interval, the signal-to-noise ratio for a pulsed system decreases as $(NN'^{1/2})$, where N is the number of pulses in a coherence interval, and N' the number of coherence intervals over which the integration extends.

For most astronomical targets, whether they are angularly resolved or not, there is some depth resolution with pulsed systems. The physical length of a pulse is $(c\tau)$, and this ranges from about 3 to 3000 km for most practical astronomical radar systems. Thus, if the depth of the target in the sense of the distance along the line of sight between the nearest and furthest points returning detectable echoes is larger than the physical pulse length, the echo will be spread over a greater time interval than the original pulse. This has both advantages and disadvantages. First, it provides information on the structure of the target with depth. If the depth information can be combined with Doppler shifts arising through the target's rotation, then the whole surface may be mappable, although there may still remain a twofold ambiguity about the equator. Second, and on the debit side, the signal-to-noise ratio is reduced in proportion approximately to the ratio of the lengths of the emitted and returned pulses.

Equation 2.77 and its variants that take account of resolved targets, etc. are often called the radar equation. Of great practical importance is its R^{-4} dependence, so that, if other things are equal, the power required to detect an object increases by a factor of 16 when the distance to the object doubles. Thus, Venus at greatest elongation needs 37 times the power required for its detection at inferior conjunction. However, the high dependence upon R can be made to work in our favor, when using spacecraft-borne radar systems. The value of R can then be reduced from tens of millions of kilometers to a few hundreds. The enormous gain in the sensitivity that results means that the small low-power radar systems that are suitable for spacecraft use are sufficient to map the target in detail. The prime example of this to date is, of course, Venus. Most of the data we have on the surface features have come from the radar carried by *Magellan* and the various *Mariner* and *Pioneer* spacecraft.

The radar on board *Magellan* was of a different type from those so far considered. It is known as a SAR and has much in common with the technique of aperture synthesis (Section 2.5.5). Such radars have also been used on board several remote sensing satellites like Seasat and ERS 1, for studying the earth.

SAR uses a single antenna that, since it is on board a spacecraft, is moving with respect to the surface of the planet below. The single antenna therefore successively occupies the positions of the multiple antennae of an array (Figure 2.48). To simulate the observation of such an array, all that is necessary therefore is to record the output from the radar when it occupies the position of one of the elements of the array, and then to add the successive recordings with appropriate shifts to counteract their various time delays. In this way, the radar can build up an image whose resolution, along the line of flight of the spacecraft, is many times better than that which the simple antenna would provide by itself. In practice, of course, the SAR operates continually, not just at the positions of the elements of the synthesized array.

The maximum array length that can be synthesized in this way is limited by the period over which a single point on the planet's surface can be kept under observation. The resolution of a parabolic dish used as a radar is approximately λ/D. If the spacecraft is at a height, h, the radar footprint of the planetary surface is thus about

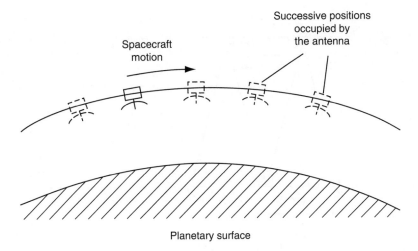

FIGURE 2.48 Synthesis of an interferometer array by a moving single antenna.

$$L = \frac{2\lambda h}{D} \qquad (2.79)$$

in diameter (Figure 2.49). Since h, in general, will be small compared with the size of the planet, we may ignore the curvature of the orbit. We may then easily see (Figure 2.50) that a given point on the surface only remains observable while the spacecraft moves a distance, L.

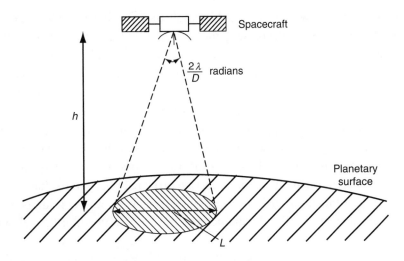

FIGURE 2.49 Radar footprint.

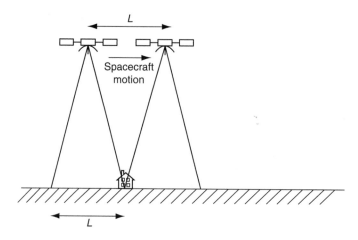

FIGURE 2.50 Maximum length of a SAR.

Now, the resolution of an interferometer (Section 2.5.2) is given by Equation 2.41. Thus, the angular resolution of a SAR of length L is

$$\text{Resolution} = \frac{\lambda}{2L} = \frac{D}{4H} \text{ radians} \tag{2.80}$$

The diameter of the synthesized radar footprint is then

$$\text{Footprint diameter} = 2\left(\frac{D}{4h}\right)h = \frac{D}{2} \tag{2.81}$$

Equation 2.81 shows the remarkable result that the linear resolution of a SAR is improved by decreasing the size of the radar dish! The linear resolution is just half the dish diameter. Other considerations such as signal-to-noise ratio (Equation 2.77) require the dish to be as large as possible. Thus, an actual SAR system has a dish size that is a compromise, and existing systems typically have 5–10 m diameter dishes.

The above analysis applies to a focused SAR, that is to say, a SAR in which the changing distance of the spacecraft from the point of observation is compensated by phase shifting the echoes appropriately. In an unfocused SAR, the point at which the uncorrelated phase shifts due to the changing distance to the observed object reach about $\lambda/4$ limits the synthesized array length. The diameter of the array footprint is then about

$$\text{Unfocused SAR footprint} = \sqrt{2\lambda h} \tag{2.82}$$

2.8.3 Equipment

Radar systems for use in studying planets, etc. have much in common with radio telescopes. Almost any filled-aperture radio antenna is usable as a radar antenna. Steerable paraboloids are particularly favored because of their convenience in use. Unfilled apertures, such as collinear arrays (Section 1.2) are not used since most of the transmitted power is wasted.

The receiver is also generally similar to those in use in radio astronomy, except that it must be very accurately tunable to provide compensation for Doppler shifts, and the stability of its frequency must be very high. Its band pass should also be adjustable so that it may be matched accurately to the expected profile of the returned pulse to minimize noise.

Radar systems, however, do require three additional components that are not found in radio telescopes. First, and obviously, there must be a transmitter, second a very high-stability master oscillator, and third an accurate timing system. The transmitter is usually second only to the antenna as a proportion of the cost of the whole system. Different targets or purposes will generally require different powers, pulse lengths, frequencies, etc. from the transmitter. Thus, it must be sufficiently flexible to cope with demands that might typically range from pulse emission at several megawatts for a millisecond burst to CW generation at several hundred kilowatts. Separate pulses must be coherent, at least over a time interval equal to the optimum pulse length for the CW mode, if full advantage of integration to improve signal-to-noise ratio is to be gained. The master oscillator is in many ways the heart of the system. In a typical observing situation, the returned pulse must be resolved to 0.1 Hz for a frequency in the region of 1 GHz, and this must be accomplished after a delay between the emitted pulse and the echo that can range from minutes to hours. Thus, the frequency must be stable to one part in 10^{12} or 10^{13}, and it is the master oscillator that provides this stability. It may be a temperature-controlled crystal oscillator, or an atomic clock. In the former case, corrections for aging will be necessary. The frequency of the master oscillator is then stepped up or down as required and used to drive the exciter for the transmitter, the local oscillators for the heterodyne receivers, the Doppler compensation system, and the timing system. The final item, the timing system, is essential if the range of the target is to be found. It is also essential to know with quite a high accuracy when to expect the echo. This arises because the pulses are not broadcast at constant intervals. If this were to be the case then there would be no certain way of relating a given echo to a given emitted pulse. Thus, instead, the emitted pulses are sent out at varying and coded intervals, and the same pattern sought amongst the echoes. Since the distance between the earth and the target is also varying, an accurate timing system is vital to allow the resulting echoes with their changing intervals to be found amongst all the noise.

2.8.4 Data Analysis

The analysis of the data to obtain the distance to the target is simple in principle, being merely half the delay time multiplied by the speed of light. In practice, the process is much more complicated. Corrections for the atmospheres of the earth and the target (should it have one) are of particular importance. The radar pulse will be refracted and delayed by the earth's atmosphere, and there will be similar effects in the target's atmosphere with in addition the possibility of reflection from an ionosphere as well as, or instead of, reflection from the surface. This applies especially to solar detection. Then reflection can often occur from layers high above the visible surface, for example, 10,000 km above the photosphere for a radar frequency of 1 GHz. Furthermore, if the target is a deep one, that is, it is resolved in depth, then the returned pulse will be spread over a greater time interval than the emitted one so that the delay length becomes ambiguous. Other effects such as refraction or delay in the interplanetary medium may also need to be taken into account.

For a deep, rotating target, the pulse is spread out in time and frequency, and this can be used to plot maps of the surface of the object (Figure 2.51). A cross section through the returned pulse at a given instant of time will be formed from echoes from all the points over an annular region such as that shown in Figure 2.52. For a rotating target, which for simplicity we assume has its rotational axis perpendicular to the line of sight, points such as A and A' will have the same approach velocity, and their echoes will be shifted to the same higher frequency. Similarly, the echoes from points such as B and B' will be shifted to the same lower frequency. Thus, the intensity of the echo at a particular instant and frequency is related to the radar properties of two points that are equidistant from the equator and are north and south of it. Hence, a map of the radar reflectivity of the target's surface may be produced, though it will have a twofold ambiguity about the equator. If the rotational axis is not perpendicular to the line of sight, then the principle is similar, but the

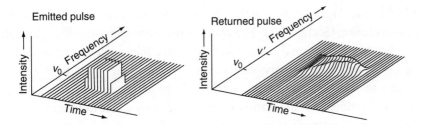

FIGURE 2.51 Schematic change in pulse shape for radar pulses reflected from a deep rotating target. The emitted pulse would normally have a Gaussian profile in frequency, but it is shown here as a square wave to enhance the clarity of the changes between it and the echo.

calculations are more complex. However, if such a target can be observed at two different angles of inclination of its rotational axis, then the twofold ambiguity may be removed and a genuine map of the surface produced. Again in practice, the process needs many corrections, and must have the

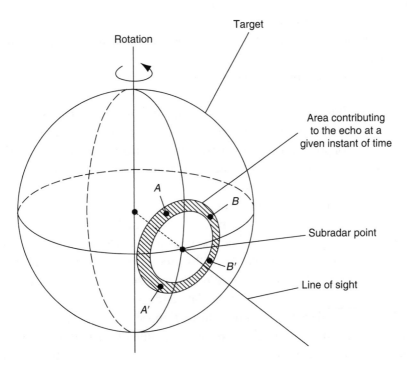

FIGURE 2.52 Region of a deep radar target contributing to the echo at a particular instant of time.

frequency and time profiles of the emitted pulse removed from the echo by deconvolution (Section 2.1). The features on radar maps are often difficult to relate to more normal surface features, since they are a conflation of the effects of surface reflectivity, small-scale structure, height, gradient, etc.

The relative velocity of the earth and the target is much simpler to obtain. Provided that the transmitter is sufficiently stable in frequency (see the earlier discussion), then it may be found from the shift in the mean frequency of the echo compared with the mean frequency of the emitted pulse, by the application of the Doppler formula. Corrections for the earth's orbital and rotational velocities are required. The former is given by Equation 4.40, while the latter may be obtained from

$$v_c = v_o + 462 \cos \delta \, \sin (\alpha - LST) \sin (LAT) \text{ ms}^{-1} \qquad (2.83)$$

where

v_c is the corrected velocity (ms^{-1})
v_o is the observed velocity (ms^{-1})
α is the right ascension of the target
δ is the declination of the target
LST is the local sidereal time at the radar station when the observation is made
LAT is the latitude of the radar station

The combination of accurate knowledge of both distance and velocity for a planet on several occasions enables its orbit to be calculated to a very high degree of precision. This then provides information on basic quantities such as the astronomical unit, and enables tests of metrical gravitational theories such as general relativity to be made, and so is of considerable interest and importance even outside the area of astronomy. However, producing a precise orbit for a planet by this method requires that the earth's and the planet's orbits both be known precisely so that the various corrections, etc. can be applied. Thus, the procedure is a bootstrap one, with the initial data leading to improved orbits, and these improved orbits then allowing better interpretation of the data and so on over many iterations.

The radar reflection contains information on the surface structure of the target. Changes in the phase, polarization and of the manner in which the radar reflectivity changes with angle, can in principle be used to determine rock types, roughness of the surface, etc. In practice, the data are rarely sufficient to allow anything more than vague possibilities to be indicated. The range of unfixed parameters describing the surface is so large, that many different models for the surface can fit the same data. This aspect of radar astronomy has therefore been of little use to date.

2.8.5   Meteors

The ionized vapor along the track of a meteor reflects radar transmissions admirably. Over the last few decades therefore many stations have been set up to detect meteors by such means. Since they can be observed equally well during the day as at night, many previously unknown daytime meteor showers have been discovered. Either pulsed or CW radar is suitable, and a wavelength of a few meters is usually used.

The echoes from a meteor trail are modulated into a Fresnel diffraction pattern (Figure 2.45), with the echoed power expressible in terms of the Fresnel integrals (Equations 2.61 and 2.62) as

$$F(t) = F(x^2 + y^2) \tag{2.84}$$

where
F is the echo strength for the fully formed trail
$F(t)$ is the echo strength at time t during the formation of the trail

The parameter l of the Fresnel integrals in this case is the distance of the meteor from the minimum range point. The modulation of the echo arises because of the phase differences between echoes returning from different parts of the trail. The distance of the meteor, and hence its height may be found from the delay time as usual. In addition to the distance however, the cross range velocity can also be obtained from the echo modulation. Several stations can combine, or a single transmitter can have several well-separated receivers, to provide further information such as the position of the radiant, atmospheric deceleration of the meteor, etc.

Exercise 2.4

If a 10 MW peak power radar is used to observe Venus from the earth at inferior conjunction, using a 300 m dish aerial, calculate the peak power required for a satellite-borne radar situated 1000 km above Venus' surface that has a 1 m dish aerial if the same signal-to-noise ratio is to be achieved. Assume that all the other parameters of the radar systems are similar, and take the distance of Venus from the earth at inferior conjunction to be 4.14×10^7 km.

If the linear resolution of the earth-based radar at Venus is 10^4 km, what is that of the satellite-based system?

2.9 ELECTRONIC IMAGES

2.9.1 Image Formats

There are various ways of storing electronic images. One currently in widespread use for astronomical images is the flexible image transport system (FITS). There are various versions of FITS, but all have a header, the image in binary form, and an end section. The header must be a multiple of 2880 bytes (this number originated in the days of punched card input to computers and represents 36 cards each of 80 bytes). The header contains information on the number of bits representing the data in the image, the number of dimensions of the image (e.g., one for a spectrum, two for a monochromatic image, three for a color image, etc.), the number of elements along each dimension, the object observed, the telescope used, details of the observation, and other comments. The header is then padded out to a multiple of 2880 with zeros. The image section of a FITS file is simply the image data in the form specified by the header. The end section then pads the image section with zeros until it too is an integer multiple of 2880 bytes.

Other formats may be encountered such as JPEG (joint photographic experts group), GIF (graphic interchange format), and TIFF (tagged image file format) that are more useful for processed images than raw data.

2.9.2 Image Compression

Most images, howsoever they may be produced and at whatever wavelength they were obtained, are nowadays stored in electronic form. This is even true of many photographic images that may be scanned using microdensitometers (Section 2.2) and fed into a computer. Electronic images may be stored by simply recording the precise value of the intensity at each pixel in an appropriate two-dimensional matrix. However, this uses large amounts of memory; for example, 2 MB for a 16-bit image from a 1000 × 1000 pixel CCD (and 32-bit, 10,000 × 10,000 CCDs, requiring 400 MB per image are being envisaged). Techniques whereby storage space for electronic images may be utilised most efficiently are therefore at a premium. The most widely used approach to the problem is to compress the images.

Image compression relies for its success upon some of the information in the image being redundant. This information may therefore be omitted or stored in shortened form without losing information from the image as a whole. Image compression that conserves all the information in the original image is known as lossless compression. For some purposes a reduction in the amount of information in the image may be acceptable (e.g., if one is only interested in certain objects or intensity ranges within the image), and this is the basis of lossy compression (see below).

There are three main approaches to lossless image compression. The first is differential compression. This stores not the intensity values for each pixel, but the differences between the values for adjacent pixels. For an image with smooth variations, the differences will normally require fewer bits for their storage than the absolute intensity values. The second approach, known as run compression, may be used when many sequential pixels have the same intensity. This is often true for the sky background in astronomical images. Each run of constant intensity values may be stored as a single intensity combined with the number of pixels in the run, rather than storing the value for every pixel. The third approach, which is widely used on the Internet and elsewhere, where it goes by various proprietary names such as ZIP and DoubleSpace, is similar to run compression except that it stores repeated sequences whatever their nature as a single code, not just constant runs. For some images, lossless compressions by a factor of 5 or 10 times are possible.

Sometimes not all the information in the original images is required. Lossy compression can then be by factors of 20, 50, or more. For example, a 32-bit image to be displayed on a computer monitor that has 256 brightness levels need only be stored to 8-bit accuracy, or only the (say) 64×64 pixels covering the image of a planet, galaxy, etc. need be stored from a 1000×1000 pixel image if that is all that is of interest within the image. However, most workers will prefer to store the original data in a lossless form.

Compression techniques do not work effectively on noisy images since the noise is incompressible. Then a lossy compression may be of use since it will primarily be the noise that is lost. An increasingly widely used lossy compression for such images is based on the wavelet transform

$$T = \frac{1}{\sqrt{|a|}} \int f(t)\psi\left(\frac{t-b}{a}\right) dt \qquad (2.85)$$

This can give high compression ratios for noisy images. However, the details of the technique are beyond the scope of this book.

2.9.3 Image Processing

Image processing is the means whereby images are produced in an optimum form to provide the information that is required from them.

Image processing divides into data reduction and image optimization. Data reduction is the relatively mechanical process of correcting for known problems and faults. Many aspects of it are discussed elsewhere, such as dark frame subtraction, flat fielding, and cosmic ray spike elimination on CCD images (Section 1.1.8), deconvolution and maximum entropy processing (Section 2.1.1) and CLEANing aperture synthesis images (Section 2.5.5), etc.

Image optimization is more of an art than a science. It is the further processing of an image with the aim of displaying those parts of the image that are of interest to their best effect for the objective for which they are required. The same image may therefore be optimized in quite different ways if it is to be used for different purposes. For example, an image of Jupiter and its satellites would be processed quite differently if the surface features on Jupiter were to be studied, compared with if the positions of the satellites with respect to the planet were of interest. There are dozens of techniques used in this aspect of image processing, and there are few rules other than experience to suggest which will work best for a particular application. The main techniques are outlined below, but the interested reader will need to look to other sources and to obtain practical experience before being able to apply them with consistent success.

2.9.3.1 Grey Scaling

This is probably the technique in most widespread use. It arises because many detectors have dynamic ranges far greater than those of the computer monitors or hard copy devices that are used to display their images. In many cases, the interesting part of the image will cover only a restricted part of the dynamic range of the detector. Grey scaling then consists of stretching that section of the dynamic range over the available intensity levels of the display device.

For example, the spiral arms of a galaxy on a 16-bit CCD image (with a dynamic range from 0 to 65,535) might have CCD level values ranging from 20,100 to 20,862. On a computer monitor with 256 intensity levels, this would be grey scaled so that CCD levels 0–20,099 corresponded to intensity 0 (i.e., black) on the monitor, levels 20,100–20,103 to intensity 1, levels 20,104–20,106 to intensity 2, and so on up to levels 20,859–20,862 corresponding to intensity 254. CCD levels 20,863–65,535 would then all be lumped together into intensity 255 (i.e., white) on the monitor.

Other types of grey scaling may aim at producing the most visually acceptable version of an image. There are many variations of this technique whose operation may best be envisaged by its effect on a histogram of the display pixel intensities. In histogram equalization, for example, the image pixels are mapped to the display pixels in such a way that there are equal numbers of image pixels in each of the display levels. For many astronomical images, a mapping that results in an exponentially decreasing histogram for the numbers of image pixels in each display level often gives a pleasing image since it produces a good dark background for the image. The number of other mappings available is limited only by the imagination of the operator.

2.9.3.2   Image Combination

Several images of the same object may be added together to provide a resultant image with lower noise. More powerful combination techniques, such as the subtraction of two images to highlight changes, or false-color displays, are, however, probably in more widespread use.

2.9.3.3   Spatial Filtering

This technique has many variations that can have quite different effects upon the image. The basic process is to combine the intensity of one pixel with those of its surrounding pixels in some manner. The filter may best be displayed by a 3×3 matrix (or 5×5, 7×7 matrices, etc.) with the pixel of interest in the center. The elements of the matrix are the weightings given to each pixel intensity when they are combined and used to replace the original value for the central pixel (Figure 2.53). Some commonly used filters are shown in Figure 2.53.

Various commercial image processing packages are available and enable anyone to undertake extensive image processing. Much of this though is biased toward commercial artwork and the production of images for advertising, etc. However, there are packages specifically designed for astronomical image processing, and many are available free of charge. The most widespread of these packages for optical images is IRAF (image reduction and analysis facility), produced and maintained by NOAO (National Optical Astronomy Observatory). IRAF is available free for downloading via the Web (see http://iraf.noao.edu/) or for a cost of a few tens of pounds/dollars on CD-ROM (see previous Web site for further

Strong smoothing filter

+1/9	+1/9	+1/9
+1/9	+1/9	+1/9
+1/9	+1/9	+1/9

Sharpening filter

0	−1/4	0
−1/4	+2	−1/4
0	−1/4	0

Weak smoothing filter

0	+1/8	0
+1/8	+1/2	+1/8
0	+1/8	0

Edge enhancement

−1	−1	−1
−1	+8	−1
−1	−1	−1

FIGURE 2.53 Examples of commonly used spatial filters.

details and also for details of IRAF's image processing packages). You will also need your computer to have a UNIX or LINUX operating system. The latter is available, also without charge for downloading onto personal computers, at http://www.linux.org/. However, you are warned that though available free of charge, these packages are not easy for an inexperienced computer user to implement successfully. If you have purchased a CCD camera designed for amateur astronomical use, then the manufacturer will almost certainly have provided at least a basic image processing package as a part of the deal. Most such manufacturers also have Web sites where other software produced by themselves and by users of the cameras may be found. For radio astronomical data, AIPS and AIPS++ are the most widely used data reduction packages (Section 2.5.5).

CHAPTER 3

Photometry

3.1 PHOTOMETRY

3.1.1 Background

3.1.1.1 Introduction

Photometry is the measurement of the energy coming from an astronomical object within a restricted range of wavelengths. On this definition just about every observation that is made is a part of photometry. It is thus customary and convenient to regard spectroscopy and imaging as separate from photometry, although the differences are blurred in the overlap regions and outside the optical part of the spectrum. Spectroscopy (Chapter 4) is the measurement of the energy coming from an object within several (usually hundreds or thousands) of adjacent wavebands and with the widths of those wavebands a hundredth of the operating wavelength or less (usually a thousandth to a hundred-thousandth of the operating wavelength). However, some narrowband photometric systems on this definition could be regarded as low-resolution spectroscopy. It is harder to separate imaging (Chapters 1 and 2) from photometry, especially with array detectors, since much of photometry is now just imaging through an appropriate filter or filters. So it is perhaps best to regard photometry as imaging with the primary purpose of measuring the brightnesses of some or all of the objects within the image and imaging as having the primary purpose of determining the spatial structure or appearance of the object. The situation is further confused by the recent development of integral field spectroscopy (Section 4.2.4) where spectra are simultaneously obtained for some or all of the objects within an image.

Since most radio receivers are narrowband instruments and many of the detectors used for far-infrared (FIR) and microwave observations operate over restricted wavebands, essentially all their observations come under photometry and have therefore been covered in Sections 1.1 and 1.2. Likewise extreme ultraviolet (EUV), x-ray, and γ-ray detectors often have some intrinsic sensitivity to wavelength (i.e., photon energy) and so automatically operate as photometers or low-resolution spectroscopes. Their properties have been covered in Section 1.3. In this section, we are thus concerned with photometry as practiced over the near ultraviolet, visible, near-infrared (NIR), and mid-infrared (MIR) regions of the spectrum (roughly 100 nm to 100 μm).

3.1.1.2 Magnitudes

The system used by astronomers to measure the visible brightnesses of stars is a very ancient one, and for that reason is a very awkward one. It originated with the earliest of the known astronomical catalogs: that due to Hipparchus in the late second century BC. The stars in this catalog were divided into six classes of brightness, the brightest being of the first class, and the faintest of the sixth class. With the invention and development of the telescope such a system proved to be inadequate, and it had to be refined and extended. William Herschel initially undertook this, but our present day system is based upon the work of Norman Pogson in 1856. He suggested a logarithmic scale that approximately agreed with the earlier measures. As we have seen in Section 1.1.3, the response of the eye is nearly logarithmic, so that the ancient system of Hipparchus was logarithmic in terms of intensity. Hipparchus' class 5 stars were about 2.5 times brighter than his class 6 stars, and were about 2.5 times fainter than class 4 stars, and so on. Pogson expressed this as a precise mathematical law in which the magnitude was related to the logarithm of the diameter of the minimum size of telescope required to see the star (i.e., the limiting magnitude, Equation 3.3). Soon after this, the proposal was adapted to express the magnitude directly in terms of the energy coming from the star, and the equation is now usually used in the form

$$m_1 - m_2 = -2.5 \log_{10} \left(\frac{E_1}{E_2} \right) \tag{3.1}$$

where
m_1 and m_2 are the magnitudes of stars 1 and 2, respectively
E_1 and E_2 are the energies per unit area at the surface of the earth for stars 1 and 2, respectively

The scale is awkward to use because of its peculiar base, leading to the ratio of the energies of two stars whose brightnesses differ by one magnitude being 2.512.* Also the brighter stars have magnitudes whose actual values are smaller than those of the fainter stars. However, it seems unlikely that astronomers will change the system, so the student must perforce become familiar with it as it stands. One aspect of the scale is immediately apparent from Equation 3.1, and that is that the scale is a relative one—the magnitude of one star is expressed in terms of that of another, there is no direct reference to absolute energy or intensity units. The zero of the scale is therefore fixed by reference to standard stars. On Hipparchus' scale, those stars that were just visible to the naked eye were of class 6. Pogson's scale is thus arranged so that stars of magnitude six are just visible to a normally sensitive eye from a good observing site on a clear, moonless night. The standard stars are chosen to be nonvariables, and their magnitudes are assigned so that the above criterion is fulfilled. The primary stars are known as the north polar sequence, and comprise stars within 2° of the pole star, so that they may be observed from almost all northern hemisphere observatories throughout the year. Secondary standards may be set up from these to allow observations to be made in any part of the sky. The stars of the north polar sequence were listed in earlier editions of this book, but in practice the secondary standards are now used almost exclusively. Since the standard stars for other regions of the spectrum, polarimetry, magnetometry, etc. are not covered at all by the north polar sequence, its details have now been dropped from this edition. If anyone should need this information, then it may be found in earlier editions of this book, or in the primary source: Leavitt, H.S., *Annals of the Harvard College Observatory*, 71(3), 49–52, 1917.

The faintest stars visible to the naked eye, from the definition of the scale, are of magnitude six; this is termed the limiting magnitude of the eye. For point sources, the brightnesses are increased by the use of a telescope by a factor, G, that is called the light grasp of the telescope (Section 1.1.18.1). Since the dark-adapted human eye has a pupil diameter of about 7 mm, G is given by

$$G \approx 2 \times 10^4 \, d^2 \tag{3.2}$$

---

* $= 10^{0.4}$. For a difference of two magnitudes, the energies differ by $\times 6.31$ (2.512^2), for three magnitudes by $\times 15.85$ (2.512^3), four magnitudes by $\times 39.81$ (2.512^4), and five magnitudes by exactly $\times 100$ (2.512^5).

where d is the telescope's objective diameter in meters. Thus, the limiting magnitude through a visually used telescope, m_L, is

$$m_L = 16.8 + 5\log_{10} d \qquad (3.3)$$

Charge-coupled devices (CCDs) and other detection techniques will generally improve on this by some 5–10 stellar magnitudes.

The magnitudes so far discussed are all apparent magnitudes and so result from a combination of the intrinsic brightness of the object and its distance. This is obviously an essential measurement for an observer, since it is the criterion determining exposure times, etc.; however, it is of little intrinsic significance for the object in question. A second type of magnitude is therefore defined that is related to the actual brightness of the object. This is called the absolute magnitude and it is "the apparent magnitude of the object if its distance from the earth were 10 pc." It is usually denoted by M, while apparent magnitude uses the lower case, m. The relation between apparent and absolute magnitudes may easily be obtained from Equation 3.1. Imagine the object moved from its real distance to 10 pc from the earth. Its energy per unit area at the surface of the earth will then change by a factor $(D/10)^2$, where D is the object's actual distance in parsecs. Thus,

$$M - m = -2.5\log_{10}\left(\frac{D}{10}\right)^2 \qquad (3.4)$$

$$M = m + 5 - 5\log_{10} D \qquad (3.5)$$

The difference between apparent and absolute magnitudes is called the distance modulus and is occasionally used in place of the distance itself

$$\text{Distance modulus} = m - M = 5\log_{10} D - 5 \qquad (3.6)$$

Equations 3.5 and 3.6 are valid so long as the only factors affecting the apparent magnitude are distance and intrinsic brightness. However, light from the object often has to pass through interstellar gas and dust clouds, where it may be absorbed. A more complete form of Equation 3.5 is therefore

$$M = m + 5 - 5\log_{10} D - AD \qquad (3.7)$$

where A is the interstellar absorption in magnitudes per parsec. A typical value for A for lines of sight within the galactic plane is 0.002 mag pc^{-1}.

Thus we may determine the absolute magnitude of an object via Equations 3.5 or 3.7 once its distance is known. More frequently, however, the equations are used in the reverse sense to determine the distance. Often the absolute magnitude may be estimated by some independent method, and then

$$D = 10^{[(m-M+5)/5]} \text{ pc} \tag{3.8}$$

Such methods for determining distance are known as standard candle methods, since the object is in effect acting as a standard of some known luminosity. The best-known examples are the classical Cepheids with their period–luminosity relationship.

$$M = -1.9 - 2.8 \log_{10} P \tag{3.9}$$

where P is the period of the variable in days. Many other types of stars such as dwarf Cepheids, RR Lyrae stars, W Virginis stars, and β Cepheids are also suitable. Also Type Ia supernovae, or the brightest novae, or the brightest globular clusters around a galaxy, or even the brightest galaxy in a cluster of galaxies can provide rough standard candles. Yet another method is due to Wilson and Bappu who found a relationship between the width of the emission core of the ionized calcium line at 393.3 nm (the Ca II K line) in late-type stars, and their absolute magnitudes. The luminosity is then proportional to the sixth power of the line width.

Both absolute and apparent magnitudes are normally measured over some well-defined spectral region. While the above discussion is quite general, the equations only have validity within a given spectral region. Since the spectra of two objects may be quite different, their relationship at one wavelength may be very different from that at another. The next section discusses the definitions of these spectral regions and their interrelationships.

3.1.2 Filter Systems

Numerous filter systems and filter-detector combinations have been devised. They may be grouped into wide, intermediate, and narrowband systems according to the bandwidths of their transmission curves. In the visible region, wideband filters typically have bandwidths of around 100 nm, intermediate band filters range from 10 to 50 nm, while narrowband filters range from 0.05 to 10 nm. The division is convenient for the purposes of discussion here, but is not of any real physical significance.

The filters used in photometry are of two main types based upon either absorption/transmission or upon interference. The absorption/transmission filters use salts such as nickel or cobalt oxides dissolved in glass or gelatin, or a suspension of colloid particles. These filters typically transmit over a 100 nm wide region. They are thus mostly used for the wideband photometric systems. Many of these filters will also transmit in the red and infrared regions, and so must be used with a red blocking filter made with copper sulfate. Short wave-blocking filters may be made using cadmium sulfide or selenide, sulfur or gold. Although not normally used for photometry, two other types of filters may usefully be mentioned here: dichroic mirrors and neutral density filters. Neutral density filters absorb by a constant amount over a wide range of wavelengths and may be needed when observing very bright objects like the Sun (Section 5.3). They are normally a thin deposit of a reflecting metal such as aluminum or stainless steel on a glass or very thin plastic substrate. Dichroic mirrors reflect over one wavelength range and transmit over another. For example, a mirror thinly plated with a gold coating is transparent to visual radiation, but reflects the infrared. They may also be produced using multilayer interference coatings. Dichroic mirrors may be used in spectroscopy to feed two spectroscopes operating over different wavelength ranges (Section 4.2.3).

Interference filters are mostly Fabry–Perot etalons (Section 4.1.4.1) with a very small separation of the mirrors. The transmission wavelength and the bandwidth of the filter can be tuned by changing the mirror separation and/or their reflectivities (see also tunable filters, Section 4.1.4.1 and solar H-α filters, Section 5.3). Combining several Fabry–Perot cavities can alter the shape of the transmission band. Interference filters are usually used for the narrower filters of a photometric system since they can be made with bandwidths ranging from a few tens of nanometers to a hundredth of a nanometer. The recently developed rugate filter can transmit or reflect simultaneously at several wavelengths. It uses interference coatings in which the refractive index varies continuously throughout the depth of the layer.

The earliest filter system was given by the response of the human eye (Figure 1.3) and peaks around 510 nm with a bandwidth of 200 nm or so. Since all normal (i.e., not color blind, etc.) eyes have roughly comparable responses, the importance of the spectral region within which a magnitude was measured was not realized until the application of the photographic plate to astronomical detection. Then it was discovered that the magnitudes of stars that had been determined from such plates often differed significantly from those found visually. The discrepancy arose from the differing sensitivities of the eye and the photographic emulsion. Early emulsions and today's unsensitized emulsions (Section 2.2.2) have a response that peaks near 450 nm; little contribution is made to the image by radiation of a wavelength longer than about 500 nm. The magnitudes determined with the two methods became known as visual and photographic magnitudes, respectively, and are denoted by m_v and m_p or M_v and M_p. These magnitudes may be encountered when reading older texts or working on archive material, but have now largely been replaced by more precise filter systems.

With the development of photoelectric methods of detection of light, and their application to astronomy by Joel Stebbins and others earlier this century, other spectral regions became available, and more precise definitions of them were required. There are now a very great number of different photometric systems. While it is probably not quite true that there is one system for every observer, there is certainly at least one separate system for every slightly differing purpose to which photometric measurements may be put. Many of these are developed for highly specialized purposes, and so are unlikely to be encountered outside their particular area of application. Other systems have a wide application and so are worth knowing about in more detail.

For a long while, the most widespread of these more general systems was the UBV system, defined by Harold Johnson and William Morgan in 1953. This is a wideband system with the B and V regions corresponding approximately to the photographic and visual responses, and with the U region in the violet and ultraviolet. The precise definition requires a reflecting telescope with aluminized mirrors and uses an RCA 1P21 photomultiplier.

The filters are as follows:

Corning 3384	for the V region
Corning 5030 plus Schott GG 13	for the B region
Corning 9863	for the U region

Filter Name	U	B	V
Central wavelength (nm)	365	440	550
Bandwidth (nm)	70	100	90

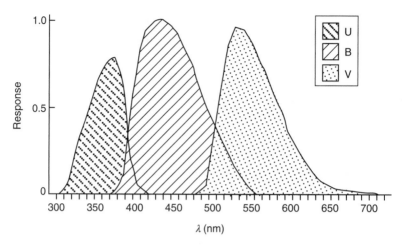

FIGURE 3.1 UBV response curves, excluding the effects of atmospheric absorption.

The response curves for the filter–photomultiplier combination are shown in Figure 3.1. The absorption in the earth's atmosphere normally cuts off the short wavelength edge of the U response. At sea level, roughly half of the incident energy at a wavelength of 360 nm is absorbed, even for small zenith angles. More importantly, the absorption can vary with changes in the atmosphere. Ideal U response curves can therefore only be approached at very high altitude, high-quality observing sites. The effect of the atmosphere on the U response curve is shown in Figure 3.2. The B and V

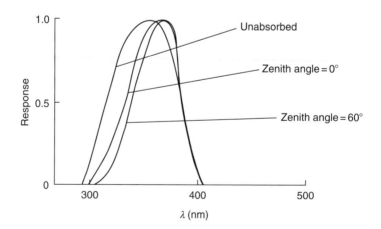

FIGURE 3.2 Effect of atmospheric absorption at sea level upon the U response curve (normalized).

response curves are comparatively unaffected by differential atmospheric absorption across their wavebands; there is only a total reduction of the flux through these filters (see Section 3.2). The scales are arranged so that the magnitudes through all three filters are equal to each other for A0 V stars such as α Lyrae (Vega). The choice of filters for the UBV system was made on the basis of what was easily available at the time. The advent of CCDs has led to several new photometric systems being devised (see below), with filters more related to the requirements of astrophysics, but the UBV system and its extensions are still in use, if only to enable comparisons to be made with measurements taken some time ago.

There are several extensions to the UBV system for use at longer wavelengths. A widely used system was introduced by Johnson in the 1960s for lead sulfide infrared detectors. With the photographic R and I bands this adds another eight regions:

Filter Name	R	I	J	K	L	M	N	Q
Central wavelength (nm)	700	900	1,250	2,200	3,400	4,900	10,200	20,000
Bandwidth (nm)		220	240	380	480	700	300	5,000	5,000

The longer wavelength regions in this system are defined by atmospheric transmission and so are very variable. Recently therefore, filters have been used to define all the passbands and the system has been adapted to make better use of the CCDs' response and the improvements in infrared detectors. This has resulted in slight differences to the transmission regions in what is sometimes called the JCG (Harold Johnson, Alan Cousins, and Ian Glass) system

Filter Name	U	B	V	R	I	Z*	J	H	K	L	M
Central wavelength (nm)[†]	367	436	545	638	797	908	1220	1630	2190	3450	4750
Bandwidth (nm)	66	94	85	160	149	96	213	307	390	472	460

---

* Added for use with CCDs.
† Data from Murdin, P. (ed.), *Encyclopedia of Astronomy and Astrophysics*, Nature and IoP Publishing, 2001, p. 1642.

Two other wideband filter systems are now used, because of the large data sets that are available. These are from the Hubble space telescope (HST) and the Sloan digital sky survey (SDSS). The HST system has six filters:

Central wavelength (nm)	336	439	450	555	675	814
Bandwidth (nm)	47	71	107	147	127	147

The SDSS, which is based upon the 2.5 m telescope at the Apache Point Observatory in New Mexico, uses five filters that cover the whole range of sensitivity of CCDs:

Filter Name	u′	g′	r′	i′	z′
Central wavelength (nm)	358	490	626	767	907
Bandwidth (nm)	64	135	137	154	147

Amongst the intermediate passband systems, the most widespread is the uvby or Strömgren system. This was proposed by Bengt Strömgren in the late 1960s and is now in fairly common use. Its transmission curves are shown in Figure 3.3. It is often used with two additional filters centered on the Hβ line, 3 and 15 nm wide, respectively, to provide better temperature discrimination for the hot stars.

Filter Name	u	v	b	y	β_n	β_w
Central wavelength (nm)	349	411	467	547	486	486
Bandwidth (nm)	30	19	18	23	3	15

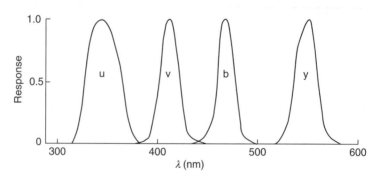

FIGURE 3.3 Normalized transmission curves for the uvby system, not including the effects of atmospheric absorption.

Narrowband work mostly concentrates on isolating spectral lines. Hα and Hβ are common choices, and their variations can be determined by measurements through a pair of filters, which are centered on the line but have different bandwidths. No single system is in general use, so that a more detailed discussion is not very profitable. Other spectral features can be studied with narrowband systems where one of the filters isolates the spectral feature, and the other is centered on a nearby section of the continuum. The reason for the invention of these and other photometric systems lies in the information that may be obtained from the comparison of the brightness of the object at different wavelengths. Many of the processes contributing to the final spectrum of an object as it is received on the earth preferentially affect one or more spectral regions. Thus, estimates of their importance may be obtained simply and rapidly by measurements through a few filters. Some of these processes are discussed in more detail in the next subsection and in Section 3.2. The most usual features studied in this way are the Balmer discontinuity and other ionization edges, interstellar absorption, and strong emission or absorption lines or bands.

There is one additional photometric system that has not yet been mentioned. This is the bolometric system, or rather bolometric magnitude, since it has only one passband. The bolometric magnitude is based upon the total energy emitted by the object at all wavelengths. Since it is not possible to observe this in practice, its value is determined by modeling calculations based upon the object's intensity through one or more of the filters of a photometric system. Although x-ray, ultraviolet, infrared, and radio data are now available to assist these calculations, many uncertainties remain, and the bolometric magnitude is still rather imprecise especially for high-temperature stars. The calculations are expressed as the difference between the bolometric magnitude and an observed magnitude. Any filter of any photometric system could be chosen as the observational basis, but in practice the V filter of the standard UBV system is normally used. The difference is then known as the bolometric correction, BC,

$$BC = m_{bol} - V \qquad (3.10)$$

$$= M_{bol} - M_V \qquad (3.11)$$

and its scale is chosen so that it is zero for main sequence stars with a temperature of about 6500 K, that is, about spectral class F5 V.

The luminosity of a star is directly related to its absolute bolometric magnitude

$$L_* = 3 \times 10^{28} \times 10^{-0.4 M_{\text{bol}}} \text{ W} \qquad (3.12)$$

Similarly, the flux of the stellar radiation just above the earth's atmosphere, f_*, is related to the apparent bolometric magnitude

$$f_* = 2.5 \times 10^{-8} \times 10^{-0.4 m_{\text{bol}}} \text{ W m}^{-2} \qquad (3.13)$$

The bolometric corrections are plotted in Figure 3.4.

Measurements in one photometric system can sometimes be converted to another system. This must be based upon extensive observational calibrations or upon detailed calculations using specimen spectral distributions. In either case, the procedure is very much second best to obtaining the data directly in the required photometric system, and requires great care in its application. Its commonest occurrence is for the conversion of data obtained with a slightly nonstandard UBV system to the standard system.

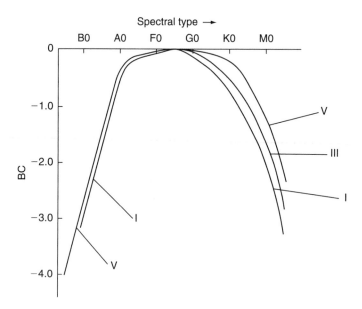

FIGURE 3.4 Bolometric corrections for main sequence stars (type V), giants (type III), and supergiants (type I).

3.1.3 Stellar Parameters

Filters may be used when observing gravitational lensing events to separate them from changes arising from the star being a variable. Gravitational lensing affects all wavelengths equally, whereas variables almost always have different amplitudes at different wavelengths. However, the usual purpose of making measurements of stars in the various photometric systems is to determine some aspect of the star's spectral behavior by a simpler and more rapid method than that of actually obtaining the spectrum. The simplest approach to highlight the desired information is to calculate one or more color indices. The color index is just the difference between the star's magnitudes through two different filters, such as the B and V filters of the standard UBV system. The color index, C, is then just

$$C = B - V \qquad (3.14)$$

where B and V are the magnitudes through the B and V filters, respectively. Similar color indices may be obtained for other photometric systems such as $439 - 555$ for the HST system and $g' - r'$ for the SDSS. Color indices for the uvby intermediate band system are discussed below. It should be noted though that a fairly common alternative usage is for C to denote the so-called international color index, which is based upon the photographic and photovisual magnitudes. The interrelationship is

$$m_p - m_{pv} = C = B - V - 0.11 \qquad (3.15)$$

The $B - V$ color index is closely related to the spectral type (Figure 3.5) with an almost linear relationship for main sequence stars. This arises from the dependence of both spectral type and color index upon temperature. For most stars, the B and V regions are located on the long-wavelength side of the maximum spectral intensity. In this part of the spectrum, the intensity then varies approximately in a black body fashion for many stars. If we assume that the B and V filters effectively transmit at 440 and 550 nm, respectively, then using the Planck equation

$$F_\lambda = \frac{2\pi hc^2}{\lambda^5 (e^{hc/\lambda kT} - 1)} \qquad (3.16)$$

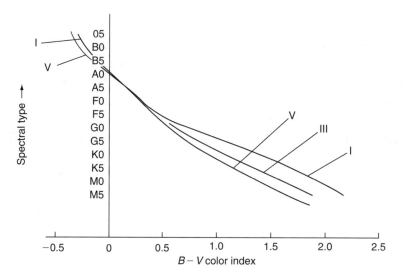

FIGURE 3.5 Relationship between spectral type and $B - V$ color index.

we obtain

$$B - V = -2.5 \log_{10} \left\{ \frac{(5.5 \times 10^{-7})^5 \left[\exp\left(\frac{6.62 \times 10^{-34} \times 3 \times 10^8}{5.5 \times 10^{-7} \times 1.38 \times 10^{-23} \times T} \right) - 1 \right]}{(4.4 \times 10^{-7})^5 \left[\exp\left(\frac{6.62 \times 10^{-34} \times 3 \times 10^8}{4.4 \times 10^{-7} \times 1.38 \times 10^{-23} \times T} \right) - 1 \right]} \right\}$$

(3.17)

which simplifies to

$$B - V = -2.5 \log_{10} \left[3.05 \frac{\exp(2.617 \times 10^4 / T) - 1}{\exp(3.27 \times 10^4 / T) - 1} \right]$$

(3.18)

which for $T < 10{,}000$ K is approximately

$$B - V \approx -2.5 \log_{10} \left[3.05 \frac{\exp(2.617 \times 10^4 / T)}{\exp(3.27 \times 10^4 / T)} \right]$$

(3.19)

$$= -1.21 + \frac{7090}{T}$$

(3.20)

Now the magnitude scale is an arbitrary one, as we have seen, and it is defined in terms of standard stars, with the relationship between the B and the V magnitude scales such that $B - V$ is zero for stars of spectral type A0

(Figure 3.5). Such stars have a surface temperature of about 10,000 K, and so a correction term of +0.5 must be added to Equation 3.20 to bring it into line with the observed relationship. Thus, we get

$$B - V = -0.71 + \frac{7090}{T} \tag{3.21}$$

giving

$$T = \frac{7090}{(B - V) + 0.71} \; K \tag{3.22}$$

Equations 3.21 and 3.22 are still only poor approximations because the filters are very broad, and so the monochromatic approximation used in obtaining the equations is a very crude one. Furthermore, the effective wavelengths (i.e., the average wavelength of the filter, taking account of the energy distribution within the spectrum) of the filters change with the stellar temperature (Figure 3.6). However, an equation of a similar nature may be fitted empirically to the observed variation (Figure 3.7), with great success for the temperature range 4,000–10,000 K, and this is

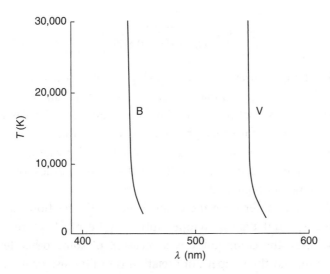

FIGURE 3.6 Change in the effective wavelengths of the standard B and V filters with changing black body temperature.

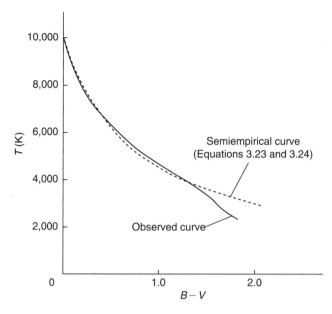

FIGURE 3.7 Observed and semiempirical $(B - V)/T$ relationships over the lower part of the main sequence.

$$B - V = -0.865 + \frac{8540}{T} \tag{3.23}$$

$$T = \frac{8540}{(B - V) + 0.865} \text{ K} \tag{3.24}$$

At higher temperatures, the relationship breaks down, as would be expected from the approximations made to obtain Equation 3.19. The complete observed relationship is shown in Figure 3.8. Similar relationships may be found for other photometric systems that have filters near the wavelengths of the standard B and V filters. For example, the relationship between spectral type and the $b - y$ color index of the uvby system is shown in Figure 3.9.

For filters that differ from the B and V filters, the relationship of color index with spectral type or temperature may be much more complex. In many cases, the color index is a measure of some other feature of the spectrum, and the temperature relation is of little use or interest. The $U - B$ index for the standard system is an example of such a case, since

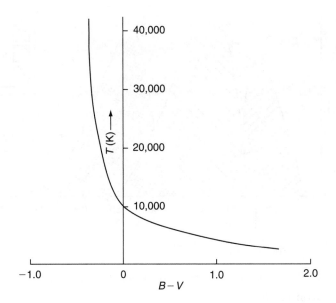

FIGURE 3.8 Observed $B - V$ versus T relationship for the whole of the main sequence.

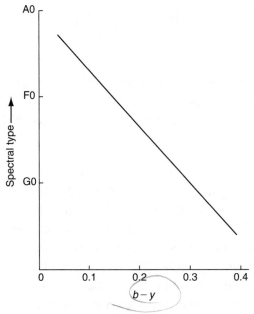

FIGURE 3.9 Relationship between spectral type and $b - y$ color index.

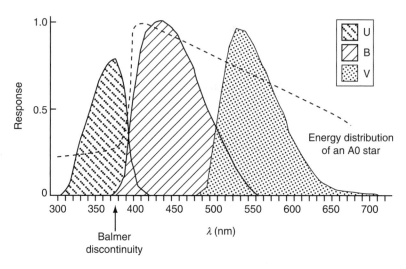

FIGURE 3.10 Position of the standard UBV filters with respect to the Balmer discontinuity.

the U and B responses bracket the Balmer discontinuity (Figure 3.10). The extent of the Balmer discontinuity is measured by a parameter, D, defined by

$$D = \log_{10} \left(\frac{I_{364+}}{I_{364-}} \right) \tag{3.25}$$

where

I_{364+} is the spectral intensity at wavelengths just longer than the Balmer discontinuity (which is at or near 364 nm)

I_{364-} is the spectral intensity at wavelengths just shorter than the Balmer discontinuity

The variation of D with the $U - B$ color index is shown in Figure 3.11. The relationship is complicated by the overlap of the B filter with the Balmer discontinuity and the variation of the effective position of that discontinuity with spectral and luminosity class. The discontinuity reaches a maximum at about A0 for main sequence stars and at about F0 for supergiants. It almost vanishes for very hot stars and for stars cooler than about G0, corresponding to the cases of too high and too low temperatures for the existence of significant populations in the hydrogen $n = 2$ level. The color index is also affected by the changing absorption coefficient of the negative hydrogen ion (Figure 3.12), and by line

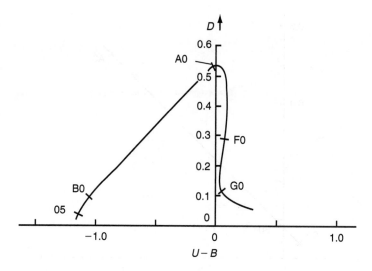

FIGURE 3.11 Variation of D with $U-B$ for main sequence stars.

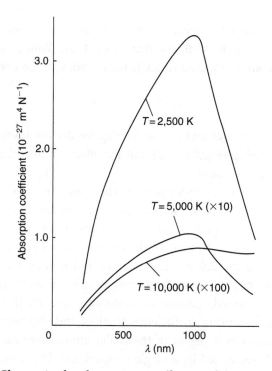

FIGURE 3.12 Change in the absorption coefficient of the negative hydrogen ion with temperature.

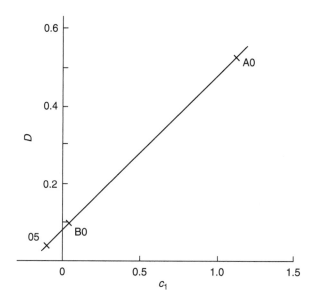

FIGURE 3.13 Variation of D with the c_1 index of the uvby system for main sequence stars.

blanketing in the later spectral types. A similar relationship may be obtained for any pair of filters that bracket the Balmer discontinuity, and another commonly used index is the c_1 index of the uvby system

$$c_1 = u + b - 2v \qquad (3.26)$$

The filters are narrower and do not overlap the discontinuity, leading to a simpler relationship (Figure 3.13), but the effects of line blanketing must still be taken into account.

Thus, the $B - V$ color index is primarily a measure of stellar temperature, while the $U - B$ index is a more complex function of both luminosity and temperature. A plot of one against the other provides a useful classification system analogous to the Hertzsprung–Russell diagram. It is commonly called the color–color diagram and is shown in Figure 3.14. The deviations of the curves from that of a black body arise from the effects just mentioned: Balmer discontinuity, line blanketing, negative hydrogen ion absorption coefficient variation, and also because the radiation originates over a region of the stellar atmosphere rather than in a single layer characterized by a single temperature. The latter effect is due to limb darkening, and because the thermalization length (the distance between successive absorptions and reemissions) normally corresponds to

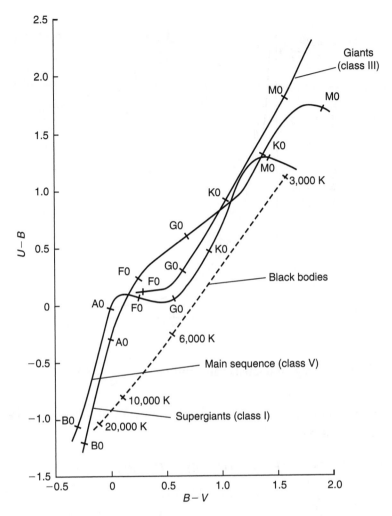

FIGURE 3.14 $(U - B)$ versus $(B - V)$ color-color diagram.

a scattering optical depth many times unity. Thus the final spectral distribution contains contributions from regions ranging from perhaps many thousands of optical depths below the visible surface to well above the photosphere and so cannot be assigned a single temperature.

Figure 3.14 is based upon measurements of nearby stars. More distant stars are affected by interstellar absorption, and since this is strongly inversely dependent upon wavelength (Figure 3.15), the $U - B$ and the $B - V$ color indices are altered. The star's spectrum is progressively weakened at shorter wavelengths and so the process is often called

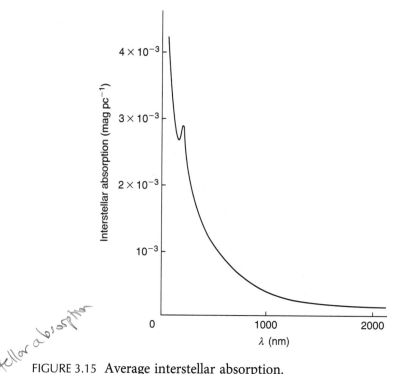

FIGURE 3.15 Average interstellar absorption.

interstellar absorption

more for short λ ...

(light λ passes through "redder")

interstellar reddening.* The color excesses measure the degree to which the spectrum is reddened

$$E_{U-B} = (U - B) - (U - B)_0 \qquad (3.27)$$

$$E_{B-V} = (B - V) - (B - V)_0 \qquad (3.28)$$

where the subscript 0 denotes the unreddened quantities. These are called the intrinsic color indices and may be obtained from the spectral type and Figures 3.5 and 3.16. Interstellar absorption (Figure 3.15) varies over most of the optical spectrum in a manner that may be described by the semiempirical relationship

$$A_\lambda = \frac{6.5 \times 10^{-10}}{\lambda} - 2.0 \times 10^{-4} \text{ mag pc}^{-1} \qquad (3.29)$$

---

* This is not the same as the redshift observed for distant galaxies. This results from the galaxies' velocities away from us and changes the observed wavelengths of the spectrum lines. With interstellar reddening, the spectrum lines' wavelengths are unchanged.

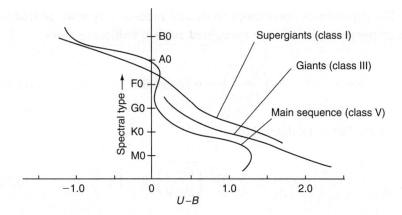

FIGURE 3.16 Relationship between spectral type and $U - B$ color index.

where λ is in nanometers. Hence, approximating U, B, and V filters by monochromatic responses at their central wavelengths, we have, for a distance of D parsecs,

$$\frac{E_{U-B}}{E_{B-V}} = \frac{(U - U_0) - (B - B_0)}{(B - B_0) - (V - V_0)} \tag{3.30}$$

$$= \frac{\left(\frac{6.5 \times 10^{-10}}{3.65 \times 10^{-7}} - 2.0 \times 10^{-4}\right)D - \left(\frac{6.5 \times 10^{-10}}{4.4 \times 10^{-7}} - 2.0 \times 10^{-4}\right)D}{\left(\frac{6.5 \times 10^{-10}}{4.4 \times 10^{-7}} - 2.0 \times 10^{-4}\right)D - \left(\frac{6.5 \times 10^{-10}}{5.5 \times 10^{-7}} - 2.0 \times 10^{-4}\right)D} \tag{3.31}$$

$$= \frac{\left(\frac{1}{365} - \frac{1}{440}\right)}{\left(\frac{1}{440} - \frac{1}{550}\right)} \tag{3.32}$$

$$= 1.027 \tag{3.33}$$

Thus, the ratio of the color excesses is independent of the reddening. This is an important result and the ratio is called the reddening ratio. Its actual value is somewhat different from that given in Equation 3.33, because of our monochromatic approximation, and furthermore, it is slightly temperature dependent and not quite independent of the reddening. Its best empirical values are

$$\frac{E_{U-B}}{E_{B-V}} = (0.70 \pm 0.10) + (0.045 \pm 0.015)E_{B-V} \quad \text{at } 30{,}000 \text{ K} \tag{3.34}$$

$$\frac{E_{U-B}}{E_{B-V}} = (0.72 \pm 0.06) + (0.05 \pm 0.01)E_{B-V} \quad \text{at } 10{,}000 \text{ K} \tag{3.35}$$

$$\frac{E_{U-B}}{E_{B-V}} = (0.82 \pm 0.12) + (0.065 \pm 0.015)E_{B-V} \quad \text{at } 5000 \text{ K} \tag{3.36}$$

The dependence upon temperature and reddening is weak, so that for many purposes we may use a weighted average reddening ratio

$$\overline{\left(\frac{E_{U-B}}{E_{B-V}}\right)} = 0.72 \pm 0.03 \tag{3.37}$$

The color factor, Q, defined by

$$Q = (U - B) - \overline{\left(\frac{E_{U-B}}{E_{B-V}}\right)}(B - V) \tag{3.38}$$

is then independent of reddening, as we may see from its expansion

$$Q = (U - B)_0 + E_{U-B} - \overline{\left(\frac{E_{U-B}}{E_{B-V}}\right)}[(B - V)_0 + E_{B-V}] \tag{3.39}$$

$$= (U - B)_0 - \overline{\left(\frac{E_{U-B}}{E_{B-V}}\right)}(B - V)_0 \tag{3.40}$$

$$= (U - B)_0 - 0.72(B - V)_0 \tag{3.41}$$

and provides a precise measure of the spectral class of the B-type stars (Figure 3.17). Since the intrinsic $B - V$ color index is closely related to the spectral type (Figure 3.5), we therefore also have the empirical relation for the early spectral type stars

$$(B - V)_0 = 0.332Q \tag{3.42}$$

shown in Figure 3.18. Hence, we have

$$E_{B-V} = (B - V) - 0.332Q = 1.4E_{U-B} \tag{3.43}$$

Thus, simple UBV photometry for hot stars results in determinations of temperature, Balmer discontinuity, spectral type, and reddening. The latter may also be translated into distance when the interstellar absorption in the star's direction is known. Thus, we have potentially a very high return of information for a small amount of observational effort, and the reader may see why the relatively crude methods of UBV and other wideband photometry still remain popular.

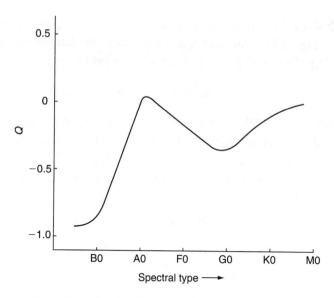

FIGURE 3.17 Variation of color factor with spectral type.

Similar or analogous relationships may be set up for other filter systems. Only one further example is briefly discussed here. The reader is referred to Appendix D for further information on these and other systems. The uvby system plus filters centered on Hβ (often called the uvbyβ system) has

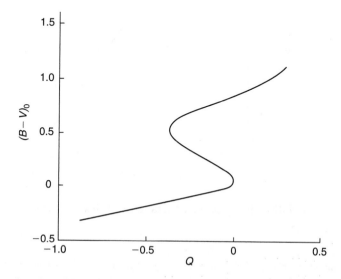

FIGURE 3.18 Relationship between color factor and the $B - V$ intrinsic color index.

several indices. We have already seen that $b - y$ is a good temperature indicator (Figure 3.9), and that c_1 is a measure of the Balmer discontinuity for hot stars (Figure 3.13). A third index is labeled m_1 and is given by

$$m_1 = v + y - 2b \tag{3.44}$$

It is sometimes called the metallicity index, for it provides a measure of the number of absorption lines in the spectrum and, hence, for a given temperature the abundance of the metals* (Figure 3.19). The fourth commonly used index of the system uses the wide and narrow Hβ filters, and is given by

$$\beta = m_n - m_w \tag{3.45}$$

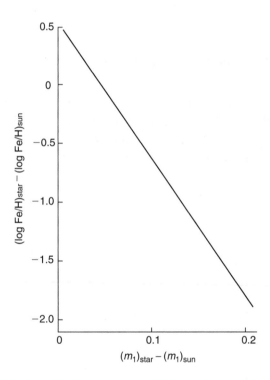

FIGURE 3.19 Relationship between metallicity index and iron abundance for solar type stars.

---

* Astronomically speaking, all elements heavier than helium are called metals, even the non-metallic ones. The reason for this is that some metals such as iron have so many spectrum lines (iron has over 20,000 in the visible part of the solar spectrum alone) that their lines dominate most stellar optical spectra.

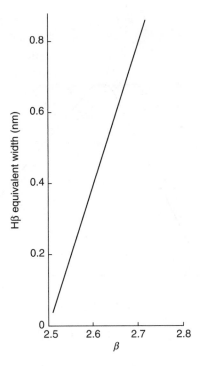

FIGURE 3.20 Relationship between β and the equivalent width of Hβ.

where m_n and m_w are the magnitudes through the narrow and wide Hβ filters, respectively. β is directly proportional to the equivalent width of Hβ (Figure 3.20), and therefore acts as a guide to luminosity and temperature among the hotter stars (Figure 3.21). Color excesses may be defined as before

$$E_{b-y} = (b - y) - (b - y)_0 \qquad (3.46)$$

and so on. Correction for interstellar absorption may also be accomplished in a similar manner to the UBV case and results in

$$(c_1)_0 = c_1 - 0.20(b - y) \qquad (3.47)$$

$$(m_1)_0 = m_1 + 0.18(b - y) \qquad (3.48)$$

$$(u - b)_0 = (c_1)_0 + 2(m_1)_0 \qquad (3.49)$$

By combining the photometry with model atmosphere work, much additional information such as effective temperatures and surface gravities may be deduced.

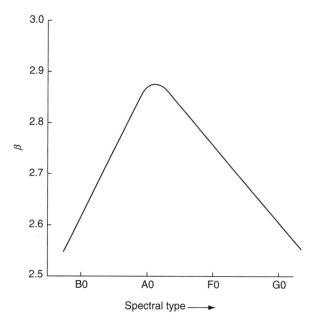

FIGURE 3.21 Relationship of β with spectral type.

Exercise 3.1

Calculate the absolute magnitude of Jupiter from its apparent magnitude at opposition, -2.6, and the absolute magnitude of $M\,31$, from its apparent magnitude, $+3.5$ (assuming it to be a point source). Their distances are 4.2 AU and 670 kpc, respectively.

Exercise 3.2

Calculate the distance of a Cepheid whose apparent magnitude is $+13^m$ on average, and whose absolute magnitude is -4^m on average, assuming that it is affected by interstellar absorption at a rate of 1.5×10^{-3} mag pc^{-1}.

Exercise 3.3

Standard UBV measures of a main sequence star give the following results: $U = 3.19$, $B = 4.03$, and $V = 4.19$. Hence, calculate or find the following: $(U - B)$, $(B - V)$; Q; $(B - V)_0$, E_{B-V}; E_{U-B}, $(U - B)_0$; spectral type; temperature; distance (assuming that the interstellar absorption shown in Figure 3.15 may be applied); U_0, B_0, V_0; and M_V.

3.2 PHOTOMETERS

3.2.1 Instruments

3.2.1.1 Introduction

We may distinguish three basic methods of measuring stellar brightness, based upon the eye, upon the photographic plate, or upon a variety of photoelectric devices. Only the latter is now of any significance, although a few amateur astronomers may still make eye-estimates of brightness as part of long-term monitoring of variable stars. Forthcoming developments like the large synoptic survey telescope (LSST) (Section 1.7.2) that will monitor tens of millions of stars every few days will soon make even this usage obsolete. Photographic magnitudes may be encountered when working with archive material. The HST *Guide Star Catalog 2* (*GSC 2*) containing details of 500 million objects, for example, is based upon photographic plates obtained by the 1.2 m Schmidt cameras at Mount Palomar and the Anglo-Australian observatory. Precision photometry, however, is now exclusively undertaken by means of photoelectric devices, with the CCD predominating in the optical region, and the solid-state devices in the infrared region.

3.2.1.2 Photographic Photometry

Apart from being used with Schmidt cameras and automatic plate measuring machines to produce relatively low accuracy ($\pm 0.01^{\text{m}}$) measurements of the brightnesses of large numbers of stars and other objects, photography has been replaced by the CCD, even for most amateur work. Only a brief overview of the method is therefore included here. The emulsion's characteristic curve (Figure 2.7) must first be found. This may be via a separate calibration exposure on the edge of the image, or if sufficient standard stars of known magnitudes are on the image, as is likely to be the case with large area Schmidt plates, then they may be measured and used instead. The images of stars of different magnitudes will be of different sizes. This arises through the image of the star not being an ideal point. It is spread out by diffraction, scintillation, scattering within the emulsion, and by reflection off the emulsion's support (halation). The actual image structure will therefore have an intensity distribution that probably approaches a Gaussian pattern. However, the recorded image will have a size governed by the point at which the illumination increases the photographic density above the gross fog

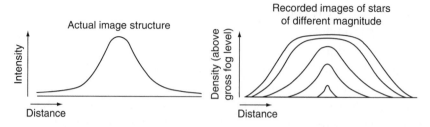

FIGURE 3.22 Schematic variation in image size and density for stars of differing magnitudes recorded on photographic emulsion.

level. Images of stars of differing magnitudes will then be recorded as shown schematically in Figure 3.22.

The measurement that is made of the stellar images may simply be that of its diameter, when a semiempirical relationship of the following form can often be fitted to the calibration curve.

$$D = A + B \log_{10} I \tag{3.50}$$

where
D is the image diameter
I is the intensity of the star
A and B are constants to be determined from the data

A more precise measurement than the actual diameter would be the distance between points in the image, which are some specified density greater than that of the gross fog, or between the half-density points. Alternatively, a combination of diameter and density may be measured. Automatic measuring machines scan the image (Section 5.1.10), so that the total density may be found by integration. Another system is based upon a variable diaphragm. This is set over the image and adjusted until it just contains that image. The transmitted intensity may then be measured and a calibration curve plotted from known standard stars as before.

The magnitudes obtained from a photograph are of course not the same as those of the standard photometric systems (Section 3.1). Usually, the magnitudes are photographic or photovisual. However, with careful selection of the emulsion and the use of filters, a response that is close to that of some filter systems may be obtained (Figure 3.23). Alternatively, a combination of emulsion types and filters may be used that gives a useful set of bands, but without trying to imitate any of the standard systems.

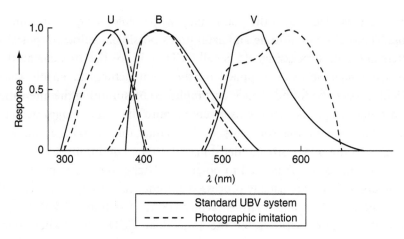

FIGURE 3.23 Comparison of the standard UBV system and its photographic simulation.

The Palomar digital sky survey (DPOSS), for example, used blue, red, and NIR bands, obtained using the now discontinued IIIaJ, IIIaF, and IVN plates combined with GG395, RG610, and RG9 filters, respectively.

3.2.1.3 CCD and Photoelectric Photometers

The most precise photometry relies upon photoelectric devices of various types. These are primarily the CCD in the visual and NIR regions, and infrared array detectors (Section 1.1) at longer wavelengths. Photomultipliers continue to be used occasionally, as do p-i-n photodiodes in small photometers aimed at the amateur market. Photometry with a CCD differs little from ordinary imaging, except that several images of the same area need to be obtained through appropriate filters (Section 3.1). Anti-blooming CCDs should be avoided since their response is nonlinear for objects nearing saturation. The normal data-reduction techniques such as dark signal subtraction and flat fielding (Section 1.1.8) need to be applied to the images. The brightness of a star is obtained directly by adding together the intensities of the pixels covered by the star image and subtracting the intensities in a similar number of pixels covering a nearby area of sky background. Conversion to stellar magnitudes then requires the brightnesses of several standard stars also to be measured. With mosaics of several large area CCDs, there are likely to be several standard stars on any image. However, if this is not the case, or with detectors covering smaller areas of the sky, one or more separate calibration images will be needed.

These are as identical to the main image as possible—as close in time as possible, as close in altitude and azimuth as possible, identical exposures, filters and data processing, etc. Small CCDs sold for the amateur market normally have software supplied by the manufacturer to enable stars' brightnesses to be found. This is accomplished by summing the intensities in a square of pixels that the user centers onto the stars' images one at a time. The size of the square is usually adjustable to take account of changes to the image sizes under different seeing conditions. Clearly, the same size of square is needed to obtain the background reading. Images from larger CCDs are usually processed automatically using software such as image reduction and analysis facility (IRAF) (Section 2.9.3.3) or specialized programs written for the individual CCD. Only when star's images overlap may the observer need to intervene in this process. Integrated magnitudes for extended objects such as galaxies are obtained in a similar fashion, though more input from the observer may be required to ensure that the object is delineated correctly before the pixel intensities are added together. From a good site under good observing conditions, photometry to an accuracy of $\pm 0.001^m$ is now achievable with CCDs.

At NIR and MIR wavelengths, the procedure is similar to that for CCDs. But in the long-wavelength infrared regions, array detectors are still small (Section 1.1), and so only one or two objects are likely to be on each image, making separate calibration exposures essential. In many cases, the object being observed is much fainter than the background noise. For the MIR and FIR regions, special observing techniques such as chopping rapidly between the source and the background, subtracting the background from the signal, and integrating the result over a long period need to be used. Most telescopes designed specifically for infrared work have secondary mirrors that can be oscillated to achieve this switching.

Photomultipliers (Section 1.1.10) continue to be used when individual photons need to be detected as in the neutrino (Section 1.5) and cosmic ray Čerenkov (Section 1.4.3.3) detectors or when very rapid responses are required as in the observation of occultations (Section 2.7). They may also be used on board spacecraft for ultraviolet measurements in the 10–300 nm region where CCDs are insensitive.

3.2.2 Observing Techniques

Probably, the single most important 'technique' for successful photometry lies in the selection of the observing site. Only the clearest and most consistent of skies are suitable for precision photometry. Haze, dust, clouds, excessive scintillation, light pollution, etc. and the variations in these, all render a site unsuitable for photometry. Furthermore, for infrared work, the amount of water vapor above the site should be as low as possible. For these reasons, good photometric observing sites are rare. Generally, they are at high altitudes and are located where the weather is particularly stable. Oceanic islands and mountain ranges with a prevailing wind from an ocean, with the site above the inversion layer, are fairly typical of the best choices. Sites that are less than ideal can still be used for photometry, but their number of good nights will be reduced. Restricting the observations to near the zenith is likely to improve the results obtained at a mediocre observing site.

The second most vital part of photometry lies in the selection of the comparison star(s). This must be nonvariable, close to the star of interest, of similar apparent magnitude and spectral class, and have its own magnitude known reliably on the photometric system that is in use. Amongst the brighter stars, there is usually some difficulty in finding a suitable standard that is close enough. With fainter stars, the likelihood that a suitable comparison star exists is higher, but the chances of its details being known are much lower. Thus, ideal comparison stars are found only rarely. The best-known variables already have lists of good comparison stars, which may generally be found from the literature. But for less well-studied stars, and those being investigated for the first time, there is no such useful information to hand. It may even be necessary to undertake extensive prior investigations, such as checking back through archive records to find nonvariable stars in the region, and then measuring these to obtain their magnitudes before studying the star itself. Also an additional star may need to be observed several times throughout an observing session to supply the data for the correction of atmospheric absorption (see Section 3.2.3).

A single observatory can clearly only observe objects for a fraction of the time, even when there are no clouds. However, rapidly changing objects may need to be monitored continuously over 24 h. A number of observatories distributed around the globe have therefore instituted various cooperative programs so that such continuous photometric observations can be made. The whole earth telescope (WET), for example, is a consortium of (currently) 28 observatories studying variable stars (see also Birmingham solar oscillations network (BISON) and global oscillation network group (GONG), Section 5.3.7). Similarly, the Liverpool telescope on La Palma and the two Faulkes telescopes in Hawaii and Australia are 2 m robotic telescopes that can respond to γ-ray burst (GRB) alerts in less than 5 min.

3.2.3 Data Reduction and Analysis

The reduction of the data is performed in three stages: correction for the effects of the earth's atmosphere, correction to a standard photometric system, and correction to heliocentric time.

The atmosphere absorbs and reddens the star's light, and its effects are expressed by Bouguer's law

$$m_{\lambda,0} = m_{\lambda,z} - a_\lambda \sec z \tag{3.51}$$

where
$m_{\lambda,z}$ is the magnitude at wavelength λ and at zenith distance z
a_λ is a constant that depends on λ

The law is accurate for zenith distances up to about 60°, which is usually sufficient since photometry is rarely carried out on stars whose zenith distances are greater than 45°. Correction for atmospheric extinction may be simply carried out once the value of the extinction coefficient a_λ is known. Unfortunately, a_λ varies from one observing site to another, with the time of year, from day to day, and even throughout the night. Thus, its value must be found on every observing occasion. This is done by observing a standard star at several different zenith distances, and by plotting its observed brightness against sec z. Now for zenith angles less than 60° or 70°, to a good approximation, sec z is just a measure of the air mass along the line of sight, so that we may reduce the observations to above the atmosphere by extrapolating them back to an air mass of zero (ignoring the question of the meaning of a value of sec z of zero, Figure 3.24). For the same standard star, this brightness should always be the same on all nights, so that an additional point, which is the average of all the previous determinations of the above-atmosphere magnitude of the star, is available to add to the observations on any given night. The extinction coefficient is simply obtained from the slope of the line in Figure 3.24. The coefficient is strongly wavelength dependent (Figure 3.25), and so it must be determined separately for every filter that is being used. Once the extinction coefficient has been found, the above-atmosphere magnitude of the star, m_λ, is given by

$$m_\lambda = m_{\lambda,z} - a_\lambda(1 + \sec z) \tag{3.52}$$

Thus, the observations of the star of interest and its standards must all be multiplied by a factor k_λ,

$$k_\lambda = 10^{0.4a_\lambda(1+\sec z)} \tag{3.53}$$

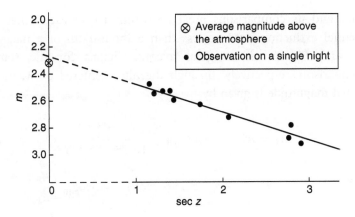

FIGURE 3.24 Schematic variation in magnitude of a standard star with zenith distance.

to correct them to their unabsorbed values. When the unknown star and its comparisons are very close together in the sky, the differential extinction will be negligible and this correction need not be applied. But the separation must be very small; for example, at a zenith distance of 45°,

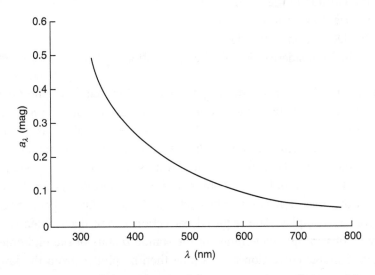

FIGURE 3.25 A typical dependence of the extinction coefficient with wavelength for a good observing site.

the star and its comparisons must be within $10'$ of each other if the differential extinction is to be less than a thousandth of a magnitude. If E_λ and E'_λ are the original average signals for the star and its comparison, respectively, through the filter centered on λ, then the corrected magnitude is given by

$$m_\lambda = m'_\lambda - 2.5 \log_{10} \left(\frac{E_\lambda 10^{0.4a_\lambda(1+\sec z)}}{E'_\lambda 10^{0.4a_\lambda(1+\sec z')}} \right) \tag{3.54}$$

$$= m'_\lambda + a_\lambda(\sec z' - \sec z) - 2.5 \log_{10} \left(\frac{E_\lambda}{E'_\lambda} \right) \tag{3.55}$$

where

m_λ and m'_λ are the magnitudes of the unknown star and its comparison, respectively

z and z' are similarly their zenith distances

The zenith angle is given by

$$\cos z = \sin\phi \sin\delta + \cos\phi \cos\delta \cos(\text{LST} - \alpha) \tag{3.56}$$

where

ϕ is the latitude of the observatory
α is the right ascension of the star
δ is the declination of the star
LST is the local sidereal time of the observation

If the photometer is working in one of the standard photometric systems, then the magnitudes obtained through Equation 3.6 may be used directly, and color indices, color excesses, etc. obtained as discussed in Section 3.1.3. Often, however, the filters or the detector may not be of the standard type. Then the magnitudes must be corrected to a standard system. This can only be attempted if the difference is small, since absorption lines, ionization edges, etc. will make any large correction exceedingly complex. The required correction is best determined empirically by observing a wide range of the standard stars of the photometric system. Suitable correction curves may then be plotted from the known magnitudes of these stars.

Finally, and especially for variable stars, the time of the observation should be expressed in heliocentric Julian days. The geocentric Julian date is tabulated in Appendix A, and if the time of the observation is expressed on this scale, then the heliocentric time is obtained by correcting to the travel time of the light to the Sun,

$$T_{Sun} = T + 5.757 \times 10^{-3}[\sin\delta_* \sin\delta_{Sun} - \cos\delta_* \cos\delta_{Sun} \cos(\alpha_{Sun} - \alpha_*)]$$

$$(3.57)$$

where

T is the actual time of observation in geocentric Julian days

T_{Sun} is the time referred to the Sun

α_{Sun} is the right ascension of the Sun at the time of the observation

δ_{Sun} is the declination of the Sun at the time of the observation

α_* is the right ascension of the star

δ_* is the declination of the star

For very precise work, the varying distance of the earth from the Sun may need to be taken into account as well.

Further items in the reduction and analysis of the data, such as the corrections for interstellar reddening, calculation of color index, temperature, etc. were covered in Section 3.1.

Exercise 3.4

Show that if the differential extinction correction is to be less than Δm magnitudes, then the zenith distances of the star and its comparison must differ by less than Δz, where

$$\Delta z = \frac{\Delta m}{a_\lambda} \cot z \operatorname{cosec} z \text{ radians}$$

CHAPTER 4

Spectroscopy

4.1 SPECTROSCOPY

4.1.1 Introduction

Practical spectroscopes are usually based upon one or other of two quite separate optical principles: interference and differential refraction. The former produces instruments based upon diffraction gratings or interferometers, while the latter results in prism-based spectroscopes. There are also some hybrid designs. The details of the spectroscopes themselves are considered in Section 4.2; here we discuss the basic optical principles that underlie their designs.

4.1.2 Diffraction Gratings

The operating principle of diffraction gratings relies upon the effects of diffraction and interference of light waves. Logically therefore, it might seem that they should be included within the section on interference-based spectroscopes below. However, diffraction gratings are in such widespread and common use in astronomy that they merit a section to themselves.

We have already seen in Section 2.5.2 the structure of the image of a single source viewed through two apertures (Figure 2.21). The angular distance of the first fringe from the central maximum is λ/d, where d is the separation of the apertures. Hence, the position of the first fringe, and of course the positions of all the other fringes, is a function of wavelength. If such a pair of apertures were illuminated with white light, all the fringes apart from the central maximum would thus be short spectra with the longer wavelengths furthest from the central maximum. In an image such as that of Figure 2.21,

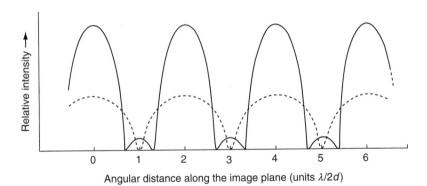

FIGURE 4.1 Small portion of the image structure for a single point source viewed through two apertures (broken curve) and three apertures (full curve).

the spectra would be of little use since the fringes are so broad that they would overlap each other long before a useful dispersion could be obtained. However, if we add a third aperture in line with the first two, and separated from the nearer of the original apertures by a distance, d, again, then we find that the fringes remain stationary, but become narrower and more intense. Weak secondary maxima also appear between the main fringes (Figure 4.1). The peak intensities are of course modulated by the pattern from a single slit when looked at on a larger scale, in the manner of Figure 2.21. If further apertures are added in line with the first three and with the same separations, then the principal fringes continue to narrow and intensify, and further weak maxima appear between them (Figure 4.2). The intensity of the pattern at some angle, θ, to the optical axis is given by (cf. Equation 1.6)

$$I(\theta) = I(0) \left[\frac{\sin^2\left(\frac{\pi D \sin \theta}{\lambda}\right)}{\left(\frac{\pi D \sin \theta}{\lambda}\right)^2} \right] \left[\frac{\sin^2\left(\frac{N\pi d \sin \theta}{\lambda}\right)}{\sin^2\left(\frac{\pi d \sin \theta}{\lambda}\right)} \right] \tag{4.1}$$

where
D is the width of the aperture
N the number of apertures

The term

$$\left[\frac{\sin^2\left(\frac{\pi D \sin \theta}{\lambda}\right)}{\left(\frac{\pi D \sin \theta}{\lambda}\right)^2} \right] \tag{4.2}$$

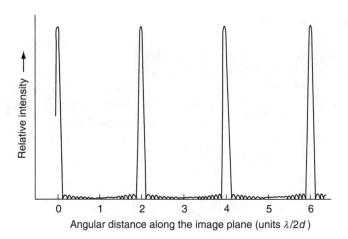

FIGURE 4.2 Small portion of the image structure for a single point source viewed through 20 apertures.

represents the modulation of the image by the intensity structure for a single aperture, while the term

$$\left[\frac{\sin^2\left(\frac{N\pi d \sin\theta}{\lambda}\right)}{\sin^2\left(\frac{\pi d \sin\theta}{\lambda}\right)}\right] \tag{4.3}$$

represents the result of the interference between N apertures. We may write

$$\Delta = \left(\frac{\pi D \sin\theta}{\lambda}\right) \tag{4.4}$$

and

$$\delta = \left(\frac{\pi d \sin\theta}{\lambda}\right) \tag{4.5}$$

and Equation 4.1 then becomes

$$I(\theta) = I(0)\frac{\sin^2\Delta}{\Delta^2}\frac{\sin^2(N\delta)}{\sin^2\delta} \tag{4.6}$$

Now consider the interference component as δ tends to $m\pi$, where m is an integer. Putting

$$P = \delta - m\pi \tag{4.7}$$

we have

$$\lim_{\delta \to m\pi} \left[\frac{\sin (N\delta)}{\sin \delta} \right] = \lim_{P \to 0} \left\{ \frac{\sin [N(P + m\pi)]}{\sin (P + m\pi)} \right\} \tag{4.8}$$

$$= \lim_{P \to 0} \left[\frac{\sin (NP) \cos (Nm\pi) + \cos (NP) \sin (Nm\pi)}{\sin P \cos (m\pi) + \cos P \sin (m\pi)} \right] \tag{4.9}$$

$$= \lim_{P \to 0} \left[\pm \frac{\sin (NP)}{\sin P} \right] \tag{4.10}$$

$$= \pm N \lim_{P \to 0} \left[\frac{\sin (NP)}{NP} \frac{P}{\sin P} \right] \tag{4.11}$$

$$= \pm N \tag{4.12}$$

Hence, integer multiples of π give the values of δ for which we have a principal fringe maximum. The angular positions of the principal maxima are given by

$$\theta = \sin^{-1} \left(\frac{m\lambda}{d} \right) \tag{4.13}$$

and m is usually called the order of the fringe. The zero intensities in the fringe pattern will be given by

$$N\delta = m'\pi \tag{4.14}$$

where m' is an integer, but excluding the cases where $m' = mN$ that are the principal fringe maxima. Their positions are given by

$$\theta = \sin^{-1} \left(\frac{m'\lambda}{Nd} \right) \tag{4.15}$$

The angular width of a principal maximum, W, between the first zeros on either side of it is thus given by

$$W = \frac{2\lambda}{Nd \cos \theta} \tag{4.16}$$

The width of a fringe is therefore inversely proportional to the number of apertures, while its peak intensity, from Equations 4.6 and 4.12, is

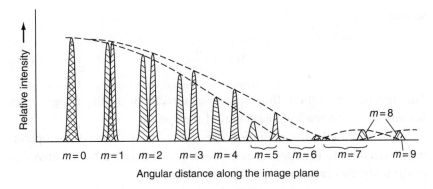

FIGURE 4.3 A portion of the image structure for a single bichromatic point source viewed through several apertures.

proportional to the square of the number of apertures. Thus, for a bichromatic source observed through a number of apertures, we obtain the type of image structure shown in Figure 4.3. The angular separation of fringes of the same order for the two wavelengths, for small values of θ, can be seen from Equation 4.13 to be proportional to both the wavelength and to the order of the fringe, while the fringe width is independent of the order (Equation 4.16). For a white light source, by a simple extension of Figure 4.3, we may see that the image will consist of a series of spectra on either side of a white central image. The Rayleigh resolution within this image is obtained from Equation 4.16

$$W' = \frac{\lambda}{Nd \cos \theta} \qquad (4.17)$$

and is independent of the fringe order. The ability of a spectroscope to separate two wavelengths is called the spectral resolution and is denoted by W_λ, and it may now be found from Equation 4.17

$$W_\lambda = W' \frac{d\lambda}{d\theta} \qquad (4.18)$$

but from Equation 4.13

$$\frac{d\lambda}{d\theta} = \frac{d}{m} \cos \theta \qquad (4.19)$$

so that

$$W_\lambda = \frac{\lambda}{Nm} \qquad (4.20)$$

The spectral resolution thus improves directly with the fringe order because of the increasing dispersion of the spectra.

More commonly, the resolution is expressed as the ratio of the operating wavelength to the spectral resolution, and is denoted by R (often and confusingly also called the spectral resolution)

$$R = \frac{\lambda}{W_\lambda} = Nm \qquad (4.21)$$

The resolution for a series of apertures is thus just the product of the number of apertures and the order of the spectrum. It is independent of the width and spacing of the apertures.

From Figure 4.3, we may see that at higher orders the spectra are overlapping. This occurs at all orders when white light is used. The difference in wavelength between two superimposed wavelengths from adjacent spectral orders is called the free spectral range, Σ. From Equation 4.13, we may see that if λ_1 and λ_2 are two such superimposed wavelengths, then

$$\sin^{-1}\left(\frac{m\lambda_1}{d}\right) = \sin^{-1}\left[\frac{(m+1)\lambda_2}{d}\right] \qquad (4.22)$$

that is for small angles

$$\Sigma = \lambda_1 - \lambda_2 \approx \frac{\lambda_2}{m} \qquad (4.23)$$

For small values of m, Σ is therefore large, and the unwanted wavelengths in a practical spectroscope may be rejected by the use of filters. However, some spectroscopes, such as those based on Fabry–Perot etalons and echelle gratings, operate at very high spectral orders and both of the overlapping wavelengths may be needed. Then it is necessary to use a cross-disperser so that the final spectrum consists of a two-dimensional (2D) array of short sections of the spectrum (see the discussion later in this section).

A practical device for producing spectra by diffraction uses a large number of closely spaced, parallel, narrow slits or grooves, and is called a diffraction grating. Typical gratings for astronomical use have between 100 and 1000 grooves per millimeter, and 1000 and 50,000 grooves in total. They are used at orders ranging from 1 up to 200 or so. Thus, the resolutions range from 10^3 to 10^5. Although the earlier discussion was based upon the use of clear apertures, a narrow plane mirror can replace each aperture without altering the results. Thus, diffraction gratings can be used either in transmission or reflection modes. Most astronomical spectroscopes are in fact based upon reflection gratings. Often the grating is inclined to the incoming beam of light, but this changes the discussion only marginally. There is a constant term, $d \sin i$, added to the path differences, where i is the angle made by the incoming beam with the normal to the grating. The whole image (Figure 4.3) is shifted an angular distance i along the image plane. Equation 4.13 then becomes

$$\theta = \sin^{-1}\left[\left(\frac{m\lambda}{d}\right) - \sin i\right] \tag{4.24}$$

and in this form is often called the grating equation.

Volume-phase holographic gratings (VPHGs) are currently starting to be used within astronomical spectroscopes. These have a grating in the form of a layer of gelatin within which the refractive index changes (Section 4.2.3), with the lines of the grating produced by regions of differing refractive indices. VPHGs operate through Bragg diffraction (Section 1.3.6.2, Figure 1.82, and Equation 1.96). Their efficiencies can thus be up to 95% in the first order. They can be used either as transmission or reflection gratings and replace conventional gratings in spectroscopes at the appropriate Bragg angle for the operating wavelength.

To form a part of a spectroscope, a grating must be combined with other optical elements. The basic layout is shown in Figure 4.4; practical designs are discussed in Section 4.2. The grating is illuminated by parallel light that is usually obtained by placing a slit at the focus of a collimating lens, but sometimes may be obtained simply by allowing the light from a very distant object to fall directly onto the grating. After reflection from the grating, the light is focused by the imaging lens to form the required spectrum, and this may then be recorded, observed through an eyepiece, projected onto a screen, etc. as desired. The collimator and imaging lenses

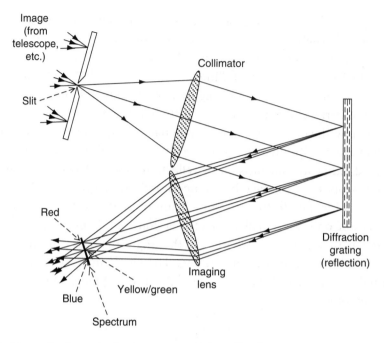

FIGURE 4.4 Basic optical arrangement of a reflection grating spectroscope.

may be simple lenses as shown, in which case the spectrum will be tilted with respect to the optical axis because of chromatic aberration, or they may be achromats or mirrors (see the next section).

The angular dispersion of a grating is not normally used as a parameter of a spectroscopic system. Instead, it is combined with the focal length of the imaging element to give either the linear dispersion or the reciprocal linear dispersion. If x is the linear distance along the spectrum from some reference point, then we have for an achromatic imaging element of focal length, f_2,

$$\frac{dx}{d\lambda} = f_2 \frac{d\theta}{d\lambda} \tag{4.25}$$

where θ is small and is measured in radians. From Equation 4.24, the linear dispersion within each spectrum is thus given by

$$\frac{dx}{d\lambda} = \pm \frac{mf_2}{d\cos\theta} \tag{4.26}$$

Now since θ varies very little over an individual spectrum, we may write

$$\frac{dx}{d\lambda} \approx \text{constant} \qquad (4.27)$$

The dispersion of a grating spectroscope is thus roughly constant compared with the strong wavelength dependence of a prism spectroscope (Equation 4.69). More commonly, the reciprocal linear dispersion, $\frac{d\lambda}{dx}$, is quoted and used. For practical astronomical spectrometers, this usually has values in the range

$$10^{-7} < \frac{d\lambda}{dx} < 5 \times 10^{-5} \qquad (4.28)$$

The commonly used units are nanometers change of wavelength per millimeter along the spectrum so that the above range is from 0.1 to 50 nm mm^{-1}. The use of angstroms per millimeter is still fairly common practice amongst astronomers; the magnitude of the dispersion is then a factor of 10 larger than the standard measure. The advent of electronic detectors has also made the use of nanometers or angstroms per pixel a measure of dispersion.

The resolving power of a spectroscope is limited by the spectral resolution of the grating, by the resolving power of its optics (Section 1.1), and by the projected slit width. The spectrum is formed from an infinite number of monochromatic images of the entrance slit. It is easy to see that the width of one of these images, S, is given by

$$S = s\frac{f_2}{f_1} \qquad (4.29)$$

where
s is the slit width
f_1 is the collimator's focal length
f_2 is the imaging element's focal length

In wavelength terms, the slit width, $S\frac{d\lambda}{dx}$, is sometimes called the spectral purity of the spectroscope. The entrance slit must have a physical width of s_{max} or less, if it is not to degrade the spectral resolution, where

$$s_{max} = \frac{\lambda f_1}{Nd \cos \theta} \qquad (4.30)$$

(cf. Equation 4.77).

If the optics of the spectroscope are well corrected, then we may ignore their aberrations and consider only the diffraction limit of the system. When the grating is fully illuminated, the imaging element will intercept a rectangular beam of light. The height of the beam is just the height of the grating, and has no effect upon the spectral resolution. The width of the beam, D, is given by

$$D = L \cos \theta \tag{4.31}$$

where
L is the length of the grating
θ is the angle of the exit beam to the normal to the plane of the grating

The diffraction limit is just that of a rectangular slit of width D. So that from Figure 1.26, the linear Rayleigh limit of resolution, W'', is given by

$$W'' = \frac{f_2 \lambda}{D} \tag{4.32}$$

$$= \frac{f_2 \lambda}{L \cos \theta} \tag{4.33}$$

If the beam is limited by some other element of the optical system, and/or is of circular cross section, then D must be evaluated as may be appropriate, or the Rayleigh criterion for the resolution through a circular aperture (Equation 1.55) must be used in place of that for a rectangular aperture. Optimum resolution occurs when

$$S = W'' \tag{4.34}$$

that is,

$$s = \frac{f_1 \lambda}{D} \tag{4.35}$$

$$= \frac{f_1 \lambda}{L \cos \theta} \tag{4.36}$$

The major disadvantage of a grating as a dispersing element is immediately obvious from Figure 4.3; the light from the original source is spread over a large number of spectra. The grating's efficiency in terms of the fraction of light concentrated into the spectrum of interest is therefore

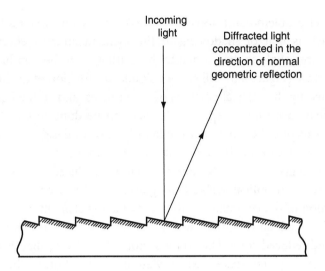

FIGURE 4.5 Enlarged section through a blazed reflection grating.

very low. However, this disadvantage may be largely overcome with reflection gratings through the use of the technique of blazing the grating. In this technique, the individual mirrors that comprise the grating are angled so that they concentrate the light into a single narrow solid angle (Figure 4.5). For instruments based upon the use of gratings at low orders, the angle of the mirrors is arranged so that the light is concentrated into the spectrum to be used, and by this means up to 90% efficiency can be achieved. In terms of the interference patterns, the grating is designed so that central peak due to an individual aperture just extends over the width of the desired spectrum. The blaze angle then shifts that peak along the array of spectra until it coincides with the desired order.

For those spectroscopes that use gratings at high orders, the grating can still be blazed, but then the light is concentrated into short segments of many different orders of spectra. By a careful choice of parameters, these short segments can be arranged so that they overlap slightly at their ends, and so coverage of a much wider spectral region may be obtained by producing a montage of the segments. Transmission gratings can also be blazed, although this is less common. Each of the grooves then has the cross section of a small prism, the apex angle of which defines the blaze angle. Blazed transmission gratings for use at infrared wavelengths can be produced by etching the surface of a block of silicon in a similar manner to the way in which integrated circuits are produced.

Another problem that is intrinsically less serious, but which is harder to counteract, is that of shadowing. If the incident and/or reflected light makes a large angle to the normal to the grating, then the step-like nature of the surface (Figure 4.5) will cause a significant fraction of the light to be intercepted by the vertical portions of the grooves, and so the light is lost to the final spectrum. There is little that can be done to eliminate this problem except either to accept the light loss, or to design the system so that large angles of incidence or reflection are not needed.

Curved reflection gratings are frequently produced. By making the curve that of an optical surface, the grating itself can be made to fulfill the function of the collimator and/or the imaging element of the spectroscope, thus reducing light losses and making for greater simplicity of design and reduced costs. The grooves should be ruled so that they appear parallel and equally spaced when viewed from infinity. The simplest optical principle employing a curved grating is that due to Rowland. The slit, grating, and spectrum all lie on a single circle that is called the Rowland circle (Figure 4.6). This has a diameter equal to the radius of

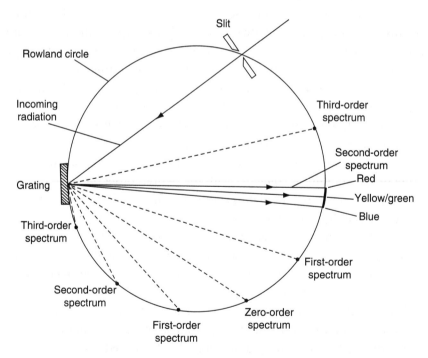

FIGURE 4.6 Schematic diagram of a spectroscope based upon a Rowland circle, using a curved grating blazed for the second order.

curvature of the grating. The use of a curved grating at large angle to its optical axis introduces astigmatism, and spectral lines may also be curved due to the varying angles of incidence for rays from the center and ends of the slit (cf. Figure 4.14). Careful design, however, can reduce or eliminate these defects, and there are several practical designs for spectroscopes based upon the Rowland circle (Section 4.2.3). Aspherical curved gratings are also possible and can be used to provide very highly corrected designs with few optical components.

Higher dispersion can be obtained by using immersed reflection gratings. The light in these interacts with the grating within a medium other than air (or a vacuum). The refractive index of the medium shortens the wavelength so that the groove spacing is effectively increased relative to the wavelength and so the dispersion is augmented (Equation 4.19). One approach to producing an immersion grating is simply to illuminate the grating from the back, that is, through the transparent substrate. A second approach is to flood the grating with a thin layer of oil kept in place by a cover sheet.

A grating spectrum generally suffers from unwanted additional features superimposed upon the desired spectrum. Such features are usually much fainter than the main spectrum and are called ghosts. They arise from a variety of causes. They may be due to overlapping spectra from higher or lower orders, or to the secondary maxima associated with each principal maximum (Figure 4.2). The first of these is usually simple to eliminate by the use of filters since the overlapping ghosts are of different wavelengths from the overlapped main spectrum. The second source is usually unimportant since the secondary maxima are very faint when more than a few tens of apertures are used, though they still contribute to the wings of the point spread function (PSF). Of more general importance are the ghosts that arise through errors in the grating. Such errors most commonly take the form of a periodic variation in the groove spacing. A variation with a single period gives rise to Rowland ghosts that appear as faint lines close to and on either side of strong spectrum lines. Their intensity is proportional to the square of the order of the spectrum. Thus, echelle gratings (see below) must be of very high quality since they may use spectral orders of several hundred. If the error is multiperiodic, then Lyman ghosts of strong lines may appear. These are similar to the Rowland ghosts, except that they can be formed at large distances from the line that is producing them. Some compensation for these errors can be obtained through deconvolution of the PSF (Section 2.1.1), but for critical work, the only real solution is to use a grating without periodic errors, such as a holographically produced grating.

Wood's anomalies may also sometimes occur. These do not arise through grating faults, but are due to light that should go into spectral orders behind the grating (were that to be possible) reappearing within lower order spectra. The anomalies have a sudden onset and a slower decline toward longer wavelengths and are almost 100% plane polarized. They are rarely important in efficiently blazed gratings.

By increasing the angle of a blazed grating, we obtain an echelle grating (Figure 4.7). This is illuminated more or less normally to the groove surfaces and therefore at a large angle to the normal to the grating. It is usually a very coarse grating—10 lines mm^{-1} is not uncommon—so that the separation of the apertures, d, is very large. The reciprocal linear dispersion

$$\frac{d\lambda}{dx} = \pm \frac{d \cos^3 \theta}{m f_2} \tag{4.37}$$

is therefore also very large. Such gratings concentrate the light into many overlapping high-order spectra, and so from Equation 4.21, the resolution is very high. A spectroscope that is based upon an echelle grating requires a second low-dispersion grating or prism whose dispersion is perpendicular to that of the echelle and so is called a cross-disperser to separate out each of the orders (Section 4.2.3).

A quantity known variously as throughput, etendu, or light gathering power is useful as a measure of the efficiency of the optical system. It is the amount of energy passed by the system when its entrance aperture is

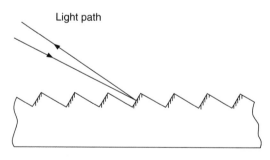

FIGURE 4.7 Enlarged view of an echelle grating.

illuminated by unit intensity per unit area per unit solid angle, and it is denoted by u

$$u = \tau A \, \Omega \qquad (4.38)$$

where
Ω is the solid angle accepted by the instrument
A is the area of its aperture
τ is the fractional transmission of its optics, incorporating losses due to scattering, absorption, imperfect reflection, etc.

For a spectroscope, Ω is the solid angle subtended by the entrance slit at the collimator, or for slitless spectroscopes, the solid angle is accepted by the telescope–spectroscope combination. A is the area of the collimator or the effective area of the dispersing element, whichever is smaller, τ will depend critically upon the design of the system, but as a reasonably general rule it may be taken to be the product of the transmissions for all the surfaces. These will usually be in the region of 0.8–0.9 for each surface, so that τ for the design illustrated in Figure 4.4 will have a value of about 0.4. Older spectroscope designs often had very low throughputs, less than 10% was not uncommon. Even today a throughput of 40% such as has been achieved by the bench-mounted high-resolution optical spectrometer (bHROS) instrument for the Gemini south telescope is considered excellent. Thus, much light is lost in spectroscopy compared with direct imaging and since the remaining light is then spread out over the spectrum, the exposures needed for spectroscopy are typically a hundred or more times those required for imaging.

The product of resolution and throughput, P, is a useful figure for comparing the performances of different spectroscope systems

$$P = R \, u \qquad (4.39)$$

Normally, it will be found that, other things being equal, P will be largest for Fabry–Perot spectroscopes (see below), of intermediate values for grating-based spectroscopes, and lowest for prism-based spectroscopes.

4.1.3 Prisms

Pure prism-based spectroscopes will rarely be encountered today, except within instruments constructed some time ago, though as an exception to that, a low-resolution prism-based mid-infrared (MIR) spectroscope is currently planned for use on the James Webb Space Telescope (JWST). However, they are commonly used in conjunction with gratings in some modern instruments. The combination is known as a grism and the deviation of the light beam by the prism is used to counteract that of the grating, so that the light passes straight through the instrument. The spectroscope can then be used for direct imaging just by removing the grism and without having to move the camera. Prisms are also often used as cross-dispersers for high spectral order spectroscopes based upon echelle gratings or etalons, and may be used nonspectroscopically for folding light beams.

When monochromatic light passes through an interface between two transparent isotropic media at a fixed temperature, then we can apply the well-known Snell's law relating the angle of incidence, i, to the angle of refraction, r, at that interface

$$\mu_1 \sin i = \mu_2 \sin r \tag{4.40}$$

where μ_1 and μ_2 are constants that are characteristic of the two media. When $\mu_1 = 1$, which strictly only occurs for a vacuum, but which holds to a good approximation for most gases, including air, we have

$$\frac{\sin i}{\sin r} = \mu_2 \tag{4.41}$$

and μ_2 is known as the refractive index of the second medium. Now we have already seen that the refractive index varies with wavelength for many media (Section 1.1.17.2, Equation 1.31). The manner of this variation may, over a restricted wavelength interval, be approximated by the Hartmann dispersion formula, in which A, B, and C are known as the Hartmann constants (see also the Cauchy formula, Equation 4.147):

$$\mu_\lambda = A + \frac{B}{\lambda - C} \tag{4.42}$$

If the refractive index is known at three different wavelengths, then we can obtain three simultaneous equations for the constants, from Equation 4.42, giving

$$C = \frac{\left[\left(\frac{\mu_1 - \mu_2}{\mu_2 - \mu_3}\right)\lambda_1(\lambda_2 - \lambda_3) - \lambda_3(\lambda_1 - \lambda_2)\right]}{\left[\left(\frac{\mu_1 - \mu_2}{\mu_2 - \mu_3}\right)(\lambda_2 - \lambda_3) - (\lambda_2 - \lambda_3)\right]} \tag{4.43}$$

$$B = \frac{\mu_1 - \mu_2}{\left(\dfrac{1}{\lambda_1 - C} - \dfrac{1}{\lambda_2 - C}\right)} \tag{4.44}$$

$$A = \mu_1 - \frac{B}{\lambda_1 - C} \tag{4.45}$$

The values for the constants for the optical region for typical optical glasses are as follows:

	A	B	C
Crown glass	1.477	3.2×10^{-8}	-2.1×10^{-7}
Dense flint glass	1.603	2.08×10^{-8}	1.43×10^{-7}

Thus, white light refracted at an interface is spread out into a spectrum with the longer wavelengths refracted less than the shorter ones. This phenomenon was encountered in Section 1.1.17 as chromatic aberration, and there we were concerned with eliminating or minimizing its effects; for spectroscopy by contrast we are interested in maximizing the dispersion.

Consider, then, a prism with light incident upon it as shown in Figure 4.8. For light of wavelength λ, the deviation, θ, is given by

$$\theta = i_1 + r_2 - \alpha \tag{4.46}$$

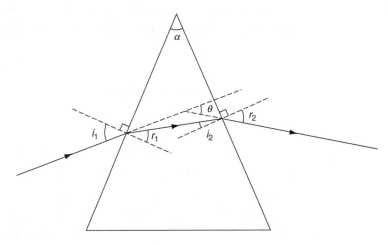

FIGURE 4.8 Optical path in a prism.

So that using Equation 4.42 and using the relations

$$\mu_\lambda = \frac{\sin i_1}{\sin r_1} = \frac{\sin r_2}{\sin i_2} \tag{4.47}$$

and

$$\alpha = r_1 + i_2 \tag{4.48}$$

we get

$$\theta = i_1 - \alpha + \sin^{-1}\left(\left(A + \frac{B}{\lambda - C}\right)\sin\left\{\alpha - \sin^{-1}\left[\frac{\sin i_1}{\left(A + \frac{B}{\lambda - C}\right)}\right]\right\}\right) \tag{4.49}$$

Now we wish to maximize $\frac{\partial\theta}{\partial\lambda}$, which we could study by differentiating Equation 4.49, but which is easier to obtain from

$$\frac{\Delta\theta}{\Delta\lambda} = \frac{\theta_{\lambda_1} - \theta_{\lambda_2}}{\lambda_1 - \lambda_2} \tag{4.50}$$

so that

$$\frac{\Delta\theta}{\Delta\lambda} = \frac{\left(\sin^{-1}\left\{\mu_{\lambda_1}\sin\left[\alpha - \sin^{-1}\left(\frac{\sin i_1}{\mu_{\lambda_1}}\right)\right]\right\} - \sin^{-1}\left\{\mu_{\lambda_2}\sin\left[\alpha - \sin^{-1}\left(\frac{\sin i_1}{\mu_{\lambda_2}}\right)\right]\right\}\right)}{(\lambda_1 - \lambda_2)} \tag{4.51}$$

The effect of altering the angle of incidence or the prism angle is now most simply followed by an example. Consider a dense flint prism for which

$$\lambda_1 = 4.86 \times 10^{-7} \text{ m} \quad \mu_{\lambda_1} = 1.664 \tag{4.52}$$

$$\lambda_2 = 5.89 \times 10^{-7} \text{ m} \quad \mu_{\lambda_2} = 1.650 \tag{4.53}$$

and

$$\frac{\Delta\theta}{\Delta\lambda} = \frac{\left(\sin^{-1}\left\{1.664\sin\left[\alpha - \sin^{-1}\left(\frac{\sin i_1}{1.664}\right)\right]\right\} - \sin^{-1}\left\{1.650\sin\left[\alpha - \sin^{-1}\left(\frac{\sin i_1}{1.650}\right)\right]\right\}\right)}{(1.03 \times 10^{-7})} \text{° m}^{-1} \tag{4.54}$$

Figure 4.9 shows the variation of $\frac{\Delta\theta}{\Delta\lambda}$ with angle of incidence for a variety of apex angles as given by Equation 4.54. From this figure, it is reasonably convincing to see that the maximum dispersion of 1.02×10^8 ° m^{-1} occurs

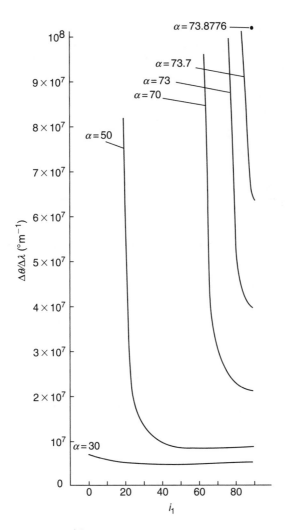

FIGURE 4.9 Variation of $\frac{\Delta\theta}{\Delta\lambda}$ with angle of incidence for a dense flint prism, for various apex angles, α.

for an angle of incidence of 90° and an apex angle of 73.8776°. This represents the condition of glancing incidence and exit from the prism (Figure 4.10), and the ray passes symmetrically through the prism.

The symmetrical passage of the ray through the prism is of importance apart from being one of the requirements for maximum $\frac{\Delta\theta}{\Delta\lambda}$. It is only when this condition applies that the astigmatism introduced by the prism is minimized. The condition of symmetrical ray passage for any prism is more normally called the position of minimum deviation. Again an

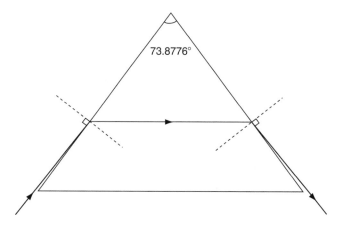

FIGURE 4.10 Optical path for maximum dispersion in a dense flint prism.

example quickly illustrates why this is so. For a dense flint prism with an apex angle of 30°, and at a wavelength of 500 nm, Figure 4.11 shows the variation of the deviation with angle of incidence. The minimum value of θ occurs for $i_1 = 25.46°$, from which we rapidly find that $r_1 = 15°$, $i_2 = 15°$,

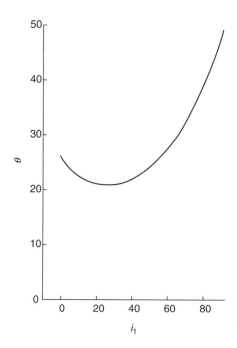

FIGURE 4.11 Deviation of dense flint prism with an apex angle of 30° at a wavelength of 500 nm.

and $r_2 = 25.46°$, and so the ray is passing through the prism symmetrically when its deviation is a minimum. More generally, we may write θ in terms of r_1,

$$\theta = \sin^{-1}(\mu_\lambda \sin r_1) + \sin^{-1}[\mu_\lambda \sin(\alpha - r_1)] - \alpha \qquad (4.55)$$

so that

$$\frac{\partial \theta}{\partial r_1} = \frac{\mu_\lambda \cos r_1}{\sqrt{1 - \mu_\lambda^2 \sin^2 r_1}} - \frac{\mu_\lambda \cos(\alpha - r_1)}{\sqrt{1 - \mu_\lambda^2 \sin^2(\alpha - r_1)}} \qquad (4.56)$$

Since $\frac{\partial \theta}{\partial r_1} = 0$ for θ to be an extremum, we obtain at this point after some manipulation

$$[\cos(2r_1) - \cos(2\alpha - 2r_1)](1 - 2\mu_\lambda^2) = 0 \qquad (4.57)$$

giving

$$r_1 = \alpha/2 \qquad (4.58)$$

and so also

$$i_2 = \alpha/2 \qquad (4.59)$$

and the minimum deviation always occurs for a symmetrical passage of the ray through the prism.

In practice, the maximum dispersion conditions of glancing incidence and exit are unusable because most of the light will be reflected at the two interactions and not refracted through the prism. Antireflection coatings (Section 1.1) cannot be employed because these can only be optimized for a few wavelengths and so the spectral distribution would be disturbed. Thus, the apex angle must be less than the optimum, but the minimum deviation condition must be retained to minimize astigmatism. A fairly common compromise, then, is to use a prism with an apex angle of 60°, which has advantages from the manufacturer's point of view in that it reduces the amount of waste material if the initial blank is cut from a billet of glass. For an apex angle of 60°, the dense flint prism considered earlier has a dispersion of 1.39×10^7 ° m^{-1} for the angle of incidence of 55.9° that is required to give minimum deviation ray passage. This is almost a factor of 10 lower than the maximum possible value.

With white light, it is obviously impossible to obtain the minimum deviation condition for all the wavelengths, and the prism is usually adjusted so that this condition is preserved for the central wavelength of the region of interest. To see how the deviation varies with wavelength, we consider the case of a prism with a normal apex angle of 60°. At minimum deviation, we then have

$$r_1 = 30° \tag{4.60}$$

Putting these values into Equation 4.55 and using Equation 4.42, we get

$$\theta = 2 \sin^{-1} \left[\frac{1}{2} \left(A + \frac{B}{\lambda - C} \right) \right] - 60 \tag{4.61}$$

where θ is in degrees, so that

$$\frac{d\theta}{d\lambda} = \frac{-180B}{\pi(\lambda - C)^2 \left[1 - \frac{1}{4} \left(A + \frac{B}{\lambda - C} \right)^2 \right]^{1/2}} \, °\,m^{-1} \tag{4.62}$$

Now

$$A + \frac{B}{\lambda - C} = \mu_\lambda \approx 1.5 \tag{4.63}$$

and so

$$\left[1 - \frac{1}{4} \left(A + \frac{B}{\lambda - C} \right)^2 \right]^{-1/2} \approx A + \frac{B}{\lambda - C} \tag{4.64}$$

Thus,

$$\frac{d\theta}{d\lambda} \approx \frac{-180AB}{\pi(\lambda - C)^2} - \frac{180B^2}{(\lambda - C)^3} \tag{4.65}$$

Now, the first term on the right-hand side of Equation 4.65 has a magnitude about 30 times larger than that of the second term for typical values of λ, A, B, and C. Hence,

$$\frac{d\theta}{d\lambda} \approx \frac{-180AB}{\pi(\lambda - C)^2} \, °\,m^{-1} \tag{4.66}$$

and hence

$$\frac{d\theta}{d\lambda} \propto (\lambda - C)^{-2} \tag{4.67}$$

and so the dispersion of a prism increases rapidly toward shorter wavelengths. For the example we have been considering involving a dense flint prism, the dispersion is nearly five times larger at 400 nm than at 700 nm.

To form a part of a spectrograph, a prism, like a diffraction grating, must be combined with other elements. The basic layout is shown in Figure 4.12; practical designs are discussed in Section 4.2.

The linear dispersion of a prism is obtained in a similar way to that for the grating. If x is the linear distance along the spectrum from some reference point, then we have for an achromatic imaging element of focal length f_2,

$$\frac{dx}{d\lambda} = f_2 \frac{d\theta}{d\lambda} \tag{4.68}$$

where θ is small and is measured in radians. Thus, from Equation 4.66

$$\frac{dx}{d\lambda} = \frac{-180ABf_2}{\pi(\lambda - C)^2} \tag{4.69}$$

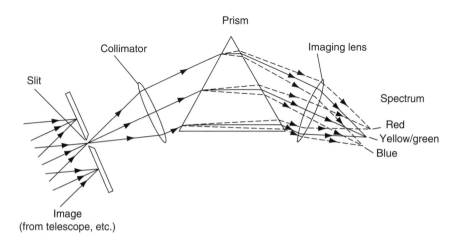

FIGURE 4.12 Basic optical arrangement of a prism spectroscope.

The reciprocal linear dispersion, $\frac{d\lambda}{dx}$, is thus

$$\frac{d\lambda}{dx} = \frac{\pi(\lambda - C)^2}{-180ABf} \tag{4.70}$$

In a similar way to the grating spectroscope, the resolving power of a prism spectroscope is limited by the spectral resolution of the prism, by the resolving power of its optics (Section 1.1), and by the projected slit width. The spectrum is formed from an infinite number of monochromatic images of the entrance slit. The width of one of these images, S, is again given by

$$S = s\frac{f_2}{f_1} \tag{4.71}$$

where
s is the slit width
f_1 is the collimator's focal length
f_2 is the imaging element's focal length

If the optics of the spectroscope are well corrected then we may ignore their aberrations and consider only the diffraction limit of the system. When the prism is fully illuminated, the imaging element will intercept a rectangular beam of light. The height of the beam is just the height of the prism, and has no effect upon the spectral resolution. The width of the beam, D, is given by

$$D = L\left[1 - \mu_\lambda^2 \sin^2\left(\frac{\alpha}{2}\right)\right]^{1/2} \tag{4.72}$$

where L is the length of a prism face, and the diffraction limit is just that of a rectangular slit of width D. So that from Figure 1.26, the linear Rayleigh limit of resolution, W'', is given by

$$W'' = \frac{f_2\lambda}{D} \tag{4.73}$$

$$= \frac{f_2\lambda}{L\left[1 - \mu_\lambda^2 \sin^2\left(\frac{\alpha}{2}\right)\right]^{1/2}} \tag{4.74}$$

If the beam is limited by some other element of the optical system, and/or is of circular cross section, then D must be evaluated as may be appropriate, or the Rayleigh criterion for the resolution through a circular aperture (Equation 1.55) must be used in place of that for a rectangular aperture. Optimum resolution occurs when

$$S = W'' \tag{4.75}$$

that is,

$$s = \frac{f_1 \lambda}{D} \tag{4.76}$$

$$= \frac{f_1 \lambda}{L\left[1 - \mu_\lambda^2 \sin^2\left(\frac{\alpha}{2}\right)\right]^{1/2}} \tag{4.77}$$

The spectral resolution may now be found from Equations 4.68 and 4.73

$$W_\lambda = W'' \frac{d\lambda}{dx} \tag{4.78}$$

$$= \frac{\lambda}{d} \frac{d\lambda}{d\theta} \tag{4.79}$$

$$\approx \frac{\lambda(\lambda - C)^2}{ABL\left[1 - \left(A + \frac{B}{\lambda - C}\right)^2 \sin^2\left(\frac{\alpha}{2}\right)\right]^{1/2}} \tag{4.80}$$

and the resolution is

$$R = \frac{\lambda}{W_\lambda} \tag{4.81}$$

$$\approx \frac{ABL\left[1 - \left(A + \frac{B}{\lambda - C}\right)^2 \sin^2\left(\frac{\alpha}{2}\right)\right]^{1/2}}{(\lambda - C)^2} \tag{4.82}$$

For a dense flint prism with an apex angle of 60° and a side length of 0.1 m, we then obtain in the visible region

$$R \approx 1.5 \times 10^4 \tag{4.83}$$

and this is a fairly typical value for the resolution of a prism-based spectroscope. We may now see another reason why the maximum dispersion (Figure 4.9) is not used in practice. Working back through the equations, we find that the term $L\left[1 - \left(A + \frac{B}{\lambda - C}\right)^2 \sin^2\left(\frac{\alpha}{2}\right)\right]^{1/2}$ that is involved in the numerator of the right-hand side of Equation 4.82 is just the width of the emergent beam from the prism. Now for maximum dispersion, the beam emerges at 90° to the normal to the last surface of the prism. Thus, however large the prism may be, the emergent beam width is zero, and thus R is zero as well. The full variation of R with α is shown in Figure 4.13. A 60° apex angle still preserves 60% of the maximum resolution and so is a reasonable compromise in terms of resolution as well as dispersion. The truly optimum apex angle for a given type of material will generally be close to but not exactly 60°. Its calculation will involve a complex trade-off between resolution, dispersion, and throughput of light, assessed in terms of the final amount of information available in the spectrum, and is not usually attempted unless absolutely necessary.

The resolution varies slightly across the width of the spectrum, unless cylindrical lenses or mirrors are used for the collimator, and these have

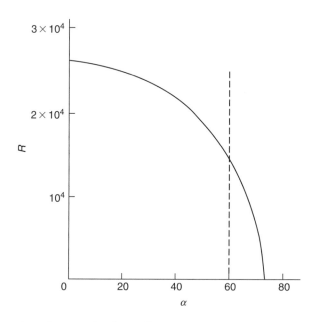

FIGURE 4.13 Resolution of a dense flint prism with a side length of 0.1 m at a wavelength of 500 nm, with changing apex angle α.

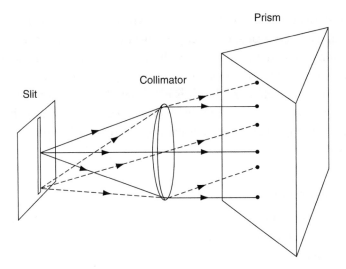

FIGURE 4.14 Light paths of rays from the center and edge of the slit of a spectroscope.

severe disadvantages of their own. The variation arises because the light rays from the ends of the slit impinge on the first surface of the prism at an angle to the optical axis of the collimator (Figure 4.14). The peripheral rays therefore encounter a prism whose effective apex angle is larger than that for the paraxial rays. The deviation is also increased for such rays, and so the ends of the spectrum lines are curved toward shorter wavelengths. Fortunately, astronomical spectra are mostly so narrow that both these effects can be neglected. Although the wider, beams encountered in integral field and multi-object spectroscopes (Sections 4.2.4 and 4.2.5, respectively) may be affected in this way.

The material used to form prisms depends upon the spectral region that is to be studied. In the visual region, the normal types of optical glass may be used, but these mostly start to absorb in the near-ultraviolet region. Fused silica and crystalline quartz can be formed into prisms to extend the limit down to 200 nm. Crystalline quartz, however, is optically active (Section 5.2), and therefore must be used in the form of a Cornu prism. This has the optical axis parallel to the base of the prism so that the ordinary and extraordinary rays coincide, and is made in two halves cemented together. The first half is formed from a right-handed crystal, and the other half from a left-handed crystal, the deviations of left- and right-hand circularly polarized beams are then similar. If required, calcium fluoride or lithium fluoride can extend the limit down to 140 nm or so, but astronomical spectroscopes working at such short wavelengths are normally based upon gratings. In the infrared region, quartz can again be used for wavelengths out to about 3.5 μm. Rock salt can be used at even longer wavelengths, but it is extremely hygroscopic that makes it difficult to use. More commonly, Fourier spectroscopy (see below) is applied when high-resolution spectroscopy is required in the far-infrared (FIR) region.

4.1.4 Interferometers

We only consider here in any detail the two main types of spectroscopic interferometry that are of importance in astronomy: the Fabry–Perot interferometer or etalon, and the Michelson interferometer or Fourier-transform spectrometer. Other systems exist but at present are of little importance for astronomy.

4.1.4.1 Fabry–Perot Interferometer

Two parallel, flat, partially reflecting surfaces are illuminated at an angle θ (Figure 4.15). The light undergoes a series of transmissions and reflections

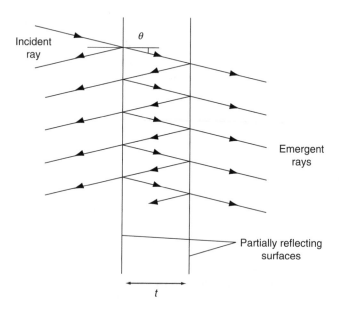

FIGURE 4.15 Optical paths in a Fabry–Perot interferometer.

as shown, and pairs of adjoining emergent rays differ in their path lengths by ΔP, where

$$\Delta P = 2t \cos \theta \tag{4.84}$$

Constructive interference between the emerging rays will then occur at those wavelengths for which

$$\mu \, \Delta P = m\lambda \tag{4.85}$$

where m is an integer, that is,

$$\lambda = \frac{2t\mu \, \cos \theta}{m} \tag{4.86}$$

If such an interferometer is used in a spectroscope in place of the prism or grating (Figure 4.16), then the image of a point source is still a point. However, the image is formed from only those wavelengths for which Equation 4.86 holds. If the image is then fed into another spectroscope, it will be broken up into a series of monochromatic images. If a slit is now used as the source, the rays from the different points along the slit will meet the etalon at differing angles, and the image will then consist of a series of superimposed short spectra. If the second spectroscope is then set so that its dispersion is perpendicular to that of the etalon (a cross-disperser), then the final image will be a rectangular array of short parallel spectra. The widths of these spectra depend upon the reflectivity of the surfaces. For a high reflectivity, we get many multiple reflections, while

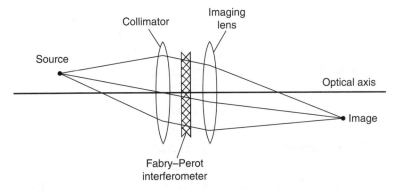

FIGURE 4.16 Optical paths in a Fabry–Perot spectroscope.

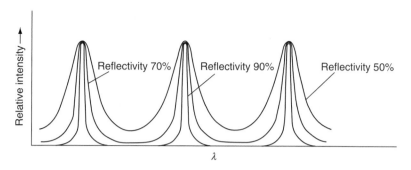

FIGURE 4.17 Intensity versus wavelength in the image of a white-light point source in a Fabry–Perot spectroscope, assuming negligible absorption.

with a low reflectivity the intensity becomes negligible after only a few reflections. The monochromatic images of a point source are therefore not truly monochromatic but are spread over a small wavelength range in a similar, but not identical, manner to the intensity distributions for several collinear apertures (Figures 4.1 and 4.2). The intensity distribution varies from that of the multiple apertures since the intensities of the emerging beams decrease as the number of reflections required for their production increases. Examples of the intensity distribution with wavelength are shown in Figure 4.17. In practice, reflectivities of about 90% are usually used.

In the absence of absorption, the emergent intensity at a fringe peak is equal to the incident intensity at that wavelength. This often seems a puzzle to readers, who on inspection of Figure 4.15 might expect there to be radiation emerging from the left as well as on the right, so that at the fringe peak the total emergent intensity would appear to be greater than the incident intensity. If we examine the situation more closely, however, we find that when at a fringe peak for the light emerging on the right, there is zero intensity in the beam emerging on the left. If the incident beam has intensity I, and amplitude $a(I = a^2)$, then the amplitudes of the successive beams on the left in Figure 4.15 are

$$-aR^{1/2}, \ aR^{1/2}T, \ aR^{3/2}T, \ aR^{5/2}T, \ aR^{7/2}T, \ \ldots$$

where the first amplitude is negative because it results from an internal reflection. It has therefore an additional phase delay of 180° compared with the other reflected beams. T is the fractional intensity transmitted and R is the fractional intensity reflected by the reflecting surfaces (note that $T + R = 1$ in the absence of absorption). Summing these terms (assumed to go on to infinity) gives zero amplitude and therefore zero intensity for the left-hand emergent beam. Similarly, the beams emerging on the right have the following amplitudes:

$$aT, \ aTR, \ aTR^2, \ aTR^3, \ aTR^4, \ \ldots \tag{4.87}$$

Summing these terms to infinity gives, at a fringe peak, the amplitude on the right as a. This is the amplitude of the incident beam, and so at a fringe maximum the emergent intensity on the right equals the incident intensity (see also Equation 4.95).

The dispersion of an etalon may easily be found by differentiating Equation 4.86

$$\frac{d\lambda}{d\theta} = -\frac{2t\mu}{m} \sin \theta \tag{4.88}$$

Since the material between the reflecting surfaces is usually air, and the etalon is used at small angles of inclination, we have

$$\mu \approx 1 \tag{4.89}$$

$$\sin \theta \approx \theta \tag{4.90}$$

and from Equation 4.86

$$\frac{2t}{m} \approx \lambda \tag{4.91}$$

so that

$$\frac{d\lambda}{d\theta} \approx \lambda \theta \tag{4.92}$$

Thus, the reciprocal linear dispersion for a typical system, with $\theta = 0.1°$, $f_2 = 1$ m, and used in the visible region, is 0.001 nm mm^{-1}, which is a factor of 100 or so larger than that achievable with more common dispersing elements.

The resolution of an etalon is rather more of a problem to estimate. Our usual measure—the Rayleigh criterion—is inapplicable since the minimum intensities between the fringe maxima do not reach zero, except for a reflectivity of 100%. However, if we consider the image of two equally bright point sources viewed through a telescope at its Rayleigh limit (Figure 2.18), then the central intensity is 81% of that of either of the peak intensities. We may therefore replace the Rayleigh criterion by the more general requirement that the central intensity of the envelope of the images of two equal sources falls to 81% of the peak intensities. Consider, therefore, an etalon illuminated by a monochromatic slit source perpendicular to the optical axis (Figure 4.16). The image will be a strip, also perpendicular to the optical axis, and the intensity will vary along the strip accordingly as the emerging rays are in or out of phase with each other. The intensity variation is given by

$$I(\theta) = \frac{T^2 I_\lambda}{(1 - R)^2 + 4R \sin^2\left(\frac{2\pi t\mu \cos\theta}{\lambda}\right)} \tag{4.93}$$

where I_λ is the incident intensity at wavelength λ. The image structure will resemble that shown in Figure 4.18. If the source is now replaced with a bichromatic source, then the image structure will be of the type shown in Figure 4.19. Consider just one of these fringes, its angular distance, θ_{max}, from the optical axis is, from Equation 4.86,

$$\theta_{max} = \cos^{-1}\left(\frac{m\lambda}{2t\mu}\right) \tag{4.94}$$

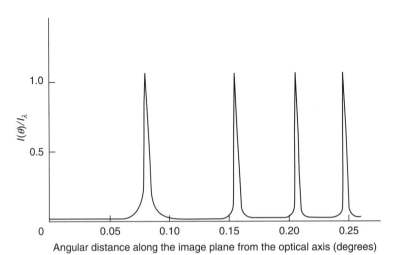

FIGURE 4.18 Image structure in a Fabry–Perot spectroscope viewing a monochromatic slit source, with $T = 0.1$, $R = 0.9$, $t = 0.1$ m, $\mu = 1$, and $\lambda = 550$ nm.

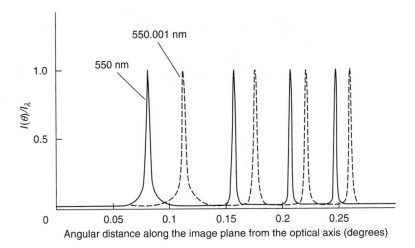

FIGURE 4.19 Image structure in a Fabry–Perot spectroscope viewing a bichromatic slit source, with $T = 0.1$, $R = 0.9$, $t = 0.1$ m, $\mu = 1$, and $\lambda = 550$ nm (full curve) and $\lambda = 550.001$ nm (broken curve).

and so the peak intensity from Equation 4.93 is

$$I(\theta_{max}) = \frac{T^2 I_\lambda}{(1 - R)^2} \tag{4.95}$$

$$= I_\lambda \text{ (when there is no absorption)} \tag{4.96}$$

Let the angular half-width of a fringe at half intensity be $\Delta\theta$, then a separation of twice the half–half width of the fringes gives a central intensity of 83% of either of the peak intensities. So that if α is the resolution by the extended Rayleigh criterion, we may write

$$\alpha \approx 2\Delta\theta = \frac{\lambda(1 - R)}{2\pi\mu t\sqrt{R}\,\theta_{max}\cos\theta_{max}} \tag{4.97}$$

Hence from Equation 4.92, we obtain the spectral resolution

$$W_\lambda = \alpha\frac{d\lambda}{d\theta} \tag{4.98}$$

$$= \frac{\lambda^2(1 - R)}{2\pi\mu t\sqrt{R}\cos\theta_{max}} \tag{4.99}$$

and so the resolution of the system (previously given the symbol R) is

$$\frac{\lambda}{\Delta\lambda} = \frac{2\pi\mu t \sqrt{R}\cos\theta_{max}}{\lambda(1-R)} \qquad (4.100)$$

or, since θ_{max} is small and μ is usually close to unity,

$$\frac{\lambda}{\Delta\lambda} \approx \frac{2\pi t\sqrt{R}}{\lambda(1-R)} \qquad (4.101)$$

Thus, for typical values of $t = 0.1$ m, $R = 0.9$ and for visible wavelengths, we have

$$\frac{\lambda}{\Delta\lambda} \approx 10^7 \qquad (4.102)$$

which is almost two orders of magnitude higher than typical values for prisms and gratings, and is comparable with the resolution for a large echelle grating while physically the device is much less bulky. An alternative measure of the resolution that may be encountered is the finesse. This is the reciprocal of the half-width of a fringe measured in units of the separation of the fringes from two adjacent orders. It is given by

$$\text{finesse} = \frac{\pi\sqrt{R}}{1-R} = \frac{\lambda}{2t} \times \text{resolution} \qquad (4.103)$$

For a value of R of 0.9, the finesse is therefore about 30.

The free spectral range of an etalon is small since it is operating at very high spectral orders. From Equation 4.94, we have

$$\Sigma = \lambda_1 - \lambda_2 = \frac{\lambda_2}{m} \qquad (4.104)$$

where λ_1 and λ_2 are superimposed wavelengths from adjacent orders (cf. Equation 4.23). Thus, the device must be used with a cross-disperser as already mentioned and/or the free spectral range must be increased. The latter may be achieved by combining two or more different etalons, then only the maxima that coincide will be transmitted through the whole system, and the intermediate maxima will be suppressed.

Practical etalons are made from two plates of glass or quartz whose surfaces are flat to 1% or 2% of their operating wavelength, they are held accurately parallel to each other by low thermal expansion spacers, with a spacing in the region of 10–200 mm. The inner faces are mirrors, which are usually produced by a metallic or dielectric coating. The outer faces are inclined by a very small angle to the inner faces so that the plates are the basal segments of very low angle prisms. Any multiple reflections other than the desired ones are then well displaced from the required image. The limit to the resolution of the instrument is generally imposed by departures of the two reflecting surfaces from absolute flatness. This limits the main use of the instrument to the visible and infrared regions. The absorption in metallic coatings also limits the short-wave use, so that 200 nm represents the shortest practicable wavelength even for laboratory usage. Etalons are commonly used as scanning instruments. By changing the air pressure by a few per cent, the refractive index of the material between the plates is changed and so the wavelength of a fringe at a given place within the image is altered (Equation 4.93). The astronomical applications of Fabry–Perot spectroscopes are comparatively few for direct observations. However, the instruments are used extensively in determining oscillator strengths and transition probabilities upon which much of the more conventional astronomical spectroscopy is based.

Another important application of etalons, and one that does have many direct applications for astronomy is in the production of narrowband filters. These are usually known as interference filters and are etalons in which the separation of the two reflecting surfaces is very small. For materials with refractive indices near 1.5, and for near-normal incidence, we see from Equation 4.86 that if t is 167 nm (say) then the maxima will occur at wavelengths of 500, 250, 167 nm, etc., accordingly, as m is 1, 2, 3, etc. While from Equation 4.97, the widths of the transmitted regions will be 8.4, 2.1, 0.9 nm, etc. for 90% reflectivity of the surfaces. Thus, a filter centered upon 500 nm with a bandwidth of 8.4 nm can be made by combining such an etalon with a simple dye filter to eliminate the shorter wavelength transmission regions (or in this example just by relying on the absorption within the glass substrates). Other wavelengths and bandwidths can easily be chosen by changing t, μ, and R. Such a filter would be constructed by evaporating a partially reflective layer onto a sheet of glass. A second layer of an appropriate dielectric material such as magnesium fluoride or cryolite is then evaporated on top of this to the desired thickness, followed by a second partially reflecting layer. A second sheet of glass is then added for protection. The reflecting layers may be silver or aluminum, or they may be formed from a double layer of two materials with very different refractive indices to improve the overall filter transmission. In the FIR region, pairs of inductive meshes can be used in a similar way for infrared filters. The band-passes of interference filters can be made squarer by using several superimposed Fabry–Perot layers. Recently, tunable filters have been developed that are especially suited to observing the emission lines of gaseous nebulae. The reflecting surfaces are mounted on stacks of piezo-electric crystals so that their separations can be altered. The TAURUS (Taylor Atherton uariable-resolution radial-uelocity system) tunable filter used at the Anglo-Australian observatory (AAO), for example, can scan between 370 and 960 nm with selectable band-passes between 6 and 0.6 nm (see also the Lyot birefringent filter, Section 5.3.4). The rather similar Maryland-Magellan tunable filter (MMTF) operates from 500 to 920 nm with a bandwidth that can be varied between 0.5 and 2.5 nm.

4.1.4.2 Michelson Interferometer

This Michelson interferometer should not be confused with the Michelson stellar interferometer that was discussed in Section 2.5.2. The instrument discussed here is similar to the device used by Michelson and Morley to try and detect the earth's motion through the aether. Its optical principles are shown in Figure 4.20. The light from the source is split into two beams by the beam splitter, and then recombined as shown. For a particular position of the movable mirror and with a monochromatic source, there will be a path difference, ΔP, between the two beams at their focus. The intensity at the focus is then

$$I_{\Delta P} = I_m \left[1 + \cos\left(\frac{2\pi\Delta P}{\lambda}\right) \right]$$

(4.105)

where I_m is a maximum intensity. If the mirror is moved, then the path difference will change, and the final intensity will pass through a series of maxima and minima (Figure 4.21). If the source is bichromatic, then two such variations will be superimposed with slightly differing periods and the final output will then have a beat frequency (Figure 4.22). The difference between the two outputs (Figures 4.20 and 4.22) gives the essential principle of the Michelson interferometer when it is used as a spectroscope. Neither output in any way resembles an ordinary spectrum,

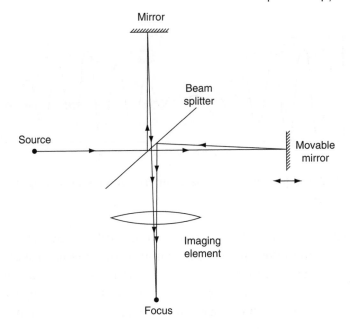

FIGURE 4.20 Optical pathways in a Michelson interferometer.

yet it would be simple to recognize the first as due to a monochromatic source, and the second as due to a bichromatic source. Furthermore, the spacing of the fringes could be related to the original wavelengths through Equation 4.105. More generally of course, sources emit a broad band of wavelengths, and the final output will vary in a complex manner. To find

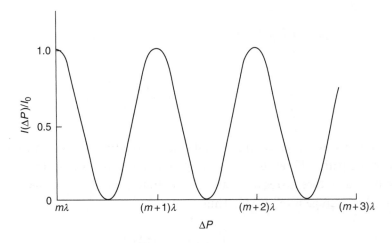

FIGURE 4.21 Variation of fringe intensity with mirror position in a Michelson interferometer.

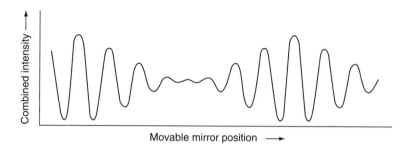

FIGURE 4.22 Output of a Michelson interferometer observing a bichromatic source.

the spectrum of an unknown source from such an output therefore requires a rather different approach than this simple visual inspection.

Let us consider therefore a Michelson interferometer in which the path difference is ΔP observing a source whose intensity at wavelength λ is I_λ. The intensity in the final image because of the light of a particular wavelength, $I'_{\Delta P}(\lambda)$, is then

$$I'_{\Delta P}(\lambda) = KI(\lambda)\left[1 + \cos\left(\frac{2\pi\Delta P}{\lambda}\right)\right] \qquad (4.106)$$

where K is a constant that takes account of the losses at the various reflections, transmissions, etc. Thus, the total intensity in the image for a given path difference is just

$$I'_{\Delta P} = \int_0^\infty I'_{\Delta P}(\lambda)\,d\lambda \qquad (4.107)$$

$$= \int_0^\infty KI(\lambda)\,d\lambda + \int_0^\infty KI(\lambda)\cos\left(\frac{2\pi\Delta P}{\lambda}\right)d\lambda \qquad (4.108)$$

Now the first term on the right-hand side of Equation 4.108 is independent of the path difference, and is simply the mean intensity of the image. We may therefore disregard it and concentrate instead on the deviations from this average level, $I(\Delta P)$. Thus,

$$I(\Delta P) = K\int_0^\infty I(\lambda)\cos\left(\frac{2\pi\Delta P}{\lambda}\right)d\lambda \qquad (4.109)$$

or in frequency terms

$$I(\Delta P) = K^{I} \int_{0}^{\infty} I(\nu) \cos\left(\frac{2\pi\Delta P\nu}{c}\right) d\nu \tag{4.110}$$

Now the Fourier transform, $F(u)$, of a function, $f(t)$ (see also Section 2.1.1), is defined by

$$F[f(t)] = F(u) = \int_{-\infty}^{\infty} f(t)e^{-2\pi iut} dt \tag{4.111}$$

$$= \int_{-\infty}^{\infty} f(t)\cos(2\pi ut) dt - i \int_{-\infty}^{\infty} f(t)\sin(2\pi ut) dt \tag{4.112}$$

Thus, we see that the output of the Michelson interferometer is akin to the real part of the Fourier transform of the spectral intensity function of the source. Furthermore, by defining

$$I(-\nu) = I(\nu) \tag{4.113}$$

we have

$$I(\Delta P) = \frac{1}{2}K^{I} \int_{-\infty}^{\infty} I(\nu) \cos\left(\frac{2\pi\Delta P\nu}{c}\right) d\nu \tag{4.114}$$

$$= K^{II} \text{Re}\left\{\int_{-\infty}^{\infty} I(\nu) \exp\left[-i\left(\frac{2\pi\Delta P}{c}\right)\nu\right] d\nu\right\} \tag{4.115}$$

where K^{II} is an amalgam of all the constants. Now by inverting the transformation

$$F^{-1}[F(u)] = f(t) = \int_{-\infty}^{\infty} F(u)e^{2\pi iut} du \tag{4.116}$$

and taking the real part of the inversion, we may obtain the function that we require—the spectral energy distribution, or as it is more commonly known, the spectrum. Thus,

$$I(\nu) = K^{\mathrm{III}} \mathrm{Re} \left\{ \int_{-\infty}^{\infty} I\left(\frac{2\pi\Delta P}{c}\right) \exp\left[i\left(\frac{2\pi\Delta P}{c}\right)\nu\right] d\left(\frac{2\pi\Delta P}{c}\right) \right\} \quad (4.117)$$

or

$$I(\nu) = K^{\mathrm{IV}} \int_{-\infty}^{\infty} I\left(\frac{2\pi\Delta P}{c}\right) \cos\left(\frac{2\pi\Delta P\nu}{c}\right) d(\Delta P) \quad (4.118)$$

where K^{III} and K^{IV} are again amalgamated constants. Finally, by defining

$$I\left(\frac{-2\pi\Delta P}{c}\right) = I\left(\frac{2\pi\Delta P}{c}\right) \quad (4.119)$$

we have

$$I(\nu) = 2K^{\mathrm{IV}} \int_{0}^{\infty} I\left(\frac{2\pi\Delta P}{c}\right) \cos\left(\frac{2\pi\Delta P\nu}{c}\right) d(\Delta P) \quad (4.120)$$

and so the spectrum is obtainable from the observed output of the interferometer as the movable mirror scans through various path differences. We may now see why a Michelson interferometer when used as a scanning spectroscope is often called a Fourier transform spectroscope. The inversion of the Fourier transform is carried out on computers using the fast Fourier transform algorithm.

In practice of course, it is not possible to scan over path differences from zero to infinity, and also measurements are usually made at discrete intervals rather than continuously, requiring the use of the discrete Fourier transform equations (Section 2.1.1). These limitations are reflected in a reduction in the resolving power of the instrument. To obtain an expression for the resolving power, we may consider the Michelson interferometer as equivalent to a two-aperture interferometer (Figure 4.1) since its image is the result of two interfering beams of light. We may therefore write

Equation 4.20 for the resolution of two wavelengths by the Rayleigh criterion as

$$W_\lambda = \frac{\lambda^2}{2\Delta P} \qquad (4.121)$$

However, if the movable mirror in the Michelson interferometer moves a distance x, then ΔP ranges from 0 to $2x$, and we must take the average value of ΔP rather than the extreme value for substitution into Equation 4.121. Thus, we obtain the spectral resolution of a Michelson interferometer as

$$W_\lambda = \frac{\lambda^2}{2x} \qquad (4.122)$$

so that the system's resolution is

$$\frac{\lambda}{W_\lambda} = \frac{2x}{\lambda} \qquad (4.123)$$

Since x can be as much as 2 m, we obtain a resolution of up to 4×10^6 for such an instrument used in the visible region.

The sampling intervals must be sufficiently frequent to preserve the resolution, but not more frequent than this, or time and effort will be wasted. If the final spectrum extends from λ_1 to λ_2, then the number of useful intervals, n, into which it may be divided, is given by

$$n = \frac{\lambda_1 - \lambda_2}{W_\lambda} \qquad (4.124)$$

so that if λ_1 and λ_2 are not too different then

$$n \approx \frac{8x(\lambda_1 - \lambda_2)}{(\lambda_1 + \lambda_2)^2} \qquad (4.125)$$

However, the inverse Fourier transform gives both $I(\nu)$ and $I(-\nu)$, so that the total number of separate intervals in the final inverse transform is $2n$. Hence, we must have at least $2n$ samples in the original transformation, and therefore the spectroscope's output must be sampled $2n$ times.

Thus the interval between successive positions of the movable mirrors, Δx, at which the image intensity is measured, is given by

$$\Delta x = \frac{(\lambda_1 + \lambda_2)^2}{16(\lambda_1 - \lambda_2)} \qquad (4.126)$$

A spectrum between 500 and 550 nm therefore requires step lengths of 1 μ, while between 2000 and 2050 nm a spectrum would require step lengths of 20 μ. This relaxation in the physical constraints required on the accuracy of the movable mirror for longer wavelength spectra, combined with the availability of other methods of obtaining visible spectra, has led to major applications of Fourier transform spectroscopy to date being in the infrared and FIR regions. Even there though, the increasing size of infrared arrays is now leading to the use of more conventional diffraction grating spectroscopes.

The basic PSF of the Fourier transform spectroscope is of the form $\frac{\sin \Delta\lambda}{\Delta\lambda}$ (Figure 4.23), where $\Delta\lambda$ is the distance from the central wavelength, λ, of a monochromatic source. This is not a particularly convenient form for the profile, and the secondary maxima may be large enough to be significant especially where the spectral energy undergoes abrupt changes, such as at ionization edges or over molecular bands. The effect of the PSF may be reduced at some cost to the theoretical resolution by a technique known as apodization (see also Sections 1.1.18.2, 1.2.3, 5.1.9, and 5.3.5). The transform is weighted by some function, ω, known as the apodization function. Many functions can be used, but the commonest is probably a triangular weighting function, that is,

$$\omega(\Delta P) = 1 - \frac{\Delta P}{2x} \qquad (4.127)$$

FIGURE 4.23 Basic instrumental profile (PSF) of a Fourier transform spectroscope.

The resolving power is halved, but the PSF becomes that for a single rectangular aperture (Figure 1.26, Equation 1.6), and has much reduced secondary maxima.

A major advantage of the Michelson interferometer over the etalon when the latter is used as a scanning instrument lies in its comparative rapidity of use when the instrument is detector-noise limited. Not only is the total amount of light gathered by the Michelson interferometer higher (the Jacquinot advantage), but even for equivalent image intensities the total time required to obtain a spectrum is much reduced. This arises because all wavelengths are contributing to every reading in the Michelson interferometer, whereas a reading from the etalon gives information for just a single wavelength. The gain of the Michelson interferometer is called the multiplex or Fellget advantage, and is similar to the gain of the Hadamard masking technique over simple scanning (Section 2.4). If t is the integration time required to record a single spectral element, then the etalon requires a total observing time of nt. The Michelson interferometer, however, requires a time of only $t/\sqrt{n/2}$ to record each sample, since it has contributions from n spectral elements. It must obtain $2n$ samples so the total observing time for the same spectrum is therefore $2nt/\sqrt{n/2}$, and it has an advantage over the etalon of a factor of $\sqrt{n/8}$ in observing time.

Michelson interferometers have another advantage in that they require no entrance slit to preserve their spectral resolution. This is of considerable significance for large telescopes where the stellar image may have a physical size of a millimeter or more because of atmospheric turbulence, while the spectroscope's slit may be only a few tenths of a millimeter wide. Thus, either much of the light is wasted or complex image dissectors (Section 4.2) must be used.

4.1.5 Fiber Optics

Fiber optic cables are now widely used to connect spectroscopes to telescopes, enabling the spectroscope to be mounted separately from the telescope. This reduces the problems of flexure that occur within telescope-mounted spectroscopes, since the gravitational loads no longer change with the different telescope positions. It also enables the spectroscope to be kept in a temperature-controlled room and in some cases cooled and inside a vacuum chamber as with, for example, the Anglo-Australian telescope's (AAT's) AAOmega that is fed by 800 fibers each 39 m in length and Gemini South's bHROS that is connected to the telescope by 35 m long fiber optics. Fiber optics can also be used to reformat stellar images so that all the light enters the spectroscope and to enable extended objects or multiple objects to be observed efficiently. Specific examples are discussed in Section 4.2. Here we are concerned with the optics of fiber optic cables.

Fiber optic cables usually consist of a thin (10–500 μm) filament of glass encased in a cladding of another glass with a lower refractive index. One or more (sometimes thousands) of these strands make up the cable as a whole. Light entering the core is transmitted through the core by multiple internal reflections off the interface between the two glasses, provided that its angle of incidence exceeds the critical angle for total internal reflection. The critical angle, θ_c, is given by

$$\theta_c = \sin^{-1}\sqrt{\left(\mu_{core}^2 - \mu_{cladding}^2\right)} \qquad (4.128)$$

where μ_{core} and $\mu_{cladding}$ are the refractive indices of the core and cladding, respectively. Fiber optic cables are usually characterized by their numerical aperture (NA) and this is simply equal to $\sin\theta_c$. The minimum focal ratio, f_{min}, that will be transmitted by the core is then

$$f_{min} = \frac{\sqrt{1 - NA}}{2NA} \qquad (4.129)$$

Commercially produced fiber optics have NAs ranging from about 0.15 to 0.35, giving minimum focal ratios ranging from $f3$ to $f0.6$.

Silica glass is used as the core of the fiber for the spectral region 400–2000 nm and is sufficiently transparent that cables can be tens of meters long without significant absorption. Further into the infrared region, specialist glasses such as zirconium fluoride need to be used. Imperfections in the walls of the fibers and internal stresses lead to focal ratio degradation. This is a decrease in the focal ratio that can lead to light loss if the angle of incidence exceeds the critical angle. It may also cause problems in matching the output from the cable to the focal ratio of the instrument being used. Focal ratio degradation affects long focal ratio light beams worst, so short (faster) focal ratios are better for transmitting light into the fibers.

Cables comprising many individual strands will not transmit all the light that they receive, because of the area occupied by the cladding. This is typically 40% of the cross-sectional area of the cable. Many astronomical applications therefore use single-strand cables with a core diameter sufficient to accept a complete stellar image. Multistrand cables can be coherent or noncoherent. In coherent cables, the individual strands have the same relative positions at the input end of the cable as at the output. In noncoherent cables, the relationship between strand positions at the input and output ends of the cable is random. For some special applications, such as reformatting stellar images to match the shape of the spectroscope's entrance slit, the input and output faces of the fiber optic cable are of different shapes. Such cables need to be coherent in the sense that the positions of the strands at the output face are related in a simple and logical manner to those at the input face.

Optical fibers also have other potential astronomical applications that use them in rather more sophisticated ways than just simply getting light to where it is needed. We have already seen (Section 1.1) that by varying the refractive index within the fiber, a Bragg grating can be formed that cuts out the atmospheric OH lines forming the main source of background noise in the near-infrared (NIR) region. Another possibility may result in a new type of spectroscope. The integrated photonic spectroscope uses a number of fibers of different lengths to form a phased array (see Section 1.2 for an account of a radio analogue). The fibers are fed by a 2D waveguide that acts as a multiplexor and their outputs are recombined to form the spectrum in a second waveguide. A working demonstration model of this device has recently been constructed at the AAO. Integrated photonic spectroscopes hold out the possibility of replacing the massive and cumbersome spectroscopes often used today with an instrument that is just a few centimeters in size.

Exercise 4.1

Calculate the heliocentric radial velocity for a star in which the Hα line (laboratory wavelength 656.2808 nm) is observed to have a wavelength of 656.1457 nm. At the time of the observation, the solar celestial longitude was 100°, and the star's right ascension and declination were 13^h 30^m, and +45°, respectively. The obliquity of the ecliptic is 23°27′.

Exercise 4.2

A prism spectroscope is required with a reciprocal linear dispersion of 1 nm mm^{-1} or better at a wavelength of 400 nm. The focal length of the imaging element is limited to 1 m through the need for the spectroscope to fit onto the telescope. Calculate the minimum number of 60° crown glass prisms required to achieve this, and the actually resulting reciprocal linear dispersion at 400 nm. Note: Prisms can be arranged in a train so that the emergent beam from one forms the incident beam of the next in line. This increases the dispersion in proportion to the number of prisms, but does not affect the spectral resolution, which remains at the value appropriate for a single prism.

Exercise 4.3

Show that the reciprocal linear dispersion of a Fabry–Perot etalon is given by

$$\frac{d\lambda}{dx} = \frac{\lambda l}{2f_2^2}$$

where l is the length of the slit, and the slit is symmetrical about the optical axis.

4.2 SPECTROSCOPES

4.2.1 Basic Design Considerations

The specification of a spectroscope usually begins from just three parameters. One is the focal ratio of the telescope upon which the spectroscope is to operate, the second is the required spectral resolution, and the third is the required spectral range. Thus, in terms of the notation used in Section 4.1, we have f', W_λ, and λ specified (where f' is the effective focal ratio of the telescope at the entrance aperture of the spectroscope), and we require the values of f_1, f_2, s, R, $\frac{d\lambda}{d\theta}$, L, and D. We may immediately write down the resolution required of the dispersion element

$$R = \frac{\lambda}{W_\lambda} \tag{4.130}$$

Now for a 60° prism from Equation 4.82, we find that at 500 nm

$$R = 6 \times 10^4 \ L \ \text{(crown glass)} \tag{4.131}$$

$$R = 15 \times 10^4 \ L \ \text{(flint glass)} \tag{4.132}$$

where L is the length of a side of a prism, in meters. So we may write

$$R \approx 10^5 \, L \qquad (4.133)$$

when the dispersing element is a prism. For a grating, the resolution depends upon the number of lines and the order of the spectrum (Equation 4.21). A reasonably typical astronomical grating (ignoring echelle gratings, etc.) will operate in its third order and will have some 500 lines mm^{-1}. Thus, we may write the resolution of a grating as

$$R \approx 1.5 \times 10^6 \, L \qquad (4.134)$$

where L is the width of the ruled area of the grating, in meters. The size of the dispersing element will thus be given approximately by Equations 4.133 and 4.134. It may be calculated more accurately in second and subsequent iterations through this design process, using Equations 4.21 and 4.82.

The diameter of the exit beam from the dispersing element, assuming that it is fully illuminated, can then be found from

$$D = L \cos \phi \qquad (4.135)$$

where ϕ is the angular deviation of the exit beam from perpendicular to exit face of the dispersing element. For a prism, ϕ is typically $60°$, while for a grating used in the third order with equal angles of incidence and reflection, it is $25°$. Thus, we get

$$D = 0.5 \, L \; (\text{prism}) \qquad (4.136)$$

$$D = 0.9 \, L \; (\text{grating}) \qquad (4.137)$$

The dispersion can now be determined by setting the angular resolution of the imaging element equal to the angle between two just resolved wavelengths from the dispersing element

$$\frac{\lambda}{D} = W_\lambda \frac{d\theta}{d\lambda} \qquad (4.138)$$

giving

$$\frac{d\theta}{d\lambda} = \frac{R}{D} \qquad (4.139)$$

Since the exit beam is rectangular in cross section, we must use the resolution for a rectangular aperture of the size and shape of the beam (Equation 1.7) for the resolution of the imaging element, and not its actual resolution, assuming that the beam is wholly intercepted by the imaging element, and that its optical quality is sufficient not to degrade the resolution below the diffraction limit.

The final parameters now follow easily. The physical separation of two just resolved wavelengths on the charge-coupled device (CCD) or other imaging detector must be greater than or equal to the separation of two pixels. CCD pixels are typically around 15–20 μm in size, so the focal length in meters of the imaging element must be at least

$$f_2 = \frac{2 \times 10^{-5}}{W_\lambda} \frac{d\theta}{d\lambda} \tag{4.140}$$

The diameter of the imaging element, D_2, must be sufficient to contain the whole of the exit beam. Thus, for a square cross-section exit beam,

$$D_2 = \sqrt{2} \, D \tag{4.141}$$

The diameter of the collimator, D_1, must be similar to that of the imaging element in general if the dispersing element is to be fully illuminated. Thus again,

$$D_1 = \sqrt{2} \, D \tag{4.142}$$

Now for the collimator to be fully illuminated in its turn, its focal ratio must equal the effective focal ratio of the telescope. Hence, the focal length of the collimator, f_1, is given by

$$f_1 = \sqrt{2} \, Df' \tag{4.143}$$

Finally, from Equation 4.76, we have the slit width

$$s = \frac{f_1 \lambda}{D} \tag{4.144}$$

and a first approximation has been obtained to the design of the spectroscope.

The low light levels involved in astronomy usually require the focal ratio of the imaging elements to be small, so that it is fast in imaging terms. Satisfying this requirement usually means a compromise in some other part of the design, so that an optimally designed system is rarely achievable in practice. The slit may also need to be wider than specified to use a reasonable fraction of the star's light.

The limiting magnitude of a telescope–spectroscope combination is the magnitude of the faintest star for which a useful spectrum may be obtained. This is a very imprecise quantity, for it depends upon the type of spectrum and the purpose for which it is required, as well as the properties of the instrument and the detector. For example, if strong emission lines in a spectrum are the features of interest, then fainter stars may be studied than if weak absorption lines are desired. Similarly, spectra of sufficient quality to determine radial velocities may be obtained for fainter stars than if line profiles are wanted. However, a guide to the limiting magnitude may be gained through the use of Bowen's formula

$$m = 12 + 2.5 \log_{10} \left[\frac{sD_1 T_D gqt\left(\frac{d\lambda}{d\theta}\right)}{f_1 f_2 \alpha H} \right] \tag{4.145}$$

where

m is the faintest B magnitude that will give a usable spectrum in t seconds of exposure

T_D is the telescope objective's diameter

g is the optical efficiency of the system, that is, the ratio of the usable light at the focus of the spectroscope to that incident upon the telescope. Typically, it has a value of 0.2. Note, however, that g does not include the effect of the curtailment of the image by the slit

q is the quantum efficiency of the detector, typical values are 0.4–0.9 for CCDs (Section 1.1.8)

α is the angular size of the stellar image at the telescope's focus, typically 5×10^{-6} to 2×10^{-5} radians

H is the height of the spectrum

This formula gives quite good approximations for spectroscopes in which the star's image is larger than the slit and it is trailed along the length of the slit to broaden the spectrum. This used to be the commonest mode of use for astronomical spectroscopes, but now many spectroscopes are fed by optical fibers that may or may not intercept the whole of the star's light.

Other situations such as an untrailed image, or an image smaller than the slit, require the formula to be modified. Thus, when the slit is wide enough for the whole stellar image to pass through it, the exposure varies inversely with the square of the telescope's diameter, while for extended sources, it varies inversely with the square of the telescope's focal ratio (cf. Equations 1.67 and 1.70, etc.).

The slit is quite an important part of the spectroscope since in most astronomical spectroscopes it fulfills two functions. First, it acts as the entrance aperture of the spectroscope. For this purpose, its sides must be accurately parallel to each other and perpendicular to the direction of the dispersion. It is also usual for the slit width to be adjustable or slits of different widths provided, so that alternative detectors may be used and/or changing observing conditions catered for. Although we have seen how to calculate the optimum slit width, it is usually better in practice to find the best slit width empirically. The slit width is optimized by taking a series of images of a sharp emission line in the comparison spectrum through slits of different widths. As the slit width decreases, the image width should also decrease at first, but should eventually become constant. The change-over point occurs when some part of the spectroscope system other than the slit starts to limit the resolution, and the slit width at changeover is the required optimum value. As well as allowing the desired light through, the slit must reject unwanted radiation. The jaws of the slit are therefore usually of a knife-edge construction with the chamfering on the inside of the slit, so that light is not scattered or reflected into the spectroscope from the edges of the jaws. On some instruments, a secondary purpose of the slit is to assist in the guiding of the telescope on the object. When the stellar image is larger than the slit width, it will overlap onto the slit jaws. By polishing the front of the jaws to an optically flat mirror finish, these overlaps can be observed via an auxiliary detector, and the telescope driven and guided to keep the image bisected by the slit.

However, guiding in this manner is wasteful of the expensively collected light from the star of interest. Most modern instruments therefore use another star within the field of view to guide on. If the stellar image is then larger than the slit, it can be reformatted so that all its light enter the spectroscope. There are several ways of reformatting the image. Early approaches such as that owing to Bowen are still in use. The Bowen image slicer consists of a stack of overlapped mirrors (Figure 4.24) that section the image and then rearrange the sections end to end to form a linear image suitable for matching a spectroscope slit. The Bowen–Walraven

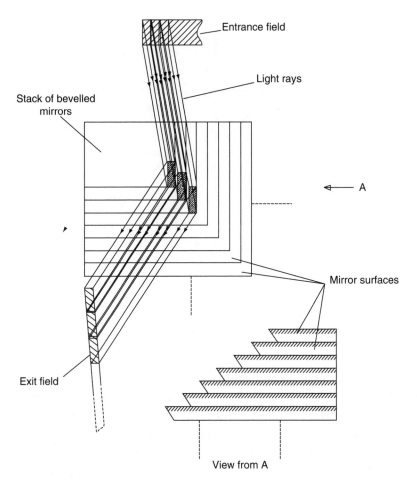

FIGURE 4.24 Bowen image slicer.

image slicer uses multiple internal reflections. A prism with a chamfered side is used and has a thin plate attached to it. Because of the chamfered side, the plate only touches the prism along one edge. A light beam entering the plate is repeatedly internally reflected wherever the plate is not in contact with the prism, but is transmitted into the prism along the contact edge (Figure 4.25). The simplest concept is a bundle of optical fibers whose cross section matches the slit at one end and the star's image at the other. Thus, apart from the reflection and absorption losses within the fibers, all the star's light is conducted into the spectroscope. Disadvantages of fiber optics are mainly the degradation of the focal ratio of the beam owing to imperfections in the walls of the fiber so that not all the light are

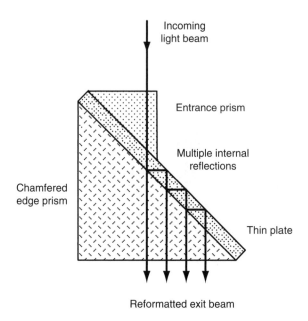

FIGURE 4.25 Bowen–Walraven image slicer.

intercepted by the collimator, and the multilayered structure of the normal commercially available units leads to other light losses since only the central core of fiber transmits the light. Thus, specially designed fiber optic cables are usually required and these are made in-house at the observatory needing them. They are usually much thicker than normal fiber optics and are formed from plastic or fused quartz. On telescopes with adaptive optics, the size of the star's image is much reduced and this allows not only the slit width to be smaller, but also a more compact design to be used for the whole spectroscope.

For extended sources, it is common practice to use long slits. Provided that the image does not move with respect to the slit (and no spectrum widener is used), then each point in that portion of the image falling onto the slit has its individual spectrum recorded at the appropriate height within the final spectrum. Several slits can be used as the entrance aperture of the spectroscope provided that the spectra do not overlap. Then, all the information that is derivable from a single-slit spectrogram is available, but the whole source can be covered in a fraction of the time. The ultimate development of this procedure, known as integral field spectroscopy or three-dimensional (3D) spectroscopy, is to obtain a spectrum for every resolved point within an extended source and this is discussed further below.

For several purposes, the slit may be dispensed with and some specific designs are considered later in this section. Apart from the Fourier transform spectroscope (Section 4.1.4.2), they fall into two main categories. In the first, the projected image size on the spectrum is smaller than some other constraint on the system's resolution. The slit and the collimator may be discarded and parallel light from the source allowed to impinge directly onto the dispersing element. In the second type of slitless spectroscope, the source is producing a nebular type of spectrum (i.e., a spectrum consisting almost entirely of emission lines with little or no continuum). If the slit alone is then eliminated from the telescope–spectroscope combination, the whole of the image of the source passes into the spectroscope. The spectrum then consists of a series of monochromatic images of the source in the light of each of the emission lines. Slitless spectroscopes are difficult to calibrate so that radial velocities can be found from their spectra, but they may be much more optically efficient than a slit spectroscope. In the latter, perhaps 1%–10% of the incident light is eventually used in the image, but some types of slitless spectroscope can use as much as 75% of the light. Furthermore, some designs, such as the objective prism (see below), can image as many as 10^5 stellar spectra in one exposure. A system that is closely related to the objective prism places the disperser shortly before the focal point of the telescope. Although the light is no longer in a parallel beam, the additional aberrations that are produced may be tolerable if the focal ratio is long. Using a zero-deviation grism in combination with correcting optics enables a relatively wide field to be covered, without needing the large sizes required for objective prisms. With suitable blazing for the grating part of the grism, the zero-order images provide wavelength reference points for the spectra. Grisms used on the wide-field imager of European Southern Observatory's (ESO's) 2.2 m telescope enable spectra to be obtained in this way over an unvignetted area $19'$ across. Similarly, the visible multi-object spectrograph (VIMOS) for the very large telescope (VLT) uses a VPH transmission grating in combination with two prisms to produce a zero-deviation disperser covering a $7' \times 8'$ field of view and a 500–1050 nm spectral region at a spectral resolution of 2500.

Spectroscopes, as we have seen, contain many optical elements that may be separated by large distances and arranged at large angles to each other. For the spectroscope to perform as expected, the relative positions of these various components must be stable to within very tight limits. The two major problems in achieving such stability arise through flexure and thermal expansion. Flexure primarily affects the smaller spectroscopes that are attached to telescopes at Cassegrain foci and so move around with the telescope. Their changing attitudes as the telescope moves cause the stresses within them to alter, so that if in correct adjustment for one position, they will be out of adjustment in other positions. The light beam from the telescope is often folded so that it is perpendicular to the telescope's optical axis. The spectroscope is then laid out in a plane that is parallel to the back of the main mirror. Hence, the spectroscope components can be rigidly mounted onto a stout metal plate, which in turn is bolted flat onto the back of the telescope. Such a design can be made very rigid, and the flexure reduced to acceptable levels. In some modern spectroscopes, active supports are used to compensate for flexure along the lines of those used for telescope mirrors (Section 1.1.18.2).

Temperature changes affect all spectroscopes, but the relatively short light paths in the small instruments that are attached directly to telescopes mean that generally the effects are unimportant.

The large fixed spectroscopes that operate at Coudé and Nasmyth foci are obviously unaffected by changing flexure and usually there is little difficulty other than that of cost in making them as rigid as desired. They experience much greater problems, however, from thermal expansion. The size of the spectrographs may be very large (e.g., see Exercise 4.4) with optical path lengths measured in tens of meters. Thus, the temperature control must be correspondingly strict. A major problem is that the thermal inertia of the system may be so large that it may be impossible to stabilize the spectroscope at ambient temperature before the night has ended. Thus, many such spectroscopes are housed in temperature-controlled sealed rooms. The light then has to be piped in through an optically flat window. Many spectroscopes now have some or all of their components cooled and enclosed in a vacuum chamber. In such cases, problems owing to changing temperatures are avoided provided that the cryostat can maintain a reasonably constant temperature.

Any spectroscope except the Michelson interferometer can be used as a monochromator. That is, a device to observe the object in a very restricted range of wavelengths. Most scanning spectroscopes are in effect mono-chromators whose waveband may be varied. The most important use of the devices in astronomy, however, is in the spectrohelioscope. This builds up a picture of the Sun in the light of a single wavelength, and this is usually chosen to be coincident with a strong absorption line. Further details are given in Section 5.3.3. A related instrument for visual use on small telescopes is called a prominence spectroscope. This has the spectroscope offset from the telescope's optical axis so that the (quite wide) entrance slit covers the solar limb. A small direct-vision prism or transmission grating then produces a spectrum and a second slit isolates an image in Hα light, allowing prominences and other solar features to be discerned.

Spectroscopy is undertaken throughout the entire spectrum. In the infrared and UV regions, techniques, designs, etc. are almost identical to those for visual work except that different materials may need to be used. Some indication of these has already been given in Section 4.1. Generally, diffraction gratings and reflection optics are preferred since there is then no worry over absorption within the optical components. The technique of Fourier transform spectroscopy, as previously mentioned, has so far had its main applications in the infrared region. At very short UV, and at x-ray wavelengths, glancing optics (Section 1.3) and diffraction gratings can be used to produce spectra using the same designs as for visual spectroscopes, the appearance and layout, however, will look very different because of the off-axis optical elements, even though the optical principles are the same. Also, many of the detectors at short wavelengths have some intrinsic spectral resolution. Radio spectroscopes have been described in Section 1.2.

4.2.2 Prism-Based Spectroscopes

As noted earlier, spectroscopes using prisms as the sole dispersing element will now rarely be encountered except for some MIR instruments such as that planned for the JWST. Only a brief guide to such spectroscopes will therefore be given here—partly for historical interest, but also because many diffraction-grating-based instruments have similar designs. The basic layout of a prism-based spectroscope is shown in Figure 4.12. Many instruments have been constructed to this design with only slight modifications, the most important of which is the use of several prisms. If several identical prisms are used with the light passing through each along minimum deviation paths, then the total dispersion is just that of one of the prisms multiplied by the number of prisms. The resolution is unchanged and remains that for a single prism. Thus, such an arrangement is of use

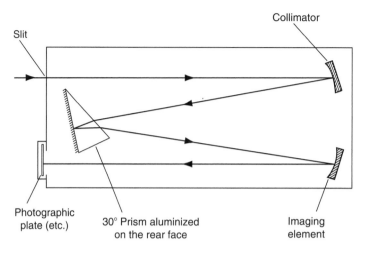

FIGURE 4.26 Compact design for the basic prism spectroscope.

when the resolution of the system is limited by some element of the spectroscope other than the prism. A rather more compact system than that shown in Figure 4.12 can be made by replacing the 60° prism by one with a 30° apex angle that is aluminized on one surface (Figure 4.26). The light therefore passes twice through the prism making its effect the equivalent of a single 60° prism. However, the minimum deviation path is no longer possible, so that some astigmatism is introduced into the image, but by careful design this can be kept lower than the resolution of the system as a whole.

Another similar arrangement, widely used for long-focus spectroscopes in laboratory and solar work, is called the Littrow spectroscope, or auto-collimating spectroscope. A single lens or mirror acts as both the collimator and imaging element (Figure 4.27), thus saving on both the cost and size of the system.

The deviation of the optical axis caused by the prism can be a disadvantage for some purposes. Direct-vision spectroscopes overcome this problem and have zero deviation for some selected wavelength. There are several designs but most consist of combinations of prisms made of two different types of glass with the deviations arranged so that they cancel out while some remnant of the dispersion remains. This is the inverse of the achromatic lens discussed in Section 1.1.17, and therefore usually uses crown and flint glasses for its prisms. The condition for

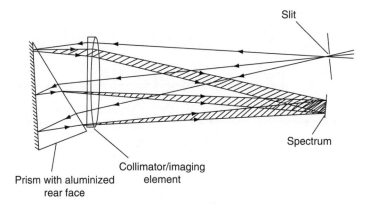

Slit

Spectrum

Prism with aluminized
rear face

Collimator/imaging
element

FIGURE 4.27 Light paths in a Littrow spectroscope.

zero deviation, assuming that the light passes through the prisms at
minimum deviation, is

$$\sin^{-1}\left[\mu_1 \sin\left(\frac{\alpha_1}{2}\right)\right] - \frac{\alpha_1}{2} = \sin^{-1}\left[\mu_2 \sin\left(\frac{\alpha_2}{2}\right)\right] - \frac{\alpha_2}{2} \qquad (4.146)$$

where
α_1 is the apex angle of prism number 1
α_2 is the apex angle of prism number 2
μ_1 is the refractive index of prism number 1
μ_2 is the refractive index of prism number 2

In practical designs, the prisms are cemented together so that the light
does not pass through all of them at minimum deviation. Nonetheless,
Equation 4.146 still gives the conditions for direct vision to a good degree of
approximation. Direct-vision spectroscopes can also be based upon grisms,
where the deviations of the prism and grating counteract each other.

Applications of most direct-vision spectroscopes are nonastronomical,
since they are best suited to visual work. They are though being increasingly used within instruments that obtain direct images as well as spectra,
such as many integral field spectroscopes (see also grisms, below). Spectra
are obtained when the direct-vision spectroscope is placed into the optical
path and direct images are obtained when it is removed, without the need
to adjust the instrument in other respects between the two operating
modes. There is also one other ingenious astronomical application that
has been used in the past, invented by Treanor, which enabled many

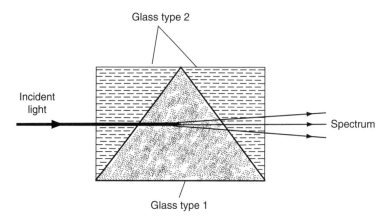

FIGURE 4.28 Treanor's direct-vision prism.

approximate stellar radial velocities to be found rapidly. The method was based upon a direct-vision prism formed from one prism with an apex angle of α, and two prisms made of a different glass with apex angles of $\alpha/2$. The combination forms a block of glass whose incident and exit faces are perpendicular to the light paths (Figure 4.28), so that there is no displacement of the undeviated ray. The two glasses are chosen so that their refractive indices are identical for the desired undeviated wavelength, but their refractive index gradients against wavelength differ. One possible example that gives an undeviated ray at 530 nm combines crown glass and borosilicate glass:

λ (nm)	μ (Crown Glass)	μ (Borosilicate Glass)
400	1.530	1.533
530	1.522	1.522
600	1.519	1.517

The prism is placed without a slit in a collimated beam of light from a telescope. After passing through the prism, the spectrum is then focused to produce an image as usual. Since no slit is used, all the stars in the field of view of the telescope are imaged as short spectra. If the collimated beam is slightly larger than the prism, then the light not intercepted by the prism will be focused to a stellar image, and this image will be superimposed upon the spectrum at a position corresponding to the undeviated wavelength. Thus, many stellar spectra may be imaged simultaneously, each

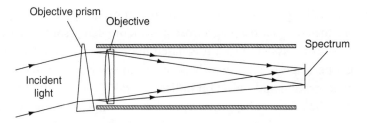

FIGURE 4.29 Objective prism spectroscope.

with a reference point from which the wavelengths of lines in the spectrum can be measured, enabling the radial velocities to be found. The dispersion of the system is low, but the efficiency is high compared with a slit spectroscope.

Another application of the prism that is also now largely of historical interest, having been supplanted by integral field spectroscopy (see below), is the simplest spectroscope of all, the objective prism. This is just a thin prism that is large enough to cover completely the telescope's objective and it is positioned immediately before the telescope's entrance aperture. The starlight is already parallel so that a collimator is unnecessary, while the scintillation disk of the star replaces the slit. The telescope acts as the imaging element (Figure 4.29). Like Treanor's prism, this has the enormous advantage that a spectrum is obtained for every star in the normal field of view. Thus, if the telescope is a Schmidt camera, up to 10^5 spectra may be obtainable in a single exposure. The system has three main disadvantages. First, the dispersion is low; second, the observed star field is at an angle to the telescope axis; and finally, there is no reference point for wavelength measurements although a number of ingenious adaptations of the device have been tried to overcome this latter difficulty.

4.2.3 Grating Spectroscopes

Most of the gratings used in astronomical spectroscopes are of the reflection type. This is because the light can be concentrated into the desired order by blazing quite easily, whereas for transmission gratings, blazing is much more difficult and costly. Transmission gratings are, however often, used in grisms, and these are finding increasing use in integral field spectroscopes, etc.

Plane gratings are most commonly used in astronomical spectroscopes and are almost invariably incorporated into one or other of two designs discussed in the previous section, with the grating replacing the prism. These are the compact basic spectroscope (Figure 4.26), sometimes called a Czerny–Turner system when it is based upon a grating and the Littrow spectroscope (Figure 4.27), which is called an Ebert spectroscope when based upon grating and reflection optics.

Most of the designs of spectroscopes that use curved gratings are based upon the Rowland circle (Figure 4.6). The Paschen–Runge mounting in fact is identical to that shown in Figure 4.6. It is a common design for laboratory spectroscopes since wide spectral ranges can be accommodated, but its size and awkward shape make it less useful for astronomical purposes. A more compact design based upon the Rowland circle is called the Eagle spectroscope (Figure 4.30). However, the vertical displacement of the slit and the spectrum (see the side view in Figure 4.30) introduces some astigmatism. The Wadsworth design abandons the Rowland circle,

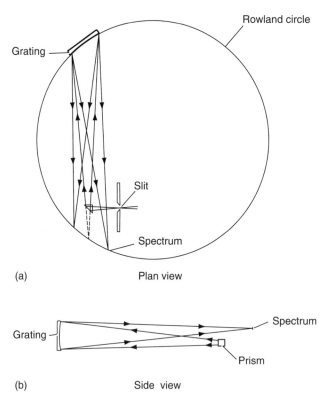

FIGURE 4.30 Optical arrangement of an Eagle spectroscope.

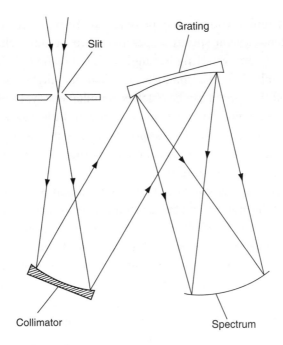

FIGURE 4.31 Wadsworth spectroscope mounting.

but still produces a stigmatic image through its use of a collimator (Figure 4.31). However, the focal surface becomes paraboloidal, and some spherical aberration and coma are introduced. Furthermore, the dispersion for a given grating, if it is mounted into a Wadsworth system, is only half what it could be if the same grating were mounted into an Eagle system, since the spectrum is produced at the prime focus of the grating and not at its radius of curvature. With designs such as the Wadsworth and its variants, the imaging element is likely to be a Schmidt camera system (Section 1.1) to obtain high-quality images with a fast system. Some recent spectroscopes though have used dioptric cameras to avoid the light loss involved with the plate holder in a Schmidt system. Exotic optical materials such as calcium fluoride often need to be used in these designs to achieve the required imaging speed, flat field, and elimination of other aberrations. Gratings can also be used as discussed earlier in various specialized applications such as nebular and prominence spectroscopes.

Spectroscopes are usually optimized for one spectral region. If spectra are needed that extend over a wider range than is covered by the spectroscope, then it may need adjusting to operate in another region, or a different instrument entirely may be required. To overcome this problem to some extent, several spectroscopes have two or three channels optimized for different wavelength ranges. The incoming light from the telescope is split into the channels by dichroic mirrors, so that (with some designs) the spectra can be obtained simultaneously. In other designs, the spectra are obtained in quick succession with little down time needed to adjust the spectroscope. ESO's ultraviolet/visual echelle spectroscope (UVES), for example, has a blue channel covering 300–500 nm and a red channel covering 420–1100 nm.

The design of a spectroscope is generally limited by the available size and quality of the grating, and these factors are of course governed by cost. The cost of a grating in turn is dependent upon its method of production. The best gratings are originals, which are produced in the following manner. A glass or other low-expansion substrate is over-coated with a thin layer of aluminum. The grooves are then scored into the surface of the aluminum by lightly drawing a diamond across it. The diamond's point is precisely machined and shaped so that the required blaze is imparted to the rulings. An extremely high-quality machine is required for controlling the diamond's movement, since not only must the grooves be straight and parallel, but also their spacings must be uniform if Rowland and Lyman ghosts are to be avoided. The position of the diamond is therefore controlled by a precision screw thread, and is monitored interferometrically. We have seen that the resolution of a grating is dependent upon the number of grooves, while the dispersion is a function of the groove spacing (Section 4.1.2). Thus, ideally, a grating should be as large as possible, and the grooves should be as close together as possible (at least until their separation is less than their operating wavelength). Unfortunately, both of these parameters are limited by the wear on the diamond point. Thus, it is possible to have large coarse gratings and small fine gratings, but not large fine gratings. The upper limits on size are about 0.5 m^2, and on groove spacing, about 1500 lines mm^{-1}. A typical grating for an astronomical spectroscope might be 0.1 m across and have 500 lines mm^{-1}.

Echelle gratings are used for many recently built spectroscopes. The rectangular format of the group of spectral segments after the cross-disperser matches well to the shape of large CCD arrays, so that the latter may be used efficiently. The UVES for ESO's VLT, for example, operates in the blue, red, and NIR regions with two 0.2 × 0.8 m echelle gratings. The gratings have 41 and 31 lines mm^{-1} and spectral resolutions of 80,000 and 115,000, respectively. Also planned for the VLT and due to receive first light in 2008 is the X-shooter. This will be a single-target echelle-based spectroscope with prism cross-dispersers and three separate arms optimized for the UV/blue, visual, and NIR regions, so that it can cover the spectral region from 300 to 2500 nm. Similarly, the Magellan Inamori Kyocera echelle instrument for the Magellan telescope that started science operations in 2003 uses two echelle gratings with prism cross-dispersers to cover the 320–1000 nm region. On a quite different scale, there is the BACHES* instrument that is designed for use on small ($\geq$0.2 m) telescopes, and thus potentially available to amateur astronomers. This instrument can obtain spectra over the 390–750 nm region for stars brighter than 5^m at a spectral resolution of 19,000 with a 15 min exposure on a 0.35 m telescope. It uses a 79 lines mm^{-1} echelle grating with a diffraction grating as the cross-disperser and is designed to monitor spectrum variables such as Be stars.

Often a replica grating will be adequate, especially for lower resolution instruments. Since many replicas can be produced from a single original, their cost is a small fraction of that of an original grating. Some loss of quality occurs during replication, but this is acceptable for many purposes. However, a replica improves on an original in one way, and that is in its reflection efficiency. This is better than that of the original because the most highly burnished portions of the grooves are at the bottoms of the grooves on the original, but are transferred to the tops of the grooves on the replicas. The original has, of course, to be the inverse of the finally desired grating. Covering the original with a thin coat of liquid plastic or epoxy resin, which is stripped off after it has set, produces the replicas. The replica is then mounted onto a substrate, appropriately curved if necessary, and then aluminized.

More recently, large high-quality gratings have been produced holographically. An intense monochromatic laser beam is collimated and used to illuminate a photoresist-covered surface. The reflections from the back of the blank interfere with the incoming radiation, and the photoresist is exposed along the nodal planes of the interference field. Etching away the surface then leaves a grating that can be aluminized and used directly, or it can have replicas formed from it as above. The wavelength of the

---

* Basic echelle spectrograph—Batches also means "pothole" in Spanish, but it is not clear, at least to the author, why this should be significant.

illuminating radiation and its inclination can be altered to give almost any required groove spacing and blaze angle.

A related technique to the last produces VPHGs (see Section 4.1.2). A layer of gelatin around 10 μm thick and containing a small proportion of ammonia or potassium dichromate is coated onto a glass substrate. It is then illuminated with an interference pattern from two laser beams. Instead of being etched however, the layer is then treated in water and alcohol baths so that the refractive index within the layer varies according to the exposure it has received. The layer thus forms a grating with the lines of the grating produced by regions of differing refractive indices. Since the gelatin is hygroscopic, it must be protected after production of the grating by a cover sheet. VPHGs can have up to 95% efficiency in their blaze region, and currently can be produced up to 300 mm in diameter and with 100–6000 lines mm^{-1}. They can be used both as transmission and reflection, and they seem likely to find increasing use within astronomical spectroscopes. It is also possible to produce two different gratings within a single element. Such gratings can be tilted with respect to each other so that the spectra are separated. Different wavelength regions can then be observed using a single spectroscope.

Most spectroscopic observations require exposures ranging from tens of seconds to hours or more. The read-out times from their detectors are therefore negligible in comparison. However, for some applications, such as observing rapidly varying or exploding stars, a series of short exposures may be needed and then the read-out and processing times can become significant. Recently, instruments optimized for high-speed spectroscopy have thus started to be developed. ULTRASPEC, for example, being developed by ESO uses a frame-transfer electron-multiplying CCD (EMCCD) that enables hundreds of spectra per second to be obtained. Clearly, such short exposures can only be used for objects with very bright apparent magnitudes; however, as the 20–50 m class telescopes that are currently being planned come on stream, high-speed spectroscopy is likely to be extended to fainter objects and become more widely used.

4.2.4 Integral Field Spectroscopy

Where the spectra of several individual parts of an extended object such as a gaseous nebula or galaxy are needed, then, as already discussed, a long slit may be used so that a linear segment of the object is covered by a single exposure. Where spectra of every resolution element of an object are required then repeated adjacent exposures using a long-slit spectroscope are one possible, but time-consuming, approach. Several techniques that come under the heading of integral field spectroscopy or 3D spectroscopy have thus been developed recently to obtain spectra for every resolved point within an extended source more efficiently.

Most simply, several long slits can be used as the entrance aperture of the spectroscope, provided that the spectra do not overlap. Then, all the information that are derivable from a single-slit spectrogram are available, but the source can be covered in a fraction of the time. Another approach is to use a scanning narrowband filter and obtain numerous images at slightly differing wavelengths. Examples of this are the TAURUS tunable filter and MMTF discussed in Section 4.1.4.1. Alternatively, imaging detectors, such as superconducting tunnel-junction detectors (STJs) (Section 1.1.11), that are also intrinsically sensitive to wavelength can be used to obtain the whole data set in a single exposure. A color photograph or image is, of course, essentially a low-resolution 3D spectrogram. A possible extension to existing color imaging techniques that may have a high-enough spectral resolution to be useful is through the use of dye-doped polymers. These are thin films of polymers such as polyvinylbutyral, containing a dye such as chlorin that has a very narrow absorption band. Changing conditions within the substrate cause the wavelength of the absorption band to scan through a few nanometers, and potentially provide high efficiency direct imaging combined with spectral resolutions of perhaps 10^5 or more.

Most integral field spectroscopy, however, relies on three other approaches. The first is to use an image slicer, though this will need to cover a larger area than that of the stellar image slicers discussed previously. A stack of long plane mirrors whose widths match that of the slit and that are slightly twisted with respect to each other around their short axes is used. The stack is placed at the focal plane of the telescope and the segments of the sliced image then reimaged into a line along the length of the slit. The spectrometer for infrared faint-field imaging (SPIFFI), for example, which forms the heart of the spectrograph for integral-field observations in the near-infrared (SINFONI) adaptive optics integral field spectroscope on the VLT, uses two sets of 32 mirrors to split fields of view of up to $8'' \times 8''$ into narrow rectangular arrays matching the shape of spectroscope's entrance aperture.

The second approach is to use a large fiber optic bundle with a square or circular distribution of the fibers at the input and a linear arrangement at the output. The arrangement of the fibers at the output must correspond in a simple and logical fashion to their arrangement at the input so that the position within the final spectrum of the individual spectrum from a point in the image can be found. As with the stellar image slicers, the cladding of the fibers would mean that significant amounts of light would be lost if the fiber optics were to be placed directly at the telescope's focus. Instead, an array of small lenses is used to feed the image into the fibers. The array of square or hexagonal lenses is placed some distance behind the telescope focus and each lens images the telescope pupil onto the end of a fiber. Providing that these sub-images are contained within the transmitting portions of the fibers, no light is lost due to cladding. The segmented pupil image-reformatting array lens (SPIRAL) on the AAT, for example, uses an array of 512 square lenses to provide spectroscopy over an $11'' \times 22''$ area of the sky feeding the AAOmega spectroscope via 18 m of fiber optic cable. The third approach also uses an array of small lenses but dispenses with the spectroscope slit. The lenses produce a grid of images that is fed directly into the spectroscope and results in a grid of spectra. By orienting the lens array at a small angle to the dispersion, the spectra can be arranged to lie side by side and not to overlap. The optically adaptive system for imaging spectroscopy (OASIS), which was originally used on the 3.6 m Canada–France–Hawaii telescope (CFHT) and is now installed at the Nasmyth focus of the 4.2 m William Herschel telescope (WHT), uses an 1100 hexagonal array of lenses in this fashion, while the supernova integral-field spectrograph (SNIFS) on the University of Hawaii's 2.2 m telescope covers a $6'' \times 6''$ area of the sky with an array of 225 lenslets.

The ideal output from an integral field spectroscope is a spectral cube (hence the alternative name of 3D spectroscopy). The spectral cube has two sides that are the x and y positional coordinates of a conventional 2D image with the third side being the wavelength or frequency. Thus, each pixel in the 2D image of the object or area of sky has its spectrum recorded—alternatively, we may regard the cube as a series of images at different wavelengths each separated by the spectral resolution of the system. Spectral cubes can in principle be obtained for any wavelength, but currently are only available in the visible, NIR and the microwave regions (where, for example, SCUBA2 on the JCMT observes a 16-pixel image over the 325–375 GHz region—Section 1.2.2).

For the future, a second-generation instrument for the VLT, the multi-unit spectroscopic explorer (MUSE), is currently being planned and is expected to cover a $1' \times 1'$ area of the sky at a spatial resolution of $0.2''$ and a spectral resolution of 3000 over the 465–930 nm region. The instrument is planned to be used with an adaptive-optics image sharpener using four guide stars and a reflective image slicer that divides the total field of view into 24 subfields, each of which will then be imaged by a 2k × 4k detector. The resulting spectral cube will thus contain some 360 million elements.

4.2.5 Multi-Object Spectroscopy

Some versions of integral field spectroscopes, especially the earlier designs, use multiple slits to admit several parts of the image of the extended object into the spectroscope simultaneously. Such an instrument can easily be extended to obtain spectra of several different objects simultaneously by arranging for the slits to be positioned over the images of those objects within the telescope's image plane. Practical devices that used this approach employ a mask with narrow linear apertures cut at the appropriate points to act as the entrance slits. However, a new mask is needed for every field to be observed, or even for the same field if different objects are selected, and cutting the masks is generally a precision job requiring several hours work for a skilled technician. This approach has therefore rather fallen out of favor, though the Gemini telescopes' multi-object spectroscopes (GMOSs) continue to use masks. Their masks are generated automatically using a laser cutter, with the slit positions determined from a direct image of the required area of the sky obtained by GMOS operating in an imaging mode. The large binocular telescope's (LBT) LUCIFER (LBT NIR utility with camera and integral-field unit for extragalactic research) can also operate with masks to provide multi-object spectroscopy over the 900–2500 nm region as well as being available in other operating modes. Similarly, one mode of operation of the Keck telescopes' deep imaging multi-object spectrograph (DEIMOS) instrument uses up to 130 slits cut into a mask for spectroscopy over the 410–1100 nm region covering an 80 square minute of arc area of the sky.

Most multi-object spectroscopy, however, is now undertaken using fiber optics to transfer the light from the telescope to the spectroscope. The individual fiber optic strands are made large enough to be able to contain the whole of the seeing disk of a stellar image or in some cases the nucleus of a distant galaxy or even the entire galaxy. Each strand then has one of its ends positioned in the image plane of the telescope so that it intercepts one of the required images. The other ends are aligned along the length of the spectroscope slit, so that hundreds of spectra may be obtained in a single exposure. Initially, the fiber optics were positioned by being plugged into holes drilled through a metal plate. But this has the same drawbacks as cutting masks for integral field spectroscopes and has been superseded. There are now two main methods of positioning the fibers, both of which are computer-controlled and allow 400 or 500 fibers to be repositioned in a matter of minutes. The first approach is to attach the input ends of the fiber optic strands to small magnetic buttons that cling to a steel back plate. The fibers are reconfigured one at a time by a robot arm that picks up a button and moves it to its new position as required. A positional accuracy for the buttons of 10–20 μm is needed. This approach was used with the first multi-object spectroscope for 2° field (2dF) of the AAT. The latter uses correcting lenses to provide an unvignetted 2° field of view, over which 400 fibers could be positioned to feed two low to medium dispersion spectroscopes. This system was replaced in 2006 with AAOmega (see above) that uses 800 fibers also attached to magnetic buttons. The OzPoz* system for ESO's VLT originally could position up to 560 fibers in the same way. Now with the FLAMES (fiber large array multielement spectrograph) instrument, it moves 15 small arrays of fiber optics each covering a 2″ × 3″ area. Thus, each fiber optic bundle represents a mini integral field feeding one of two spectroscopes and so enabling the whole of the visible spectrum to be observed. The system uses two backing plates so that one can be in use while the other is being reset. Swapping the plates takes about 5 min, thus minimizing the dead time between observations. The second approach mounts the fibers on the ends of computer-controlled arms that can be moved radially and/or from side to side. The arms may be moved by small motors, or as in the Subaru telescope's Echidna system by an electromagnetic system. The latter has 400 arms (or spines to fit in with the device's name), each of which can be positioned within a 7 mm circle to within ±10 μm. Typically, about 90% of 400 target objects can be reached in any one configuration and the light beams are fed to an NIR spectroscope. Repositioning the spines takes only 10 min. A possible ambitious instrument for the future is the kilo-aperture optical spectrograph (KAOS), currently being considered for the Gemini telescopes. This could have up to 5000 fibers positioned by Echidna-type spines and covering a 1.5° × 1.5° field of view.

---

* The device is based upon the positioner developed for the 2dF on the AAT, hence Oz for Australia and Poz for positioner.

Schmidt cameras can also be used with advantage for multi-object spectroscopy, because their wide fields of view provide more objects for observation, and their short focal ratios are well matched to transmission through optical fibers. Thus, for example, the 6° field (6dF) project developed from the fiber-linked array-image reformatter (FLAIR) on the U.K. Schmidt camera, which could obtain up to 150 spectra simultaneously over a 40 square-degree field of view with automatic positioning of the fiber optics using the magnetic button system. FLAIR has now been replaced by the 6dF multi-object spectroscope. This still uses 150 fibers positioned using magnetic buttons, but the turnaround time has been shortened so that ten times the output from FLAIR is possible. This system is currently engaged in the radial-velocity experiment (RAVE) that aims to have measured the radial velocities of a million stars by 2010.

Once the light is traveling through the fiber optic cables, it can be led anywhere. There is thus no requirement for the spectroscope to be mounted on the telescope, although this remains the case in a few systems. More frequently, the fiber optics take the light to a fixed spectroscope that can be several tens of meters away. This has the advantage that the spectroscope is not subject to changing gravitational loads and so does not flex, and it can be in a temperature-controlled room. For example, Subaru's Echidna feeds a spectroscope 50 m away from the telescope. Fiber optic links to fixed spectroscopes can also be used for single-object instruments. ESO's high-accuracy radial-velocity planet searcher (HARPS) spectroscope on the 3.6 m telescope is intended for extrasolar planet finding and so needs to determine stellar velocities to within ± 1 m s^{-1}. Thus, a very stable instrument is required and HARPS is not only mounted 38 m away from the telescope, but is also enclosed in a temperature-controlled vacuum chamber.

4.2.6 Techniques of Spectroscopy

There are several problems and techniques that are peculiar to astronomical spectroscopy and that are essential knowledge for the intending astrophysicist.

One of the greatest problems, and one which has been mentioned several times already, is that the image of the star may be broadened by atmospheric turbulence until its size is several times the slit width. Only a small percentage of the already pitifully small amount of light from a star therefore enters the spectroscope. The design of the spectroscope, and in particular the use of a large focal length of the collimator in comparison to that of the imaging element, can enable wider slits to be used. Even then, the slit is generally still too small to accept the whole stellar image, and the size of the spectroscope may have to become very large if reasonable resolutions and dispersions are to be obtained. The alternative approach to the use of a large collimator focal length lies in the use of an image slicer and/or adaptive optics as discussed earlier.

Another problem in astronomical spectroscopy concerns the width of the spectrum. If the stellar image is held motionless at one point of the slit during the exposure, then the final spectrum may only be a few microns high. Not only is such a narrow spectrum very difficult to measure, but also individual spectral features are recorded by only a few pixels and so the noise level is high. Both of these difficulties can be overcome, though at the expense of longer exposure times, by widening the spectrum artificially. There are several ways of doing this, such as introducing a cylindrical lens to make the images astigmatic, or by trailing the telescope during the exposure so that the image moves along the slit. For single-object spectroscopes, the spectrum can be widened by using a rocker block. This is just a thick piece of glass (etc.) with polished plane-parallel sides through which the light passes. The block is oscillated about an axis parallel to the spectrum and the displacement of the light beam (Figure 4.32) moves the beam up and down, so broadening the spectrum.

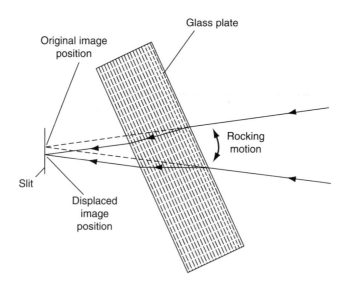

FIGURE 4.32 Image displacement by a plane-parallel glass plate.

In multi-object spectroscopes using fiber-optic links, the fiber size is generally sufficient to provide a useable spectrum, while clearly the spectra in integral field spectroscopes cannot be widened without destroying the imaging information. Astronomical spectra are typically widened to about 0.1–1 mm.

Spectrographs operating in the NIR region (1–5 μm) are of conventional designs, but need to be cooled to reduce the background noise level. Typically, instruments such as the United Kingdom infrared telescope imager spectrometer (UIST) and ESO's infrared spectrometer and array camera (ISAAC) and cryogenic infrared echelle spectrograph (CRIRES) that operate from 1 to 5 μm have their main components cooled to 70 K, while the detectors are held at 25 K to 4 K or less and the whole instrument is enclosed in a vacuum chamber. The long-slit intermediate resolution infrared spectrograph (LIRIS) on the WHT observes over the 0.9–2.4 μm region with its detectors cooled to 65 K and the rest of the instrument cooled by liquid nitrogen. The multi-object spectrometer for infrared exploration (MOSFIRE) that is due to be commissioned for the Keck I telescope in 2009 will use a mask of 46 slits to cover objects over a $6.1' \times 6.1'$ area in the NIR region. The precision radial-velocity spectrometer (PRVS) planned for the Gemini telescopes in 2010 will provide radial velocities to ± 3 m s^{-1} from 1 to 1.8 μm spectra. This will enable small (0.05–0.5 solar mass) stars and brown dwarfs, which are far more numerous than larger stars, to be investigated for the presence of extrasolar planets. At MIR wavelengths, the entire spectroscope may need to be cooled to less than 4 K. The VLT MIR imager and spectrometer (VISIR) that started operations in 2004, for example, has most of its structure and optics cooled to 33 K, the parts near the detectors to 15 K, and the detectors to 7 K.

UV spectroscopy is not possible from ground-based telescopes (see also Section 1.3), except in the most limited sense from the eye's cutoff wavelength of about 380 nm (persons having had cataract operations may be able to see further in to the UV than this, since the eye's lens absorbs short wave radiation strongly) to the point where the earth's atmosphere becomes opaque (300–340 nm depending upon altitude, state of the ozone layer, and other atmospheric conditions)—that is, the UVA and UVB regions as they are popularly known. All except the nearest UV spectroscopy therefore needs spectroscopes flown on board balloons, rockets, or spacecraft to lift them above most or all of the earth's atmosphere. Two prime examples of UV spectroscopy spacecraft—both now decommissioned—were *International Ultraviolet Explorer* (*IUE*) and *Far Ultraviolet Spectroscopic Explorer* (*FUSE*). *IUE* operated from 1978 to 1996 observing spectra from 100 to 300 nm, while *FUSE* was launched in 1999 and operated until 2007 observing the 90–120 nm spectral region. For the future, it is planned to install the *Cosmic Origins Spectrograph* (*COS*) UV spectroscope, whose design is based upon that for *FUSE*, on the *Hubble Space Telescope* (*HST*) during the fourth servicing mission, now expected in 2008. *COS* will use windowless micro-channel plates (MCPs) as its detectors and will cover the 115–320 nm region at a spectral resolution of 20,000.

Atmospheric dispersion is another difficulty that needs to be considered. Refraction in the earth's atmosphere changes the observed position of a star from its true position (Equation 5.13). But the refractive index varies with wavelength. For example, at standard temperature and pressure, we have Cauchy's formula (cf. the Hartmann formula, Equation 4.42) for the refractive index of the atmosphere:

$$\mu = 1.000287566 + \frac{1.3412 \times 10^{-18}}{\lambda^2} + \frac{3.777 \times 10^{-32}}{\lambda^4} \qquad (4.147)$$

where the wavelength is in nanometers so that the angle of refraction changes from one wavelength to another and the star's image is drawn out into a very short vertical spectrum. In a normal basic telescope system (i.e., without the extra mirrors required, for example, by the Coudé system), the long wavelength end of the spectrum will be uppermost. To avoid spurious results, particularly when undertaking spectrophotometry, the atmospheric dispersion of the image must be arranged to lie along the slit, otherwise certain parts of the spectrum may be preferentially selected from the image. Alternatively, an atmospheric dispersion corrector may be used. This is a low but variable dispersion direct-vision spectroscope that is placed before the entrance slit of the main spectroscope and whose dispersion is equal and opposite to that of the atmosphere. The problem is usually only significant at large zenith distances, so that it is normal practice to limit spectroscopic observations to zenith angles of less than 45°. The increasing atmospheric absorption, and the tendency for telescope tracking to deteriorate at large zenith angles, also contributes to the wisdom of this practice.

In many cases, it will be necessary to try and remove the degradation introduced into the observed spectrum because the spectroscope is not perfect. Many techniques and their ramifications for this process of deconvolution are discussed in Section 2.1.1.

To determine radial velocities, it is necessary to be able to measure the actual wavelengths of the lines in the spectrum, and to compare these with their laboratory wavelengths. The difference, $\Delta\lambda$, then provides the radial velocity via the Doppler shift formula

$$v = \frac{\Delta\lambda}{\lambda}c = \frac{\lambda_{\text{Observed}} - \lambda_{\text{Laboratory}}}{\lambda_{\text{Laboratory}}}c \qquad (4.148)$$

where c is the velocity of light. The observed wavelengths of spectrum lines are most usually determined by comparison with the positions of emission lines in an artificial spectrum. This comparison spectrum is normally that of an iron or copper arc or comes from a low-pressure gas-emission lamp, such as sodium or neon or combinations such as thorium and argon. The light from the comparison source is fed into the spectroscope and appears as one or two spectra on one or both sides of the main spectrum (Figure 4.33). The emission lines in the comparison spectra are at their rest wavelengths and are known very precisely. The observed wavelengths of the stellar (or other object's) spectrum lines are found by comparison with the positions of the artificial emission lines.

However, the emission lines produced by real atoms have drawbacks when used to produce comparison spectra. This problem is becoming particularly apparent with the requirement to be able to measure radial velocities to a precision of ± 0.1 m s^{-1} or even ± 0.01 m s^{-1} necessary for the detection of terrestrial-sized extrasolar planets and the need for this

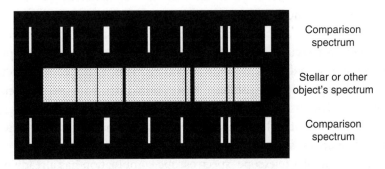

FIGURE 4.33 Wavelength comparison spectrum.

precision to be stable over several years. Thus, the lines from real atoms are not distributed uniformly and there may be long gaps between usable lines, the intensities of the lines vary widely, and also many lines are blends whose exact median wavelength will depend upon the relative strengths of the two or more individual lines contributing to the blend—and these can vary with the physical conditions within the emission lamp, etc. Finally, the wavelengths of the lines are not known a priori, but have to be measured individually in the laboratory. At the time of writing therefore, the production of a comb of close, uniformly intense, and regularly spaced emission features using femtosecond laser is being investigated. The laser-frequency comb is based upon the repetitive emissions from a mode-locked laser. Stability is ensured by synchronizing the repetitions with an atomic clock. The Fourier transform of the output of such a laser is the frequency comb and its emission features are spaced at frequencies of $1/T$ and their spectral widths are $1/D$, where T is the laser repetition interval and D is the laser pulse duration. Such a system would be as close to an ideal comparison as we are likely to get, but some practical difficulties remain to be solved before it can be used in practice.

For rapid determinations of radial velocity, the cross-correlation spectroscope, originally devised by Roger Griffin, can be used, although this device will not now be much encountered. This instrument places a mask over the image of the spectrum, and reimages the radiation passing through the mask onto a point source detector. The mask is a negative version (either a photograph or artificially generated) of the spectrum. The mask is moved across the spectrum and when the two coincide exactly, there is a very sharp drop in the output from the detector. The position of the mask at the correlation point can then be used to determine the radial velocity with in the best cases, a precision of a few meters per second. More conventional spectroscopes can provide velocities to a precision of a few meters per second for bright stars through the use of an absorption cell containing hydrogen fluoride or iodine. This is placed in the light path, and superimposes artificial absorption lines with very precisely known wavelengths onto the stellar spectrum. With this system, there is no error because of differing light paths as there may be with a conventional comparison spectrum.

The guiding of a telescope–spectroscope combination has already been mentioned. For bright objects, the reflected image from slit jaws polished to a mirror finish can be viewed through an optical inspection system.

Guiding becomes more difficult for faint objects or when an image dissector is in use. Normally, an accurate offset guiding system on some nearby brighter star must be used. Many of the 8–10 m class telescopes and some smaller instruments have drive and tracking systems sufficiently accurate that little guiding is needed. The short exposures often possible with such large instruments also helps to reduce the need for guiding.

4.2.7 Future Developments

The major foreseeable developments in spectroscopy seem likely to lie in the direction of improving the efficiency of existing systems rather than in any radically new systems or methods. The lack of efficiency of a typical spectroscope arises mainly from the loss of light within the overall system. To improve on this requires gratings of greater efficiency, reduced scattering and surface reflection, etc. from the optical components, and so on. Since these factors are already quite good, and it is mostly the total number of components in the telescope–spectroscope combination that reduces the efficiency, improvements are thus likely to be slow and gradual.

Adaptive optics is already used on most VLTs to reduce the seeing disk size of stellar images, and so allow more compact spectroscopes to be used. This approach is likely to spread to smaller instruments in the near future. Direct energy detectors such as STJs are likely to be used more extensively, although at present they have relatively low spectral resolutions, and require extremely low operating temperatures. The use of integral field and multi-object spectroscopes is likely to become more common, with wider fields of view and more objects being studied for individual instruments. The extension of high-resolution spectroscopy to longer infrared wavelengths is likely to be developed, even though this may involve cooling the fiber optic connections and large parts of the telescope as well as most of the spectroscope. Plus, of course, the continuing increase in the power and speed of computers will make real-time processing of the data from spectroscopes much more commonplace.

Exercise 4.4

Design a spectroscope (i.e., calculate its parameters) for use at a Coudé focus where the focal ratio is $f25$. A resolution of 10^{-3} nm is required over the spectral range 500–750 nm. A grating with 500 lines mm^{-1} is available, and the final recording of the spectrum is to be by photography.

Exercise 4.5

Calculate the limiting B magnitude for the system designed in Exercise 4.4 when it is used on a 4 m telescope. The final spectrum is widened to 0.2 mm, and the longest practicable exposure is 8 h.

Exercise 4.6

Calculate the apex angle of the dense flint prism required to form a direct-vision spectroscope in combination with two 40° crown glass prisms, if the undeviated wavelength is to be 550 nm. (See Section 1.1 for some of the data.)

CHAPTER 5

Other Techniques

5.1   ASTROMETRY

5.1.1   Introduction

Astrometry is probably the most ancient branch of astronomy, dating back to at least several centuries BC and possibly to a couple of millennia BC. Indeed until the late eighteenth century, astrometry was the whole of astronomy. Yet although it is such an ancient sector of astronomy, it is still alive and well today and employing the most modern techniques, plus some that William Herschel would recognize. Astrometry is the science of measuring the positions in the sky of galaxies, stars, planets, comets, asteroids, and recently, spacecraft. From these positional measurements come determinations of distance via parallax, motions in space via proper motion, orbits and hence sizes and masses within binary systems, and a reference framework that is used by the whole of the rest of astronomy and astrophysics as well as by space scientists to direct and navigate their spacecraft. Astrometry also leads to the production of catalogs of the positions, and sometimes the natures, of objects that are then used for other astronomical purposes.

From the invention of the telescope until the 1970s, absolute positional accuracies of about $0.1''$ ($= 100$ mas)* were the best that astrometry could deliver. That has now improved to better than 1 mas, and space missions

---

* The units of milliarcseconds (mas) and microarcseconds (μas) are widely used in astrometry and are generally more convenient. $1'' = 1000$ mas and 1 mas $= 1000$ μas. To put these units in perspective, a resolution of 1 μas would mean that the headlines in a newspaper held by an astronaut walking on the moon, could be read from the earth.

planned for the next couple of decades should improve that by a factor of ×100 at least. Positional accuracies of a few microarcseconds potentially allow proper motions of galaxies to be determined, though the diffuse nature of galaxy images may render this difficult. However, such accuracies will enable the direct measurement of stellar distances throughout the whole of the Milky Way galaxy, and out to the Magellanic Clouds and the Andromeda galaxy (M31).

Astrometry may be absolute (sometimes called fundamental) when the position of a star is determined without knowing the positions of other stars, or relative when the star's position is found with respect to the positions of its neighbors. Relative astrometry may be used to give the absolute positions of objects in the sky, provided that some of the reference stars have their absolute positions known. It may also be used for determinations of parallax, proper motion, binary star orbital motion, without needing to convert to absolute positions. An important modern application of relative astrometry is to enable the optical fibers of multi-object spectroscopes (Section 4.2.5) to be positioned correctly in the focal plane of the telescope to intercept the light from the objects of interest. The absolute reference frame is called the ICRS (International Celestial Reference System) and is now defined using 212 extragalactic compact radio sources (known as the defining sources) with their positions determined by radio interferometry (see Section 5.1.7) to an accuracy of about ±0.5 mas. Until 1998, the ICRS was based upon optical astrometric measurements, and when space-based interferometric systems (see Section 5.1.8) produce their results, the definition may well revert to being based upon optical measurements. The practical realization of the ICRS for optical work is to be found in the *Hipparcos* catalog (see Section 5.1.8).

5.1.2 Background

5.1.2.1 Coordinate Systems

The measurement of a star's position in the sky must be with respect to some coordinate system. There are several such systems in use by astronomers, but the commonest is that of right ascension and declination. This system, along with most of the others, is based upon the concept of the celestial sphere, that is, a hypothetical sphere, centered upon the earth and enclosing all objects observed by astronomers. The space position of an object is related to a position on the celestial sphere by a radial projection from the center of the earth. Henceforth in this section, we talk about the position of an object as its position on the celestial sphere, and we ignore

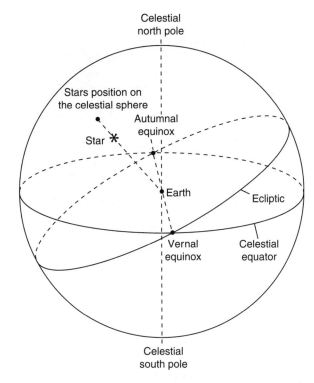

FIGURE 5.1 Celestial sphere.

the differing radial distances that may be involved. We also extend the polar axis and the equatorial and orbital planes of the earth until these too meet the celestial sphere (Figure 5.1). These intersections are called the celestial north pole, the celestial equator, etc. Usually, there is no ambiguity if the celestial qualification is omitted, so that they are normally referred to as the north pole, the equator, etc. The intersections (or nodes) of the ecliptic and the equator are the vernal and autumnal equinoxes. The former is additionally known as the first point of Aries from its position in the sky some 2000 years ago. The ecliptic is also the apparent path of the Sun across the sky during a year, and the vernal equinox is defined as the node at which the Sun passes from the southern to the northern hemisphere. This passage occurs within a day of March 21 each year. The position of a star or other object is thus given with respect to these reference points and planes.

The declination of an object, δ, is its angular distance north or south of the equator. The right ascension, α, is its angular distance around from the

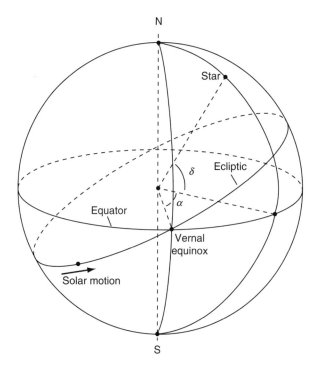

FIGURE 5.2 Right ascension and declination.

meridian (or great circle) that passes through the vernal equinox and the poles, measured in the same direction as the solar motion (Figure 5.2). By convention, declination is measured from $-90°$ to $+90°$, in units of degrees, minutes, and seconds of arc, and is positive to the north of the equator and negative to the south. Right ascension is measured from $0°$ to $360°$, in units of hours, minutes, and seconds of time where

$$1 \text{ h} = 15° \tag{5.1}$$

$$1 \text{ min} = 15' \tag{5.2}$$

$$1 \text{ s} = 15'' \tag{5.3}$$

The earth's axis moves in space with a period of about 25,750 years, a phenomenon known as precession. Hence, the celestial equator and poles also move. The positions of the stars therefore slowly change with time. Catalogs of stars thus customarily give the date, or epoch, for which the stellar positions that they list are valid. To obtain the position at some other date, the effects of precession must be added to the catalog positions

$$\delta_T = \delta_E + (\theta \sin \varepsilon \cos \alpha_E)T \qquad (5.4)$$

$$\alpha_T = \alpha_E + [\theta(\cos \varepsilon + \sin \varepsilon \sin \alpha_E \tan \delta_E)]T \qquad (5.5)$$

where

α_T and δ_T are the right ascension and declination of the object at an interval T years after the epoch E

α_E and δ_E are the coordinates at the epoch

θ is the precession constant

$$\theta = 50.40'' \text{ year}^{-1} \qquad (5.6)$$

ε is the angle between the equator and the ecliptic, more commonly known as the obliquity of the ecliptic

$$\varepsilon = 23°27'8'' \qquad (5.7)$$

Frequently used epochs are the beginnings of the years 1900, 1950, 2000, etc. with 1975 and 2025 also being encountered. Other effects upon the position, such as nutation, proper motion, etc., may also need to be taken into account in determining an up-to-date position for an object.

Two alternative coordinate systems that are in use are first celestial latitude (β) and longitude (λ), which are, respectively, the angular distances up or down from the ecliptic and around the ecliptic from the vernal equinox, and second galactic latitude (b) and longitude (l), which are, respectively, the angular distances above or below the galactic plane, and around the plane of the galaxy measured from the direction to the center of the galaxy.*

---

* The system is based upon a position for the galactic center of RA_{2000} 17 h 45 min 36 s, Dec_{2000} 28°56'18''. It is now known that this is in error by about 4', however the incorrect position continues to be used. Before about 1960, the intersection of the equator and the galactic plane was used as the zero point and this is about 30° away from the galactic center. Coordinates based upon this old system are sometimes indicated by a superscript "I," and those using the current system by a superscript "II," that is, b^I or b^{II} and l^I or l^{II}.

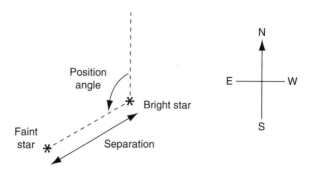

FIGURE 5.3 Position angle and separation of a visual double star (as seen on the sky directly).

5.1.2.2 Position Angle and Separation

The separation of a visual double star is just the angular distance between its components. The position angle is the angle from the north, measured in the sense, north → east → south → west → north, from 0° to 360° (Figure 5.3) of the fainter star with respect to the brighter star. The separation and position angle are related to the coordinates of the star by

$$\text{separation} = \left\{ \left[(\alpha_F - \alpha_B) \cos \delta_B \right]^2 + (\delta_F - \delta_B)^2 \right\}^{1/2} \tag{5.8}$$

$$\text{position angle} = \tan^{-1} \left[\frac{\delta_F - \delta_B}{(\alpha_F - \alpha_B) \cos \delta} \right] \tag{5.9}$$

where
α_B and δ_B are the right ascension and declination of the brighter star
α_F and δ_F are the right ascension and declination of the fainter star

with the right ascensions of both stars being converted to normal angular measures.

5.1.3 Transit Telescopes

Transit telescopes (Figure 5.4), which are also sometimes called meridian circles, historically provided the basic absolute measurements of stellar positions for the calibration of other astrometric methods, though they have now largely been superseded by the results from *Hipparcos* (see Section 5.1.8) and radio interferometers. They developed from the mural

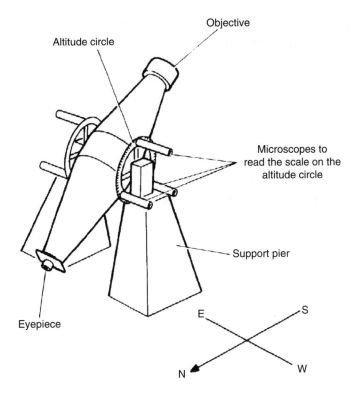

FIGURE 5.4 Transit telescope.

quadrants that were used before the invention of the telescope. These were just sighting bars, pivoted at one end, restricted to moving in just one plane and with a divided arc to measure the altitude at the other end; nonetheless, they provided surprisingly good measurements when in capable hands. The one used by Tycho Brahe (1546–1601), for example, was mounted on a north–south aligned wall and had a 2 m radius. He was able to measure stellar positions using it to a precision of about ±30″.

The principle of the transit telescope is simple, but great care was required in practice if it were to produce reliable results. The instrument was almost always a refractor, because of the greater stability of the optics, and it was pivoted only about a horizontal east–west axis. The telescope was thereby constrained to look just at points on the prime meridian. A star was observed as it crossed the center of the field of view (i.e., when it transited the prime meridian), and the precise time of the passage was noted. The declination of the star is obtainable from the altitude of the telescope, while the right ascension is given by the local sidereal time at the observatory at the instant of transit,

$$\delta = A + \phi - 90° \tag{5.10}$$

$$\alpha = \text{LST} \tag{5.11}$$

where

A is the altitude of the telescope
ϕ is the latitude of the observatory
LST is the local sidereal time at the instant of transit

To achieve an accuracy of a tenth of a second of arc in the absolute position of the star, a very large number of precautions and corrections are needed. A few of the more important ones are temperature control; use of multiple or driven cross wires in the micrometer eyepiece; reversal of the telescope on its bearings; corrections for flexure, nonparallel objective, and eyepiece focal planes; rotational axis not precisely horizontal; rotational axis not precisely east–west; errors in the setting circles; incorrect position of the micrometer cross wire; personal setting errors; etc. Then of course all the normal corrections for refraction, aberration, etc. have also to be added. Modern versions of the instrument, such as the Carlsberg meridian telescope (CMT), use charge-coupled devices (CCDs) for detecting transits. The CMT is a refractor with a 17.8 cm diameter objective and a 2.66 m focal length sited on La Palma in the Canary Islands. It is operated remotely via the Internet, and measures between 100,000 and 200,000 star positions in a single night. The CCD detector on the CMT uses charge transfer to move the accumulating electrons from pixel to pixel at the same rate as the images drift over the detector (also known as time-delayed integration [TDI], drift scanning and image tracking—see Section 1.1 [liquid mirror telescopes] and Section 2.7). This enables the instrument to detect stars down to 17^m, and also to track several stars simultaneously. The main program of the CMT is to link the positions of the bright stars measured by *Hipparcos* with those of fainter stars.

5.1.4 Photographic Zenith Tube and the Impersonal Astrolabe

Two more modern instruments that until recently have performed the same function as the transit telescope are the photographic zenith tube (PZT) and the astrolabe. Both of these use a bath of mercury to determine the zenith position very precisely. The PZT obtained photographs of stars that transit close to the zenith and provided an accurate determination of time and the latitude of an observatory, but it was restricted in its observations of stars to those that culminated within a few tens of minutes of arc of the zenith. The PZT has been superseded by very long baseline radio interferometers (Sections 2.5 and 5.1.7) and measurements from spacecraft that are able to provide positional measurements for far more stars and with significantly higher accuracies.

The astrolabe observes near an altitude of 60° and so can give precise positions for a wider range of objects than the PZT. In its most developed form (Figure 5.5), it is known as the Danjon or impersonal astrolabe since its measurements are independent of focusing errors. Two separate beams of light are fed into the objective by the 60° prism, one of the beams having been reflected from a bath of mercury. A Wollaston prism (Section 5.2.3) is used to produce two focused beams parallel to the optic axis with the other two nonparallel emergent beams being blocked off. Two images of the star are then visible in the eyepiece. The Wollaston prism is moved along the optical axis to compensate for the sidereal motion of the star. The measurement of the star's position is accomplished by moving the 60° prism along the optical axis until the two images merge into one. The position of the prism at a given time for this coincidence to occur can then be converted into the position of the star in the sky. The astrolabe has also largely fallen out of use, but one or two examples are still in operation such as the Mark III astrolabe at the Beijing National Observatory.

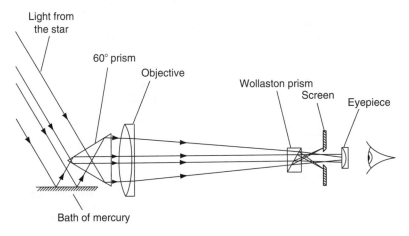

FIGURE 5.5 Optical paths in the impersonal astrolabe.

5.1.5 Micrometers

While the addition of one or more cross wires to an eyepiece can convert it into a micrometer when it is used in conjunction with a clock, it is the bifilar micrometer (sometimes called the filar micrometer) that is more normally brought to mind by the term micrometer. Although for most purposes the bifilar micrometer has been superseded by CCDs and photography, it is even now a good method for measuring close visual double stars and is thus still sometimes used by amateur double star observers. It also finds application as a means of off-setting from a bright object to a faint or invisible one. Designs for the device vary in their details, but the most usual model comprises a fixed cross wire at the focus of an eyepiece, with a third thread parallel to one of the fixed ones. The third thread is displaced slightly from the others, although still within the eyepiece's field of sharp focus, and it may be moved perpendicularly to its length by a precision screw thread. The screw has a calibrated scale so that the position of the third thread can be determined to within 1 μm. The whole assembly is mounted on a rotating turret whose angular position may also be measured (Figure 5.6).

Another type of micrometer that has the advantage of being far less affected by scintillation is the double-image micrometer. Two images of each star are produced, and the device is adjusted so that one image of one

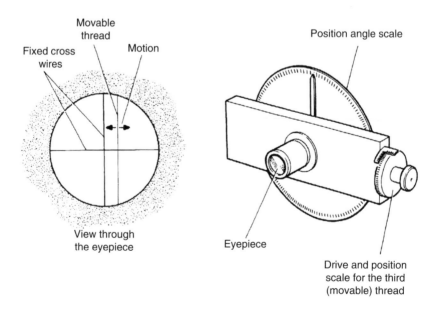

FIGURE 5.6 Bifilar micrometer.

star is superimposed upon one image of the other.* The settings of the device can be calibrated, and then the separation and position angle follow directly. Commonly used methods of splitting the images are an objective cut into two along a diameter whose two halves may be tilted with respect to each other, and a Nicol prism (Section 5.2.2.2). The advantage of systems of this type derives from the instantaneous determination of both separation and position angle, compared with the delay of several seconds or even minutes for obtaining the various readings from a bifilar micrometer, and from the reduced effect of scintillation since both images are likely to be affected by it in a similar manner. Very few examples of double-image micrometers now remain in use, but an amateur program using one to track binary star systems is still in progress at the Observatori Astronomic del Garraf in Spain.

5.1.6 Astrographs and Other Telescopes

Almost any telescope can be used to obtain images from which the positions of objects in the sky can be obtained, but some telescopes are optimized for the work. The astrograph has a wide (a degree or more) field of view so that many faint stars' positions can be determined from those of a few brighter reference stars. Astrographs designed around the beginning of the twentieth century were refractors with highly corrected multiele-ment objectives. Their apertures were in the region of 10–25 cm, and focal lengths between 2 and 4 m. Measurements using these instruments formed the basis of the ICRS. More recently, Schmidt cameras and conventional reflectors with correcting lenses have also been used.

Smaller fields of view are adequate for parallax determination. Long focus refractors are favored for this work because of their greater stability when compared with reflectors, because of their closed tubes, because their optics may be left undisturbed for many years, and because flexure is generally less deleterious in them. However, a few specially designed reflectors, especially the U.S. Naval Observatory's (USNO's) 1.55 m Strand reflector, are used successfully for astrometry. Against the advantages of the refractor their disadvantages must be set. These are primarily their residual chromatic aberration and coma, temperature effects, and slow changes in the optics such as the rotation of the elements of the objective in their mounting cells. Refractors in use for astrometry usually have focal

---

* This has close parallels with the way in which the navigator's sextant operates, although then it is usually the sun or star and the horizon that are superimposed.

ratios of f15–f20, and range in diameter from about 0.3 to 1.0 m. Their objectives are achromatic (Section 1.1.17.1), and filters are usually used so that the bandwidth is limited to a few tens of nanometers either side of their optimally corrected wavelength. Coma may need to be limited by stopping down the aperture, by introducing correcting lenses near the focal plane, or by compensating the measurements later, in the analysis stages, if the magnitudes and spectral types of the stars are well known. Observations are usually obtained within a few degrees of the meridian to minimize the effects of atmospheric refraction and dispersion and to reduce changes in flexure. The position of the focus may change with temperature and needs to be checked frequently throughout the night. Other corrections for such things as plate tilt collimation errors, astigmatism, distortion, emulsion movement, etc. need to be known and their effects allowed for when reducing the raw data.

5.1.7 Interferometers

As we have seen (Section 2.5), interferometers have the capability of providing much higher resolutions than single dish-type instruments since their baselines can be extended easily and cheaply. High resolution can be translated into high-positional accuracy, provided that the mechanical construction of the interferometer is adequate. Interferometers also provide absolute positions for objects. The disadvantage of interferometers for astrometry is that only one object can be observed at a time, and each observation requires several hours. Both radio and optical interferometers are used for this purpose.

Optical interferometers such as Cambridge optical aperture synthesis telescope (COAST) (Section 2.5.5) and the USNO's navy prototype optical interferometer (NPOI) are essentially Michelson stellar interferometers (Section 2.5.2) incorporating a delay line into one or both light beams. The delay line is simply an arrangement of several mirrors that reflects the light around an additional route. By moving one or more of the mirrors in the delay line, the delay can be varied. The observational procedure is to point both telescopes (or the flat mirrors feeding the telescopes) at the same star and adjust the delay until the fringe visibility (Section 2.5.2) is maximum. The physical length of the delay line then gives the path difference between the light beams to the two telescopes. Combining this with the baseline length gives the star's altitude. Several such observations over a night will then enable the star's position in the sky to be calculated. Currently, absolute positions may be obtained in this manner to accuracies of a few milliarcseconds, with the relative positions of close stars being determined to perhaps an order of magnitude better precision. Relative measurements for widely separated objects may be obtained from the variation of the delay over several hours. The delay varies sinusoidally over a 24 h period, and the time interval between the maxima for two stars equals the difference in their right ascensions. As with conventional interferometers, the use of more than two telescopes enables several delays to be measured simultaneously, speeding up the process and enabling instrumental effects to be corrected. The optical interferometer on Mount Wilson recently completed for CHARA (Center for High Angular Resolution Astronomy), for example, uses six 1 m telescopes in a two-dimensional (2D) array and has a maximum baseline of 350 m.

Radio interferometers are operated in a similar manner for astrometry, and very long baseline interferometry (VLBI) systems (Section 2.5.5) provide the highest accuracies at the time of writing. For some of the radio sources, this accuracy may reach 100 μas. The measurements by VLBI of some 212 compact radio sources have thus been used to define the ICRS since January 1, 1998. Recently, the very long baseline array (VLBA) has been able to measure the parallax of a star within the Orion Nebula to an uncertainty of ±0.15 mas, giving the distance to the nebula as 390 ± 23 pc, or 90 pc closer to us than had been previously thought.

5.1.8 Space-Based Systems

By operating instruments in space and so removing the effects of the atmosphere and gravitational loading, astrometry is expected to reach accuracies of a few microarcseconds in the next decade or so. Space astrometry missions, actual and planned, divide into two types: scanning (or survey) and point and stare. Scanning means that exposures are short, and so only the brighter stars can be observed and with relatively low accuracy, the accuracy also depends upon the star's brightness. However, large numbers of star positions may be measured quickly. Point and stare missions, as the name suggests, look at individual stars (or a few stars very close together) for long periods of time. Such missions can observe relatively few star positions, but do so with high accuracy and to very faint magnitudes. Hipparcos and Gaia (see below) are examples of scanning systems, while the Hubble Space Telescope (HST) and Space Interferometry Mission (SIM) are point and stare.

The first astrometric spacecraft, called Hipparcos (high-precision parallax collecting satellite), was launched by ESA in 1990. Unfortunately, the apogee boost motor failed and instead of going into a geostationary orbit, it eventually ended up in a 10 1/2 h orbit with a perigee of 6800 km and apogee of 42,000 km. However, by using more ground stations and extending the mission duration to 3 years, the original aim of measuring the positions of 120,000 stars to an accuracy of 2 mas was achieved. The *Hipparcos* telescope was fed by two flat mirrors at an angle of 29° to each other enabling two areas of the sky 58° apart to be observed simultaneously. It determined the relative phase difference between two stars viewed via the two mirrors, to obtain their angular separation. A precision grid at the focal plane modulated the light from the stars, and the modulated light was then detected from behind the grid to determine the transit times (see also the multichannel astrometric photometer [MAP] instrument in Section 5.1.9). The satellite rotated once every 2 h and measured all the stars down to 9.0^m about 120 times, observing each star for about 4 s on each occasion. Each measurement was from a slightly different angle and the final positional catalog was obtained by the processing of some 10^{12} bits of such information. The data from *Hipparcos* are to be found in the *Hipparcos*, *Tycho*, and *Tycho-2** catalogs. These may be accessed at http://archive.ast.cam.ac.uk/hipp/. The *Hipparcos* catalog contains the positions, distances, proper motions, and magnitudes of some 118,000 stars to an accuracy of ±0.7 mas. The *Tycho* catalog contains the positions and magnitudes for about one million stars and has positional accuracies ranging from ±7 to ±25 mas. The *Tycho-2* catalog utilizes ground-based data as well as that from *Hipparcos* and has positional accuracies ranging from ±10 to ±70 mas, depending upon the star's brightness. It contains positions, magnitudes, and proper motions for some two-and-a-half million stars.

The *HST* can use the WFPC2 (second wide-field and planetary camera) and FGS (fine guidance sensors) instruments for astrometry. The planetary camera provides measurements to an accuracy of ±1 mas for objects down to magnitude 26^m, but only has a 34″ field of view. Its images are also under-sampled so that dithering (Section 1.1.8, CCDs) has to be used to reach this level of accuracy. The telescope has three FGS but only needs two for guidance purposes. The third is therefore available for astrometry. The FGS are interferometers based upon Köster prisms (Section 2.5.2), and provide positional accuracies of ±5 mas.

---

* A separate instrument, the star mapper, observed many more stars, but for only 0.25 s, and its data produced the *Tycho* and *Tycho-2* catalogs.

Several other spacecraft missions have been proposed for astrometry, but have not received continuing support. Thus, NASA withdrew its sponsorship of *FAME* (*Full Sky Astrometric Mapping Explorer*) in 2004, and has postponed the launch of *SIM* indefinitely, although a launch in 2015 or 2016 is still perhaps a possibility. *FAME* would have measured the positions of 40 million stars to an accuracy of 50–500 μas using a similar approach to that used by *Hipparcos*, while *SIM* would have used a 10 m baseline Michelson optical interferometer to obtain positions to 4 μas of some 40,000 stars. ESA's *Gaia* mission is still on stream for a launch perhaps as early as 2011, but astrometry is now a subsidiary aim of the project. It will be positioned at the Sun–earth inner Lagrange point, some 1.5×10^6 km from the earth, to avoid eclipses and occultations and so provide a stable thermal environment. It will operate in a similar fashion to *Hipparcos* but using two separate telescopes set at a fixed angle and with elongated apertures 1.7×0.7 m. The long axis of the aperture in each case will be aligned along the scanning direction to provide the highest resolution. The readout will be by CCDs using TDI. The aim is to determine positions for a billion stars to ± 10 μas accuracy at visual magnitude 15^m. *Gaia* will also determine radial velocities for stars down to 17^m using a separate spectroscopic telescope. The objective for the *Gaia* mission is to establish a precise three-dimensional (3D) map of the galaxy and provide a massive database for other investigations of the Milky Way.

5.1.9 Detectors

CCD (Section 1.1.8) are now widely used as detectors for astrometry. They have the enormous advantage that the stellar or other images can be identified with individual pixels whose positions are automatically defined by the structure of the detector. The measurement of the relative positions of objects within the image is therefore vastly simplified. Software to fit a suitable point spread function (Section 2.1.1) to the image enables the position of its centroid to be defined to a fraction of the size of the individual pixels.

The disadvantage of electronic images is their small scale. Most CCD chips are only a few centimeters in size (though mosaics are larger) compared with tens of centimeters for astrographic photographic plates. They therefore cover only a small region of the sky and it can become difficult to find suitable reference stars for the measurements. However, the greater dynamic range of CCDs compared with photographic emulsion and the use of anti-blooming techniques (Section 1.1.8) enables the positions of many more stars to be usefully measured. With transit telescopes (see Section 5.1.3), TDI may be used to reach fainter magnitudes. TDI is also proposed for the *Gaia* spacecraft detection system (see Section 5.1.8). Otherwise the operation of a CCD for positional work is conventional (Section 1.1.8). In other respects, the processing and reduction of electronic images for astrometric purposes is the same as that for a photographic image.

Grid modulation is used by the MAP at the Allegheny Observatory. Light from up to 12 stars is fed into separate detectors by fiber optics as a grating with 4 lines mm^{-1} is passed across the field. Timing the disappearances of the stars behind the bars of the grating provides their relative positions to an accuracy of 3 mas. A similar technique was used by the *Hipparcos* spacecraft (see Section 5.1.8), although in this case it was the stars that moved across a fixed grid.

Photography is no longer used for astrometry. However, much material in the archives is in the form of photographs and this is still essential to long-term programs measuring visual binary star orbits or proper motion, etc. Thus, the reduction of data from photographs will continue to be needed for astrometry for some considerable time to come.

CCD imaging may usefully be used to observe double stars when the separation is greater than about 3″. Many exposures are made with a slight shift of the telescope between each exposure. Averaging the measurements can give the separation to better than a hundredth of a second of arc and the position angle to within a few minutes of arc. For double stars with large magnitude differences between their components, an objective grating can be used (see next section), or the shape of the aperture can be changed so that the fainter star lies between two diffraction spikes and has an improved noise level. The latter technique is a variation of the technique of apodization mentioned in Sections 1.1.18.2, 1.2.3, 4.1.4.2, and 5.3.5.

5.1.10 Measurement and Reduction

Transit telescopes and interferometers give absolute positions directly as discussed above. In most other cases, the position of an unknown star is obtained by comparison with reference stars that also appear on the image and whose positions are known. If absolute positions are to be determined, then the reference stars' positions must have been found via a transit instrument, interferometer or astrolabe, etc. For relative positional work, such as that involved in the determination of parallax and proper motion, any very distant star may be used as a comparison star.

It is advantageous to have the star of interest and its reference stars of similar brightnesses, although the much greater dynamic ranges of CCDs and other electronic detectors compared with that of photographic emulsions render this requirement of less importance than it was in the past. If still needed, the brightness of one or more stars may be altered through the use of variable density filters, by the use of a rotating chopper in front of the detector, or by the use of an objective grating (Section 4.2). The latter device is arranged so that pairs of images, which in fact are the first-order or higher order spectra of the star, appear on either side of it. By adjusting the spacing of the grating, the brightness of these secondary images for the brighter stars may be arranged to be comparable with the brightness of the primary images of the fainter stars. The position of the brighter star is then taken as the average of the two positions of its secondary images. The apparent magnitude, m, of such a secondary image resulting from the nth order spectrum is given by

$$m = m_0 + 5\log_{10}(N\eta \cosec \eta) \qquad (5.12)$$

where m_0 is the apparent magnitude of the star, η is given by

$$\eta = \pi Nd/D \qquad (5.13)$$

where
D is the separation of the grating bars
d is the width of the gap between the grating bars
N is the total number of slits across the objective

Once an image of a star field has been obtained, the positions of the stars' images must be measured to a high degree of accuracy and the measurements converted into right ascension and declination or separation and position angle. Although no longer used for new astrometric imaging, much archive material is in the form of photographic plates and even modern catalogs such as the Hubble *Guide Star Catalog* and the Palomar Digital Sky Survey (DPOSS) are based upon photographic data. On photographs the measuring is undertaken in a manner similar to that used to measure spectra, except that 2D positions are required. The measuring procedure is a tedious process and, as in the case of spectroscopic measurement, a number of automatic measuring machines have been developed over the last two or three decades. Since much of the photographic archive has now been scanned, many of these machines have now been decommissioned. The Royal Observatory Edinburgh's SuperCOSMOS (coordinates, sizes, magnitudes, orientations, and shapes) machine, for example, could analyze a 350 mm square Schmidt plate containing 2 GB of data in just 2 h. It was based upon a moving granite table onto which the plate was mounted and a 2048 pixel linear CCD giving 10 μm linear resolution and about 50 mas accuracy in stars' positions. The USNO's precision measuring machine (PMM) used two CCDs to scan a plate and processed the data in real time.

CCD give positions directly (see Section 5.1.9) once the physical structure of the device has been calibrated. Usually, the position will be found by fitting a suitable point spread function (Section 2.1.1) for the instrument used to obtain the overall image to the stellar images when they spread across several pixels. The stars' positions may then be determined to subpixel accuracy.

Howsoever the raw data may have been obtained, the process of converting the measurements into the required information is known as reduction. For accurate astrometry, it is a lengthy process. A number of corrections have already been mentioned, and in addition to those, there are the errors of the measuring machine itself to be corrected, and the distortion introduced by the image to be taken into account. The latter is caused by the projection of the curved sky onto the flat photographic plate or CCD (known as the tangential plane), or in the case of Schmidt cameras, its projection onto the curved focal plane of the camera, followed by the flattening of the plate after it is taken out of the plate holder. For the simpler case of the astrograph, the projection is shown in Figure 5.7, and the relationship between the coordinates on the flat image and in the curved sky is given by

$$\rho = \tan^{-1}\left(\frac{y}{F}\right) \tag{5.14}$$

$$\tau = \tan^{-1}\left(\frac{x}{F}\right) \tag{5.15}$$

where
x and y are the coordinates on the image
τ and ρ are the equivalent coordinates on the image of the celestial sphere
F is the focal length of the telescope

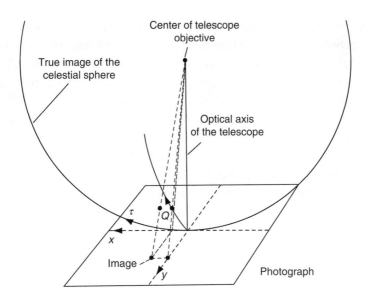

FIGURE 5.7 Projection onto a photographic plate or flat CCD.

One instrument that, while it is not strictly of itself a part of astrometry, is very frequently used to identify the objects that are to be measured is called the blink comparator. In this machine, the worker looks alternately at two aligned images of the same star field obtained sometime apart from each other, the frequency of the interchange being about 2 Hz. Stars, asteroids, comets, etc. that have moved in the interval between the images then call attention to themselves by appearing to jump backward and forward while the remainder of the stars are stationary. The device can also pick out variable stars for further study by photometry (see Chapter 3). Although these remain stationary, they appear to blink on and off (hence the name of the instrument). Early blink comparators were basically binocular microscopes that had a mechanical arrangement for swapping the view of two photographic plates. Nowadays, the images are usually viewed on a computer screen and software aligns the images and provides the alternating views. Software can also be used to identify the moving or changing objects directly. But this is still one area where the human eye–brain combination functions as efficiently as the computer.

5.1.11 Sky Surveys and Catalogs

The end result of most astrometry is a catalog of positions and other properties of (usually) a large number of objects in the sky. The *Hipparcos*,

Tycho, and *Tycho-2* (see Section 5.1.8) catalogs are just the latest examples of the process (see also Section 5.5.2). Other recent astrometric catalogs include the second issue of the USNO's *Twin Astrographic Catalog* (*TAC 2.0*) based upon photographic plates obtained with the twin astrograph and containing over 700,000 stellar positions to between ±50 and ±120 mas accuracy, the USNO's *A2.0* catalog containing 526 million entries, the USNO's *B1.0* catalog containing data on a billion stars with 200 mas positional accuracy, and the *UCAC* (*USNO CCD Astrograph Catalog*) that aims to measure the positions of some 40 million stars in the 10^m to 14^m range to ±20 mas accuracy and which is expected to be distributed in 2008. Older astrometric catalogs include the *Fundamental Katalog* series that culminated in FK5 in 1998 containing 1500 stars with positional accuracies of better than ±100 mas, the *Astrographic Catalog* (*AC*) series, the *Astronomische Gesellschaft Katalog* (*AGK*) series, plus arguably Argelander's, Flamsteed's, Brahe's and even the original Hipparchus' catalogs, since these gave state-of-the-art stellar positions for their day.

There have been many other sky surveys and catalogs and more are being produced all the time that are nonastrometric. That is to say their purpose is other than that of providing accurate positions, and the position if determined at all for the catalog is well below the current levels of astrometric accuracy. Indeed, many such catalogs just use the already-known astrometric positions. Nonastrometric catalogs are also produced for all regions of the spectrum, not just from optical and radio sources. There are tens of thousands of nonastrometric catalogs produced for almost as many different reasons, and ranging in content from a few tens to half-a-billion objects. Examples of such catalogs include the HST *Guide Star Catalog 2* (*GSC2*) containing 500 million objects, the Two-Micron All Sky Survey (2MASS) and Deep Near Infrared Survey (DENIS) of the southern sky in the infrared region, and the Sloan Digital Sky Survey (SDSS) with its million red shifts of galaxies and quasars.

Exercise 5.1

If the axis of a transit telescope is accurately horizontal, but is displaced north or south of the east–west line by E seconds of arc, then show that for stars on the equator

$$\Delta\alpha^2 + \Delta\delta^2 \approx E^2 \sin 2\phi$$

where

$\Delta\alpha$ and $\Delta\delta$ are the errors in the determinations of right ascension and declination in seconds of arc

ϕ is the latitude of the observatory

Exercise 5.2

Calculate the widths of the slits required in an objective grating for use on a 1 m telescope if the third-order image of Sirius A is to be of the same brightness as the primary image of Sirius B. The total number of slits in the grating is 99. The apparent magnitude of Sirius A is -1.5, while that of Sirius B is $+8.5$.

5.2 POLARIMETRY

5.2.1 Background

Although the discovery of polarized light from astronomical sources dates back to the beginning of the nineteenth century, when Dominique Arago detected its presence in moonlight, the extensive development of its study is a relatively recent phenomenon. This is largely due to the technical difficulties that are involved, and initially at least due to the lack of any expectation of polarized light from stars by astronomers. Many phenomena, however, can contribute to the polarization of radiation, and so, conversely, its observation can potentially provide information upon an equally wide range of basic causes.

5.2.1.1 Stokes' Parameters

Polarization of radiation is simply the nonrandom angular distribution of the electric vectors of the photons in a beam of radiation. Customarily, two types are distinguished: linear and circular polarizations. In the former, the electric vectors are all parallel and their direction is constant, while in the latter, the angle of the electric vector rotates with time at the frequency of the radiation. These are not really different types of phenomena, however, and all types of radiation may be considered to be different aspects of partially elliptically polarized radiation. This has two components, one of which is unpolarized, the other being elliptically polarized. Elliptically polarized light is similar to circularly polarized light in that the electric vector rotates at the frequency of the radiation, but in addition the magnitude varies at twice that frequency, so that plotted on a polar diagram the electric vector would trace out an ellipse

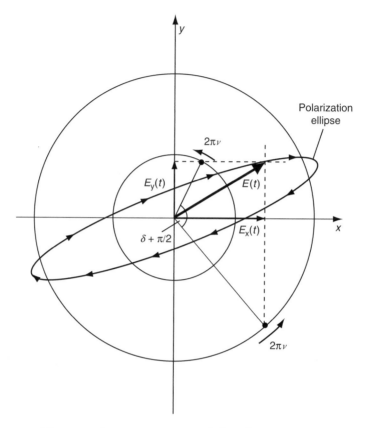

FIGURE 5.8 The x and y components of the elliptically polarized component of partially elliptically polarized light.

(e.g., Figure 5.8). The properties of partially elliptically polarized light are completely described by four parameters that are called the Stokes' parameters. These fix the intensity of the unpolarized light, the degree of ellipticity, the direction of the major axis of the ellipse, and the sense (left- or right-handed rotation) of the elliptically polarized light. If the radiation is imagined to be propagating along the z-axis of a 3D rectangular coordinate system, then the elliptically polarized component at a point along the z-axis may have its electric vector resolved into components along the x and y axes (Figure 5.8), these being given by

$$E_x(t) = e_1 \cos(2\pi\nu t) \tag{5.16}$$

$$E_y(t) = e_2 \cos(2\pi\nu t + \delta) \tag{5.17}$$

where

ν is the frequency of the radiation

δ is the phase difference between the x and y components

e_1 and e_2 are the amplitudes of the x and y components

It is then a tedious but straightforward matter to show that

$$a = \left[\frac{(e_1^2 + e_2^2)}{1 + \tan^2\left(\frac{1}{2} \sin^{-1}\left\{ \left[\frac{2e_1 e_2}{(e_1^2 + e_2^2)}\right] \sin \delta \right\}\right)} \right]^{1/2} \tag{5.18}$$

$$b = a \tan\left(\frac{1}{2} \sin^{-1}\left\{ \left[\frac{2e_1 e_2}{(e_1^2 + e_2^2)}\right] \sin \delta \right\}\right) \tag{5.19}$$

$$a^2 + b^2 = e_1^2 + e_2^2 \tag{5.20}$$

$$\psi = \frac{1}{2} \tan^{-1}\left\{ \left[\frac{2e_1 e_2}{(e_1^2 - e_2^2)}\right] \cos \delta \right\} \tag{5.21}$$

where

a and b are the semimajor and semiminor axes of the polarization ellipse

ψ is the angle between the x-axis and the major axis of the polarization ellipse

The Stokes' parameters are then defined by

$$Q = e_1^2 - e_2^2 = \frac{a^2 - b^2}{a^2 + b^2} \cos (2\psi) I_p \tag{5.22}$$

$$U = 2e_1 e_2 \cos \delta = \frac{a^2 - b^2}{a^2 + b^2} \sin (2\psi) I_p \tag{5.23}$$

$$V = 2e_1 e_2 \sin \delta = \frac{2ab}{a^2 + b^2} I_p \tag{5.24}$$

where I_p is the intensity of the polarized component of the light. From Equations 5.22 through 5.24, we have

$$I_p = (Q^2 + U^2 + V^2)^{1/2} \tag{5.25}$$

The fourth Stokes' parameter, I, is just the total intensity of the partially polarized light

$$I = I_u + I_p \qquad (5.26)$$

where I_u is the intensity of the unpolarized component of the radiation. (Note: The notation and definitions of the Stokes' parameters can vary. While that given here is probably the commonest usage, a check should always be carried out in individual cases to ensure that a different usage is not being employed.)

The degree of polarization, π, of the radiation is given by

$$\pi = \frac{(Q^2 + U^2 + V^2)^{1/2}}{I} = \frac{I_p}{I} \qquad (5.27)$$

while the degree of linear polarization, π_L, and the degree of ellipticity, π_e, are

$$\pi_L = \frac{(Q^2 + U^2)^{1/2}}{I} \qquad (5.28)$$

$$\pi_e = \frac{V}{I} \qquad (5.29)$$

When $V = 0$ (i.e., the phase difference, δ, is 0 or π radians), we have linearly polarized radiation. The degree of polarization is then equal to the degree of linear polarization, and is the quantity that is commonly determined experimentally

$$\pi = \pi_L = \frac{I_{max} - I_{min}}{I_{max} + I_{min}} \qquad (5.30)$$

where I_{max} and I_{min} are the maximum and minimum intensities that are observed through a polarizer as it is rotated. The value of π_e is positive for right-handed radiation, and negative for left-handed radiation.

When several incoherent beams of radiation are mixed, their various Stokes' parameters combine individually by simple addition. A given monochromatic partially elliptically polarized beam of radiation may therefore have been formed in many different ways, and these are indistinguishable by the intensity and polarization measurements alone. It is often customary therefore to regard partially elliptically polarized light as formed from two separate components, one of which is unpolarized and the other completely elliptically polarized. The Stokes' parameters of these components are then as follows:

	I	Q	U	V
Unpolarized component	I_u	0	0	0
Elliptically polarized component	I_p	Q	U	V

and the normalized Stokes' parameters for more specific mixtures are

	Stokes' Parameters			
Type of Radiation	I/I	Q/I	U/I	V/I
Right-hand circularly polarized (clockwise)	1	0	0	1
Left-hand circularly polarized (anticlockwise)	1	0	0	−1
Linearly polarized at an angle ψ to the x-axis	1	$\cos 2\psi$	$\sin 2\psi$	0

The Stokes' parameters are related to more familiar astronomical quantities by

$$\theta = \frac{1}{2} \tan^{-1}\left(\frac{U}{Q}\right) = \psi \qquad (5.31)$$

$$e = \left(1 - \tan^2 \left\{\frac{1}{2} \sin^{-1}\left[\frac{V}{(Q^2 + U^2 + V^2)^{1/2}}\right]\right\}\right)^{1/2} \qquad (5.32)$$

where

θ is the position angle (Section 5.1) of the semimajor axis of the polarization ellipse (when the x-axis is aligned north–south)

e is the eccentricity of the polarization ellipse and is 1 for linearly polarized radiation and 0 for circularly polarized radiation

5.2.2 Optical Components for Polarimetry

Polarimeters usually contain a number of components that are optically active in the sense that they alter the state of polarization of the radiation. They may be grouped under three headings: polarizers, converters, and depolarizers. The first produces linearly polarized light, the second converts elliptically polarized light into linearly polarized light, or vice versa, while the last eliminates polarization. Most of the devices rely upon birefringence for their effects, so we must initially discuss some of its properties before looking at the devices themselves.

5.2.2.1   Birefringence

The difference between a birefringent material and a more normal optical material may best be understood in terms of the behavior of the Huygens' wavelets. In a normal material, the refracted ray can be constructed from the incident ray by taking the envelope of the Huygens' wavelets, which spread out from the incident surface with a uniform velocity. In a birefringent material, however, the velocities of the wavelets depend upon the polarization of the radiation, and for at least one component, the velocity will also depend on the orientation of the ray with respect to the structure of the material. In some materials, it is possible to find a direction for linearly polarized radiation for which the wavelets expand spherically from the point of incidence as in normal optical materials. This ray is then termed the ordinary ray and its behavior may be described by normal geometrical optics. The ray that is polarized orthogonally to the ordinary ray is then termed the extraordinary ray, and wavelets from its point of incidence spread out elliptically (Figure 5.9). The velocity of the

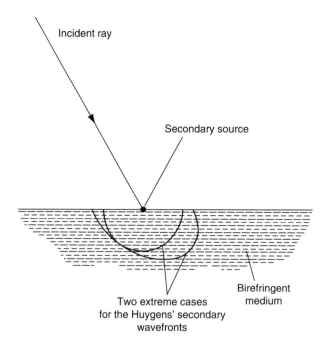

FIGURE 5.9 Huygens' secondary wavelets in a birefringent medium.

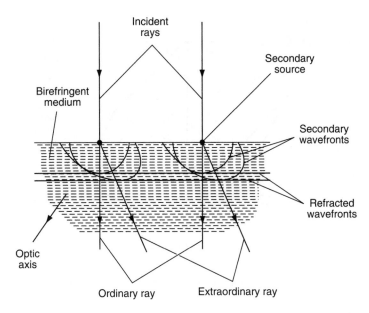

FIGURE 5.10 Formation of ordinary and extraordinary rays in a birefringent medium.

extraordinary ray thus varies with direction. We may construct the two refracted rays in a birefringent material by taking the envelopes of their wavelets as before (Figure 5.10). The direction along which the velocities of the ordinary and extraordinary rays are equal is called the optic axis of the material. When the velocity of the extraordinary ray is in general larger than that of the ordinary ray (as illustrated in Figures 5.9 and 5.10), then the birefringence is negative. It is positive when the situation is reversed. The degree of the birefringence may be obtained from the principal extraordinary refractive index, μ_E. This is the refractive index corresponding to the maximum velocity of the extraordinary ray for negative materials, and the minimum velocity for positive materials. It will be obtained from rays traveling perpendicular to the optic axis of the material. The degree of birefringence, which is often denoted by the symbol J, is then simply the difference between the principal extraordinary refractive index and the refractive index for the ordinary ray, μ_O

$$J = \mu_E - \mu_O \tag{5.33}$$

Most crystals exhibit natural birefringence and it can be introduced into many more and into amorphous substances such as glass by the presence

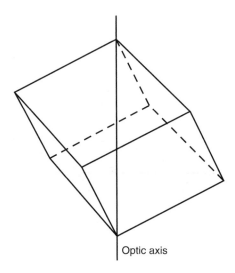

Optic axis

FIGURE 5.11 Equilateral calcite cleavage rhomb and its optic axis.

of strain in the material. One of the most commonly encountered birefringent materials is calcite. The cleavage fragments form rhombohedrons and the optic axis then joins opposite blunt corners if all the edges are of equal length (Figure 5.11). The refractive index of the ordinary ray is 1.658, while the principal extraordinary refractive index is 1.486, giving calcite the very high degree of birefringence of −0.172.

Crystals such as calcite are uniaxial and have only one optic axis. Uniaxial crystals belong to the tetragonal or hexagonal crystallographic systems. Crystals that belong to the cubic system are not usually birefringent, while crystals in the remaining systems—orthorhombic, monoclinic, or triclinic—generally produce biaxial crystals. In the latter case, normal geometrical optics breaks down completely and all rays are extraordinary.

Some crystals such as quartz that are birefringent ($J = +0.009$) are in addition optically active. This is the property whereby the plane of polarization of a beam of radiation is rotated as it passes through the material. Many substances other than crystals, including most solutions of organic chemicals, can be optically active. Looking down a beam of light, against the motion of the photons, a substance is called dextro-rotatory or right-handed if the rotation of the plane of vibration is clockwise. The other case is called laevo-rotatory or left-handed. Unfortunately and confusingly, the opposite convention is also in occasional use.

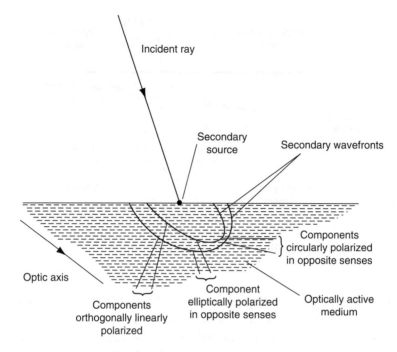

FIGURE 5.12 Huygens' secondary wavelets in an optically active medium.

The description of the behavior of an optically active substance in terms of the Huygens' wavelets is rather more difficult than was the case for birefringence. Incident light is split into two components as previously, but the velocities of both components vary with angle, and in no direction do they become equal. The optic axis therefore has to be taken as the direction along which the difference in velocities is minimized. Additionally, the nature of the components changes with angle as well. Along the optic axis they are circularly polarized in opposite senses, while perpendicular to the optic axis they are orthogonally linearly polarized. Between these two extremes the two components are elliptically polarized to varying degrees and in opposite senses (Figure 5.12).

5.2.2.2 Polarizers*

These are devices that only allow the passage of light that is linearly polarized in some specified direction. There are several varieties that are based upon birefringence, of which the Nicol prism is the best known.

---

* Also often known as analyzers

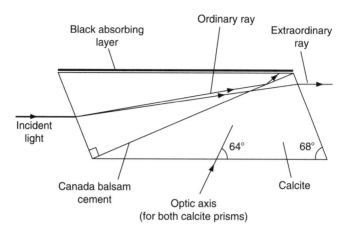

FIGURE 5.13 Nicol prism.

This consists of two calcite prisms cemented together by Canada balsam. Since the refractive index of Canada balsam is 1.55, it is possible for the extraordinary ray to be transmitted, while the ordinary ray is totally internally reflected (Figure 5.13). The Nicol polarizer has the drawbacks of displacing the light beam and of introducing some elliptical polarization into the emergent beam. Its inclined faces also introduce additional light losses by reflection. Various other designs of polarizers have therefore been developed, some of which in fact produce mutually orthogonally polarized beams with an angular separation. Examples of several such designs are shown in Figure 5.14. Magnesium fluoride and quartz can also be used to form polarizers for the visible region, while lithium niobate and sapphire are used in the infrared region.

The ubiquitous polarizing sunglasses are based upon another type of polarizer. They employ dichroic crystals that have nearly 100% absorption for one plane of polarization and less than 100% for the other. Generally, the dichroism varies with wavelength so that these polarizers are not achromatic. Usually, however, they are sufficiently uniform in their spectral behavior to be usable over quite wide wavebands. The polarizers are produced commercially in sheet form and contain many small aligned crystals rather than one large one. The two commonly used compounds are polyvinyl alcohol impregnated with iodine and polyvinyl alcohol catalyzed to polyvinylene by hydrogen chloride. The alignment is achieved by stretching the film. The use of microscopic crystals and the existence of the large commercial market mean that dichroic polarizers are far cheaper

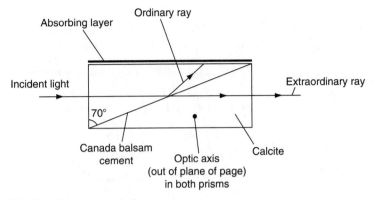

(a) Glan–Thompson polarizer

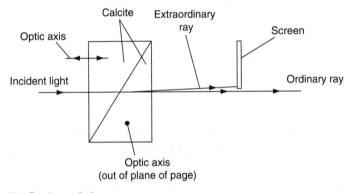

(b) Rochon polarizer

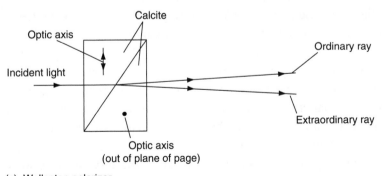

(c) Wollaston polarizer

FIGURE 5.14 Examples of birefringence polarizers.

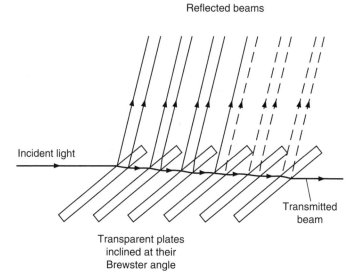

FIGURE 5.15 Reflection polarizer.

than birefringent polarizers, and so they may be used even when their performance is poorer than that of the birefringent polarizers.

Polarization by reflection can be used to produce a polarizer. A glass plate inclined at the Brewster angle will reflect a totally polarized beam. However, only a small percentage (about 7.5% for crown glass) of the incident energy is reflected. Reflection from the second surface, though, will reinforce the first reflection, and then several plates may be stacked together to provide further reflections (Figure 5.15). The transmitted beam is only partially polarized. However, as the number of plates is increased, the total intensity of the reflected beams (ignoring absorption) will approach half of the incident intensity (Figure 5.16). Hence, the transmitted beam will approach complete polarization. In practice therefore, the reflection polarizer is used in transmission, since the problem of recombining the multiple reflected beams is thereby avoided and the beam suffers no angular deviation.

Outside the optical, near-infrared, and near-ultraviolet regions, the polarizers tend to be somewhat different in nature. In the radio region, a linear dipole is naturally only sensitive to radiation polarized along its length. In the microwave region and the medium to far-infrared, wire grid polarizers are suitable. These, as their name suggests, are grids of electrically conducting wires. Their spacing is about five times their thickness,

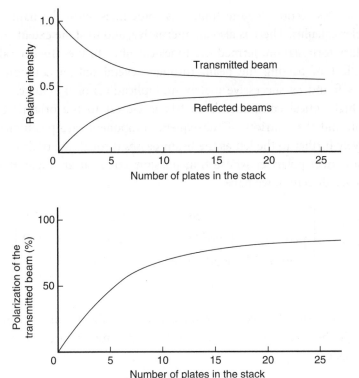

FIGURE 5.16 Properties of reflection polarizers, assuming negligible absorption and a refractive index of 1.5.

and they transmit the component of the radiation that is polarized along the length of the wires. They work efficiently for wavelengths longer than their spacing. Recently, the degree angular scale interferometer (DASI) telescope sited at the Amundsen–Scott South Polar station detected polarization in the cosmic microwave background (CMB) radiation using transition edge sensor (TES) detectors coupled with micro-strip antennas, a result subsequently confirmed by the *Wilkinson Microwave Anisotropy Probe* (*WMAP*). In the x-ray region, Bragg reflection is polarized (Section 1.3.6.2) and so a rotating Bragg spectrometer can also act as a linear polarization detector.

The behavior of a polarizer may be described mathematically by its effect upon the Stokes' parameters of the radiation. This is most easily accomplished by writing the parameters as a column vector. A matrix multiplying the vector on the left may then represent the effect of the polarizer, and also of the other optical components that we will discuss later in this section. The technique is sometimes given the name of the Mueller calculus. There is also an alternative, and to some extent complementary formulation, termed the Jones calculus. In the Mueller calculus, the effect of passing the beam through several optical components is simply found by successive matrix multiplications of the Stokes' vector. The first optical component's matrix is closest to the original Stokes' vector, and the matrices of subsequent components are positioned successively further to the left as the beam passes through the optical system. For a perfect polarizer whose transmission axis is at an angle, θ, to the reference direction, we have

$$
\begin{bmatrix} I' \\ Q' \\ U' \\ V' \end{bmatrix} = \begin{bmatrix} \frac{1}{2} & \frac{1}{2}\cos 2\theta & \frac{1}{2}\sin 2\theta & 0 \\ \frac{1}{2}\cos 2\theta & \frac{1}{2}\cos^2 2\theta & \frac{1}{2}\cos 2\theta \sin 2\theta & 0 \\ \frac{1}{2}\sin 2\theta & \frac{1}{2}\cos 2\theta \sin 2\theta & \frac{1}{2}\sin^2 2\theta & 0 \\ 0 & 0 & 0 & 0 \end{bmatrix} \begin{bmatrix} I \\ Q \\ U \\ V \end{bmatrix}
\tag{5.34}
$$

where the primed Stokes' parameters are for the beam after passage through the polarizer and the unprimed parameters are for it before such passage.

5.2.2.3 Converters
These are devices that alter the type of polarization and/or its orientation. They are also known as retarders or phase plates. We have seen earlier (Equations 5.16 and 5.17 and Figure 5.8) that elliptically polarized light may be resolved into two orthogonal linear components with a phase difference. Altering that phase difference will alter the degree of ellipticity (Equations 5.18 and 5.19) and inclination (Equation 5.21) of the ellipse. We have also seen that the velocities of mutually orthogonal linearly polarized beams of radiation will in general differ from each other when the beams pass through a birefringent material. From inspection of Figure 5.13, it will be seen that if the optic axis is rotated until it is perpendicular to the incident radiation, then the ordinary and extraordinary rays will travel in the same direction. Thus, they will pass together through a layer of a birefringent material that is oriented in this way, and will recombine upon emergence, but with an altered phase delay, because of their differing velocities. The phase delay, δ', is given to a first-order approximation by

$$
\delta' = \frac{2\pi d}{\lambda} J
\tag{5.35}
$$

where
d is the thickness of the material
J is the birefringence of the material

Now let us define the x-axis of Figure 5.8 to be the polarization direction of the extraordinary ray. We then have from Equation 5.21 the intrinsic phase difference, δ, between the components of the incident radiation

$$\delta = \cos^{-1}\left[\frac{(e_1^2 - e_2^2)}{2e_1 e_2} \tan 2\psi\right] \tag{5.36}$$

The ellipse for the emergent radiation then has a minor axis given by

$$b' = a' \tan\left(\frac{1}{2}\sin^{-1}\left\{\left[\frac{2e_1 e_2}{(e_1^2 + e_2^2)}\right]\sin(\delta + \delta')\right\}\right) \tag{5.37}$$

where the primed quantities are for the emergent beam. So

$$b' = 0 \quad \text{for } \delta + \delta' = 0 \tag{5.38}$$

$$b' = a' \quad \text{for } \delta + \delta' = \sin^{-1}\left[\frac{(e_1^2 + e_2^2)}{2e_1 e_2}\right] \tag{5.39}$$

and also

$$\psi = \frac{1}{2}\tan^{-1}\left\{\left[\frac{2e_1 e_2}{(e_1^2 - e_2^2)}\right]\cos(\delta + \delta')\right\} \tag{5.40}$$

Thus,

$$\psi' = -\psi \quad \text{for } \delta' = 180° \tag{5.41}$$

and

$$a' = a \quad \text{and} \quad b' = b \tag{5.42}$$

Thus we see that, in general, elliptically polarized radiation may have its degree of ellipticity altered and its inclination changed by passage through a converter. In particular cases, it may be converted into linearly polarized or circularly polarized radiation, or its orientation may be reflected about the fast axis of the converter.

In real devices, the value of delta is chosen to be 90° or 180° and the resulting converters are called quarter-wave plates or half-wave plates, respectively, since one beam is delayed with respect to the other by a quarter or a half of a wavelength. The quarter-wave plate is then used to convert elliptically or circularly polarized light into linearly polarized light or vice versa, while the half-wave plate is used to rotate the plane of linearly polarized light.

Many substances can be used to make converters, but mica is probably the commonest because of the ease with which it may be split along its cleavage planes. Plates of the right thickness (about 40 μm) are therefore simple to obtain. Quartz cut parallel to its optic axis can also be used, while for ultraviolet work magnesium fluoride is suitable. Amorphous substances may be stretched or compressed to introduce stress birefringence. It is then possible to change the phase delay in the material by changing the amount of stress. Extremely high-acceptance angles and chopping rates can be achieved with low power consumption if a small acoustic transducer at one of its natural frequencies drives the material. This is then commonly called a photoelastic modulator. An electric field can also induce birefringence. Along the direction of the field, the phenomenon is called the Pockels effect, while perpendicular to the field it is called the Kerr effect. Suitable materials abound and may be used to produce quarter- or half-wave plates. Generally, these devices are used when rapid switching of the state of birefringence is needed, as, for example, in Babcock's solar magnetometer (Section 5.4.2). In glass, the effects take up to a minute to appear and disappear, but in other substances the delay can be far shorter. The Kerr effect in nitrobenzene, for example, permits switching at up to 1 MHz, and the Pockels effect can be used similarly with ammonium dihydrogen phosphate or potassium dihydrogen phosphate.

All the above converters will normally be chromatic and usable over only a restricted wavelength range. Converters that are more nearly achromatic can be produced, which are based upon the phase changes that occur in total internal reflection. The phase difference between components with electric vectors parallel and perpendicular to the plane of incidence is shown in Figure 5.17. For two of the angles of incidence (approximately 45° and 60° in the example shown), the phase delay is 135°. Two such reflections therefore produce a total delay of 270°, or, as one may also view the situation, an advance of the second component with respect to the first by 90°. The minimum value of the phase difference

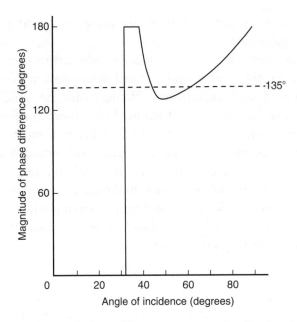

FIGURE 5.17 Differential phase delay upon internal reflection in a medium with a refractive index of 1.6.

shown in Figure 5.17 is equal to 135° when the refractive index is 1.497. There is then only one suitable angle of incidence and this is 51°47'. For optimum results, the optical design should approach as close to this ideal as possible. A quarter-wave retarder that is nearly achromatic can thus be formed using two total internal reflections at appropriate angles. The precise angles usually have to be determined by trial and error since additional phase changes are produced when the beam interacts with the entrance and exit faces of the component. Two practical designs, the Fresnel rhomb and the Mooney rhomb, are illustrated in Figure 5.18.

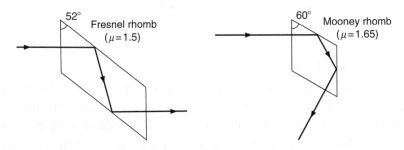

FIGURE 5.18 Quarter-wave retarders using total internal reflection.

Half-wave retarders can be formed by using two such quarter-wave retarders in succession.

Combining three retarders can make pseudo quarter- and half-wave plates that are usable over several hundred nanometers in the visual. The Pancharatnam design employs two retarders with the same delay and orientation, together with a third sandwiched between the first two. The inner retarder is rotated with respect to the other two, and possibly has a differing delay. A composite half-wave plate with actual delays in the range $180° \pm 2°$ over wavelengths from 300 to 1200 nm can, for example, be formed from three $180°$ retarders with the center one oriented at about $60°$ to the two outer ones. Quartz and magnesium fluoride are commonly used materials for the production of such super-achromatic wave plates.

The Mueller matrices for converters are

$$
\begin{bmatrix}
1 & 0 & 0 & 0 \\
0 & \cos^2 2\psi & \cos 2\psi \sin 2\psi & -\sin 2\psi \\
0 & \cos 2\psi \sin 2\psi & \sin^2 2\psi & \cos 2\psi \\
0 & \sin 2\psi & -\cos 2\psi & 0
\end{bmatrix}
\tag{5.43}
$$

for a quarter-wave plate, and

$$
\begin{bmatrix}
1 & 0 & 0 & 0 \\
0 & \cos^2 2\psi - \sin^2 2\psi & 2 \cos 2\psi \sin 2\psi & 0 \\
0 & 2 \cos 2\psi \sin 2\psi & \sin^2 2\psi - \cos^2 2\psi & 0 \\
0 & 0 & 0 & -1
\end{bmatrix}
\tag{5.44}
$$

for a half-wave plate, while in the general case, we have

$$
\begin{bmatrix}
1 & 0 & 0 & 0 \\
0 & \cos^2 2\psi + \sin^2 2\psi \cos \delta & (1 - \cos \delta) \cos 2\psi \sin 2\psi & -\sin 2\psi \sin \delta \\
0 & (1 - \cos \delta) \cos 2\psi \sin 2\psi & \sin^2 2\psi + \cos^2 2\psi \cos \delta & \cos 2\psi \sin \delta \\
0 & \sin 2\psi \sin \delta & -\cos 2\psi \sin \delta & \cos \delta
\end{bmatrix}
\tag{5.45}
$$

5.2.2.4 Depolarizers

The ideal depolarizer would accept any form of polarized radiation and produce unpolarized radiation. No such device exists, but pseudo-depolarizers can be made. These convert the polarized radiation into radiation that is unpolarized when it is averaged over wavelength, time, or area.

A monochromatic depolarizer can be formed from a rotating quarter-wave plate that is in line with a half-wave plate rotating at twice its rate. The emerging beam at any given instant will have some form of elliptical polarization, but this will change rapidly with time, and the output will average to zero polarization over several rotations of the plates.

The Lyot depolarizer averages over wavelength. It consists of two retarders with phase differences very much greater than 360°. The second plate has twice the thickness of the first and its optic axis is rotated by 45° with respect to that of the first. The emergent beam will be polarized at any given wavelength, but the polarization will vary very rapidly with wavelength. In the optical region, averaging over a waveband of a few tens of nanometers wide is then sufficient to reduce the net polarization to 1% of its initial value.

If a retarder is left with a rough surface and immersed in a liquid whose refractive index is the average refractive index of the retarder, then a beam of light will be undisturbed by the roughness of the surface because the hollows will be filled in by the liquid. However, the retarder will vary in its effect on the scale of its roughness. Thus, the polarization of the emerging beam will vary on the same scale, and a suitable choice for the parameters of the system can lead to an average polarization of zero over the whole beam.

The Mueller matrix of an ideal depolarizer is simply

$$
\begin{bmatrix}
1 & 0 & 0 & 0 \\
0 & 0 & 0 & 0 \\
0 & 0 & 0 & 0 \\
0 & 0 & 0 & 0
\end{bmatrix}
\tag{5.46}
$$

5.2.3 Polarimeters

A polarimeter is an instrument that measures the state of polarization, or some aspect of the state of polarization, of a beam of radiation. Ideally, the values of all four Stokes' parameters should be determinable, together with their variations with time, space, and wavelength. In practice, this is rarely possible, at least for astronomical sources, most of the time only the degree of linear polarization and its direction are found. Astronomical polarimeters now normally use CCDs, but photographic or photoelectric detectors have been used in the past.

5.2.3.1 Photographic Polarimeters

You are now unlikely to come across a photographic polarimeter, but many of their designs can be used for CCD-based polarimeters and they do have some historical interest. At its simplest, photographic polarimetry is just photographic photometry (Section 3.2.1.2) with the addition of a polarizer into the light path. Polaroid film is most frequently used as the polarizer. Several separate exposures are made through it with the polarizer rotated relative to the image plane by some fixed angle each time. The errors of the photographic photometric method are usually 0.01–0.05 magnitudes or more. This technique is therefore of little use for most stellar polarimetry, since the degree of polarization to be measured is normally 1% or less. It is useful, however, for highly polarized extended objects such as the Crab Nebula.

The photographic method can be improved considerably by the use of a double-beam polarizer such as the Wollaston prism. Two images are produced for each star in light that is mutually orthogonally linearly polarized. Again, several exposures with a rotation of the polarizer are needed. Multicolor observations can be made by using two Wollaston prisms with an interposed optically active element (Figure 5.19). The degree of rotation introduced by the latter is generally wavelength dependent. Hence, four

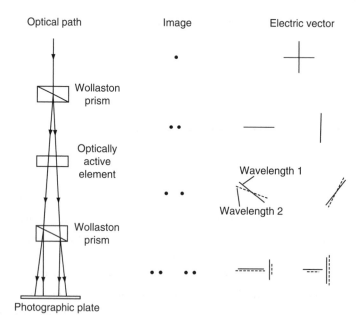

FIGURE 5.19 Öhman's multicolor polarimeter.

images result in which one pair has a similar color balance and mutually orthogonal polarizations, while the second pair are also orthogonally polarized but have a different color balance from the first pair.

Circular polarization can be studied photographically using large flat plastic quarter-wave plates. These are placed before the polarizer and convert the circular (or elliptical) radiation into linearly polarized radiation. The detection of this then proceeds along the lines just discussed. The detection threshold is currently about 10% polarization for significantly accurate measurements.

5.2.3.2 Photoelectric Polarimeters

Most of these devices bear a distinct resemblance to photoelectric photometers (Section 3.2.1.3), and indeed, many polarimeters can also function as photometers. The major differences, apart from the components that are needed to measure the polarization, arise from the necessity of reducing or eliminating the instrumentally induced polarization. The instrumental polarization originates primarily in inclined planar reflections or in the detector. Thus, Newtonian and Coudé telescopes cannot be used for polarimetry because of their inclined subsidiary mirrors. Inclined mirrors must also be avoided within the polarimeter. Cassegrain and other similar designs should not in principle introduce any net polarization because of their symmetry about the optical axis. However, problems may arise even with these telescopes if mechanical flexure or poor adjustments move the mirrors and/or the polarimeter from the line of symmetry.

Many detectors, especially photomultipliers (Section 1.1.10), are sensitive to polarization, and furthermore this sensitivity may vary over the surface of the detector. If the detector is not shielded from the earth's magnetic field, the sensitivity may vary with the position of the telescope as well. In addition, although an ideal Cassegrain system will produce no net polarization, this sensitivity to polarization arises because rays on opposite sides of the optical axis are polarized in opposite but equal ways. The detector nonuniformity may then result in incomplete cancellation of the telescope polarization. Thus, a Fabry lens and a depolarizer immediately before the detector are almost always essential components of a polarimeter.

Alternatively, the instrumental polarization can be reversed using a Bowen compensator. This comprises two retarders with delays of about $\lambda/8$ that may be rotated with respect to the light beam and to each other until their polarization effects are equal and opposite to those of the instrument. Small amounts of instrumental polarization can be corrected by the insertion of a tilted glass plate into the light beam to introduce opposing polarization. In practice, an unpolarized source is observed, and the plate tilted and rotated until the observed polarization is minimized or eliminated.

There are many detailed designs of polarimeter, but we may group them into single- and double-channel devices, with the former further subdivided into discrete and continuous systems. The single-channel discrete polarimeter is the basis of most of the other designs, and its main components are shown in Figure 5.20, although the order is not

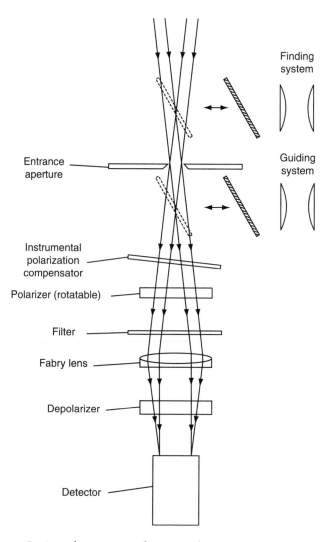

FIGURE 5.20 Basic polarimeter: schematic design and optical paths.

critical for some of them. Since stellar polarizations are usually of the order of 1% or less, it is also usually necessary to incorporate a chopper and use a phase-sensitive detector (also known as a lock-in amplifier) plus integration to obtain sufficiently accurate measurements. Switching between sky and background is also vital since the background radiation is very likely to be polarized. The polarizer is rotated either in steps or continuously, but at a rate that is slow compared with the chopping and integration times. The output for a linearly polarized source is then modulated with the position of the polarizer.

The continuous single-beam systems use a rapidly rotating polarizer. This gives an oscillating output for a linearly polarized source, and this may be fed directly into a computer or into a phase-sensitive detector that is driven at the same frequency. Alternatively, a quarter-wave plate and a photoelastic modulator can be used. The quarter-wave plate produces circularly polarized light, and this is then converted back into linearly polarized light by the photoelastic modulator, but with its plane of polarization rotated by 90° between alternate peaks of its cycle. A polarizer that is stationary will then introduce an intensity variation with time that may be detected and analyzed as before. The output of the phase-sensitive detector in systems such as these is directly proportional to the degree of polarization, with unpolarized sources giving zero output. They may also be used as photometers to determine the total intensity of the source, by inserting an additional fixed polarizer before the entrance aperture, to give in effect a source with 100% polarization and half the intensity of the actual source (provided that the intrinsic polarization is small). An early polarimeter system that is related to these is due to Dolfus. It uses a quarter-wave plate to convert the linearly polarized component of the radiation into circularly polarized light. A rotating disk that contains alternate sectors made from zero- and half-wave plates then chops the beam. The circular polarization alternates in its direction but remains circularly polarized. Another quarter-wave plate converts the circularly polarized light back to linearly polarized light, but the plane of its polarization switches through 90° at the chopping frequency. Finally, a polarizer is used to eliminate one of the linearly polarized components, so that the detector produces a square wave output.

The continuous single-beam polarimeter determines both linear polarization components quasi-simultaneously, but it still requires a separate observation to acquire the total intensity of the source. In double-beam devices, the total intensity is found with each observation, but several observations are required to determine the polarization. These instruments are essentially similar to that shown in Figure 5.20, but a double-beam polarizer such as a Wollaston prism is used in place of the single-beam polarizer shown there. Two detectors then observe both beams simultaneously. The polarizer is again rotated either in steps or continuously but slowly to measure the polarization. Alternatively, a half-wave plate can be rotated above a fixed Wollaston prism. The effects of scintillation and other atmospheric changes are much reduced by the use of dual beam instruments.

The current state of the art is represented by PlanetPol, a polarimeter designed by James Hough at the University of Hertfordshire to detect extrasolar planets. It uses a 20 kHz photoelastic modulator and a three-wedge Wollaston prism to detect the ordinary and extraordinary rays simultaneously using avalanche photodiodes and has a separate sky channel. The whole instrument rotates through $45°$ to measure the Q and U Stokes' parameters. Its sensitivity is 1 part in 10^6 or better, and on an 8 m telescope it should be able to detect the polarization induced into the light from a star by a Jupiter-sized planet for stars brighter than 6.5^m. PlanetPol has yet to detect any extrasolar planets, but the induced polarization caused by a hot Jupiter orbiting the star HD 189733 has been picked up using a conventional polarimeter on the University of Turku's (Finland) 0.6 m KVA telescope. In this latter instance though, the existence of the planet and the details of its orbit were already known from transit studies before the polarimetric work.

5.2.3.3 CCD and Other Array Detector Polarimeters

Again many designs are possible, but the dual-beam polarizer using a rotating wave plate and a Wollaston prism is in widespread use. For extended sources, the polarized images are detected directly by a CCD or infrared detector array. To avoid overlap, a mask is used at the focal plane of the telescope comprising linear obscuring bars and gaps of equal width (sometimes known as a comb dekker). The image from one of the beams then occupies the space occupied by the image of the mask. Clearly, two sets of observations, shifted by the width of an obscuring bar, are then needed to show the complete extended object. In the ultraviolet, little work has been undertaken to date. The Solar Maximum Mission (SMM) satellite, however, did include an ultraviolet spectrometer into which a magnesium fluoride retarder could be inserted, so that some measurements of the polarization in strong spectrum lines could be made. The Wisconsin ultraviolet photo-polarimeter experiment (WUPPE) was flown on the space shuttle in 1993 and the faint object camera (FOC) on board the HST can be used for polarimetry in the ultraviolet.

A few polarimeter designs have been produced, particularly for infrared work, in which a dichroic mirror is used. These are mirrors that reflect at one wavelength and are transparent at another. They are not related to the other form of dichroism that was discussed earlier in this section. For example, a thinly plated gold mirror reflects infrared radiation but transmits visible radiation. Continuous guiding on the object being studied is therefore possible, provided that it is emitting visible radiation, through the use of such a mirror in the polarimeter. Unfortunately, the reflection introduces instrumental polarization, but this can be reduced by incorporating a second 90° reflection whose plane is orthogonal to that of the first into the system.

5.2.4   Spectropolarimetry

Spectropolarimetry that provides information on the variation of polarization with wavelength (see also Öhman's multicolor polarimeter, Section 5.2.3.1) can be realized by several methods. Almost any spectroscope (Section 4.2) suited for long-slit spectroscopy can be adapted to spectropolarimetry. A quarter- or half-wave plate and a block of calcite are placed before the slit. The calcite is oriented so that the ordinary and extraordinary images lie along the length of the slit. Successive exposures are then obtained with the wave plate rotated between each exposure. For linear spectropolarimetry, four positions separated by 22.5° are needed for the half-wave plate, while for circular spectropolarimetry just two orthogonal positions are required for the quarter-wave plate. Solar spectropolarimetry is considered in Section 5.3.

5.2.5   Data Reduction and Analysis

The output of a polarimeter is usually in the form of a series of intensity measurements for varying angles of the polarizer. These must be corrected for instrumental polarization if this has not already been removed or has only been incompletely removed by the use of an inclined plate. The atmospheric contribution to the polarization must then be removed by comparison of the observations of the star and its background. Other photometric corrections, such as for atmospheric extinction (Section 3.2.3), must also be made, especially if the observations at differing orientations of the polarizer are made at significantly different times so that the position of the source in the sky has changed.

The normalized Stokes' parameters are then obtained from

$$\frac{Q}{I} = \frac{I(0) - I(90)}{I(0) + I(90)} \tag{5.47}$$

and

$$\frac{U}{I} = \frac{I(45) - I(135)}{I(45) + I(135)} \tag{5.48}$$

where $I(\theta)$ is the intensity at a position angle θ for the polarizer. Since for most astronomical sources the elliptically polarized component of the radiation is 1% or less of the linearly polarized component, the degree of linear polarization, π_L, can then be found via Equation 5.28 or more simply from Equation 5.30 if sufficient observations exist to determine I_{max} and I_{min} with adequate precision. The position angle of the polarization may be found from Equation 5.31, or again directly from the observations if these are sufficiently closely spaced. The fourth Stokes' parameter, V, is rarely determined in astronomical polarimetry. However, the addition of a quarter-wave plate to the optical train of the polarimeter is sufficient to enable it to be found if required (e.g., Section 5.4). The reason for the lack of interest is, as already remarked, the very low levels of circular polarization that are usually found. However recently, the European Southern Observatory's (ESO) 3.6 m telescope has had interchangeable linear and circular polarization detectors installed (ESO's faint object spectrograph and camera [EFOSC2]), the latter being similar to the former except for the addition of a quarter-wave plate. The first Stokes' parameter, I, is simply the intensity of the source, and so if it is not measured directly as a part of the experimental procedure, it can simply be obtained from

$$I = I(\theta) + I(\theta + 90°) \tag{5.49}$$

A flowchart to illustrate the steps in an observation and analysis sequence to determine all four Stokes' parameters is shown in Figure 5.21.

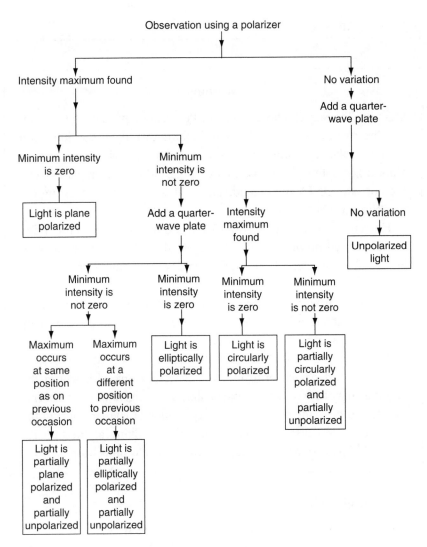

FIGURE 5.21 Flowchart of polarimetric analysis.

Exercise 5.3

Obtain Equation 5.21 from Equations 5.16 and 5.17.

Exercise 5.4

Show, using the Mueller calculus, that the effect of two ideal half-wave plates upon a beam of radiation of any form of polarization is zero.

5.3 SOLAR STUDIES

Warning

Studying the Sun can be dangerous. Permanent eye damage or even blindness can easily occur through looking directly at the Sun, even with the naked eye. Smoked glass, exposed film, CDs, space blankets, aluminized helium balloons, crossed Polaroid filters, sunglasses, etc. are not safe; ONLY filters made and sold specifically for viewing the Sun should ever be used.

Herschel wedges (see Section 5.3.2) are no longer recommended except if used on very small telescopes, also care must be taken to ensure that the rejected solar radiation does not cause burns or start fires.

Eyepiece projection is safe from the point of view of personal injury, but can damage the eyepiece and/or the telescope, especially with the shorter focal ratio instruments.

This warning should be carefully noted by all intending solar observers, and passed onto inexperienced persons with whom the observer may be working. In particular, the dangers of solar observing should be made quite clear to any groups of laymen touring the observatory, because the temptation to try looking at the Sun for themselves is very great, especially when there is an eclipse.

5.3.1 Introduction

The Sun is a reasonably typical star and as such it can be studied by many of the techniques discussed elsewhere in this book. In one major way, however, the Sun is quite different from all other stars, and that is that it may easily be studied using an angular resolution smaller than its diameter. This, combined with the vastly greater amount of energy that is available compared with that from any other star, has led to the development of many specialized instruments and observing techniques. Many of these have already been mentioned or discussed in some detail in other sections. The main such techniques include the following: solar cosmic ray and neutron detection (Section 1.4), neutrino detection (Section 1.5), radio observations (Sections 1.2 and 2.5), x-ray and γ-ray observations (Section 1.3), adaptive optics (Section 2.5.2) and radar observations

(Section 2.8), magnetometry (Section 5.4), and spectroscopy and prominence spectroscopes (Section 4.2). The other remaining important specialist instrumentation and techniques for solar work are covered in this section.

The warning at the start of this section may be quantified by the European Community directive, EN 167:

> For safe viewing of the Sun with the unaided eye, filters should have a minimum size of 35 mm × 115 mm so that both eyes are covered and have maximum transmissions of
>
> 0.003%—ultraviolet (280–380 nm)
> 0.0032%—visible (380–780 nm)
> 0.027%—near infrared (780–1400 nm)

Although these maxima are safe with regard to damage to the eye, the Sun may still be uncomfortably bright, and a transmittance of 0.0003% may be found to be better. Since the surface brightness of the Sun is not increased by the use of a telescope (Equation 1.67 et seq.), the same limits are safe for telescopic viewing, although the larger angular size of the image may increase the total energy entering the eye to the point of discomfort, so filters should generally be denser than this directive suggests. A transmission of 0.003% corresponds to an optical density for the filter of 4.5. Full aperture filters (see next section) sold for solar observing generally have optical densities of five to six.

5.3.2 Solar Telescopes

Almost any telescope may be used to image the Sun. While specialized telescopes are often built for professional work (see later discussion), small conventional telescopes are frequently used by amateurs or for teaching or training purposes. Since it is dangerous to look at the Sun through an unadapted telescope (see the warning above), special methods must be used. By far the best approach is to use a full aperture filter sold by a reputable supplier specifically for solar observing (see below), but eyepiece projection may also be possible. In this, the image is projected onto a sheet of white card held behind the eyepiece. This method may be preferred when demonstrating to a group of people, since all can see the image at the same time, and they are looking away from the Sun itself to see its image. However, telescopes such as Schmidt–Cassegrains and Maksutovs have short focal-length primary mirrors, and these can concentrate the light internally sufficiently to damage the instrument. The use of a telescope for solar projection may well invalidate the manufacturer's guarantee, and the manufacturer should always be consulted on this in advance of using the telescope for solar work. Even for other telescope designs, the heat passing through the eyepiece may be sufficient to damage it, especially with the more expensive multicomponent types. The Herschel wedge, in the past a common telescope accessory, is no longer recommended (see below). Filters placed at the eyepiece end of the telescope, even if of adequate optical density, should never be used. They can become very hot and shatter, and the unfiltered image within the telescope can cause damage to the instrument.

The Herschel wedge or solar diagonal is shown in Figure 5.22. It is a thin prism with unsilvered faces. The first face is inclined at 45° to the optical axis and thus reflects about 5% of the solar radiation into an eyepiece in the normal star diagonal position. The second face also reflects about 5% of the radiation, but its inclination to the optical axis is not 45°, and so it is easy to arrange for this radiation to be intercepted by a baffle before it reaches the eyepiece. The remaining radiation passes through the prism and can be absorbed in a heat trap, or more commonly (and dangerously) just allowed to emerge as an intense beam of radiation at the rear end of the telescope. To reduce the solar intensity to less than the recommended maximum, the wedge must be used on a very small telescope

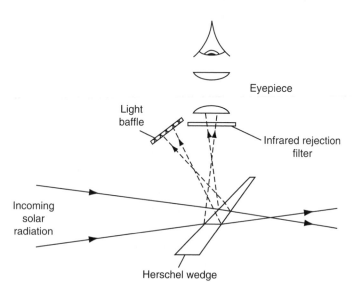

FIGURE 5.22 Optical paths in a Herschel wedge.

at a high magnification—for example, a 50 mm telescope would need a minimum magnification of ×300 to be safe. The device must also incorporate a separate infrared filter. Since modern small telescopes range from 100 to 300 mm or more in aperture, they are much too large to use with a Herschel wedge. Even if a larger telescope were to be stopped down to 50 mm (say), the Herschel wedge still has the disadvantage of producing a real image inside the telescope that may damage the telescope and/or invalidate its guarantee. The recommendation now therefore is that Herschel wedges should not be used for solar observing. If you have a Herschel wedge already, then it would be best to get the front surface of the wedge aluminized and then use it as a star diagonal.

The principle of eyepiece projection is shown in Figure 5.23, and it is the same as the eyepiece projection method of imaging (Section 2.2.6). The eyepiece is moved outward from its normal position for visual use, and a real image of the Sun is obtained that may easily be projected onto a sheet of white paper or cardboard, etc. for viewing purposes. From Equation 2.26, we have the effective focal length (EFL) of the telescope

$$\mathrm{EFL} = \frac{Fd}{e} \tag{5.50}$$

where
F is the focal length of the objective
e is the focal length of the eyepiece
d is the projection distance (see Figure 5.23)

The final image size, D, for the Sun is then simply

$$D = \frac{0.0093Fd}{e} \tag{5.51}$$

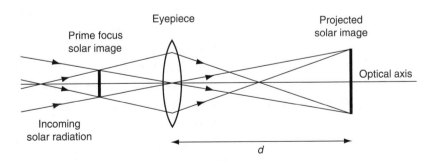

FIGURE 5.23 Projection of the solar image.

There are two drawbacks to this method. The first is that a real image of the Sun is formed at the prime focus of the objective. If this is larger than the eyepiece acceptance area, or if this image of the Sun is intercepted by a part of the telescope's structure during acquisition, then structural damage may occur through the intense heat in the beam. This is especially a problem for modern short focal ratio Schmidt–Cassegrain and Maksutov telescopes. The second drawback is that the eyepiece lenses will absorb a small fraction of the solar energy, and this may stress the glass to the point at which it may fracture. Even if this does not occur, the heating will distort the lenses and reduce the quality of the final image. The technique is thus normally limited to telescopes with apertures of 75 mm or less. Larger instruments will need to be stopped down to about this area if they are to be used. As noted above, the use of a commercially produced telescope for solar observation by eyepiece projection may invalidate its guarantee, even if the telescope is not damaged, and the manufacturer should always be consulted before using the telescope in this fashion.

Full aperture filters are produced commercially for use on small telescopes. They cover the whole aperture of the telescope as their name suggests, and may be made sufficiently opaque for use on telescopes up to 0.5 m in diameter. They are the preferred approach to converting a smallish telescope to solar observation. One of the earliest types to be produced, and which is still available today, consists of a very thin Mylar film that has been coated with a reflective layer of aluminum, stainless steel, etc. The film is so thin that very little degradation of the image occurs; however, this also makes the film liable to damage, and its useful life when in regular use can be quite short. Such filters are produced with single or double layers of reflective coating. The double layer is to be preferred since any pinholes in one layer are likely to be covered by the second layer and vice versa. Reflective coatings on optically flat glass substrates have become available recently, though they are usually somewhat more expensive than the coated plastic filters. Finally, the filter may simply be a thin sheet of absorbing plastic (do not be tempted to use any black plastic that may be lying around—it may not be sufficiently opaque in the ultraviolet or infrared—always purchase a purpose-made filter from a reliable supplier). An additional note of warning—do not forget to put a full aperture filter on the guide/finder telescope (or blank it off with an opaque screen). Without such a filter, the guide/finder telescope must not be used to align the telescope—instead the telescope's shadow should be circularized—this is usually sufficient to bring the solar image into the field of view of the (filtered) main telescope. Alternatively, if a small portable telescope's mounting is correctly aligned and has accurate setting circles, or for larger permanently mounted instruments, the Sun may be found in the telescope by setting onto its current position (listed in the *Astronomical Almanac* each year).

Larger instruments built specifically for solar observing come in several varieties. They are designed to try and overcome the major problem of the solar observer that is the turbulence within and outside the instrument, while taking advantage of the observational opportunity afforded by the plentiful supply of radiation. The most intractable of the problems is that of external turbulence. This arises from the differential heating of the observatory, its surroundings, and the atmosphere that leads to the generation of convection currents. The effect reduces with height above the ground, and so one solution is to use a tower telescope. This involves a long focal-length telescope that is fixed vertically with its objective supported on a tower at a height of several tens of meters. The solar radiation is fed into it using a coelostat (Section 1.1.18.2) or heliostat. The latter uses a single-plane mirror rather than the double mirror of the coelostat (cf. siderostat in Section 1.1.18.2), but has the disadvantage that the field of view rotates as the instrument tracks the Sun. The 150 ft (46 m) tower telescope on Mount Wilson, for example, was constructed by George Ellery Hale in 1911 with a 0.3 m lens. It has been observing the Sun every clear day since 1912 and continues to do so to this day.

The other main approach to the turbulence problem (the two approaches are not necessarily mutually exclusive) is to try and reduce the extent of the turbulence. The planting of low growing tree or bush cover of the ground is one method of doing this; another is to surround the observatory by water. The Big Bear Solar Observatory is thus built on an artificial island in Big Bear Lake, California, and houses several solar telescopes including a 0.65 m reflector. Painting the telescope and observatory with titanium-dioxide-based white paint will help to reduce heating since this reflects the solar incoming radiation, but allows the long-wave infrared radiation from the building to be radiated away. A gentle breeze also helps to clear away local turbulence and some tower telescopes such as the 0.45 m Dutch open solar telescope on La Palma are completely exposed to the elements to facilitate the effect of the wind.

Within the telescope, the solar energy can create convection currents, especially from heating the objective. Sometimes, these can be minimized by careful design of the instrument, or the telescope may be sealed and filled with helium whose refractive index is only 1.000036 compared with 1.000293 for air and whose high conductivity helps to keep the components cool and minimize convection. But the ultimate solution to them is to evacuate the optical system to a pressure of a few millibars, so that all the light paths after the objective, or a window in front of the objective, are effectively in a vacuum. Most modern solar telescopes are evacuated such as the 0.7 m German vacuum tower telescope (VTT) and the 0.9 m French THEMIS (télescope héliographique pour l'etude du magnétisme et des instabilités solaires), both on Tenerife. However, it is impractical to evacuate larger instruments, such as the proposed 4 m advanced technology solar telescope (ATST) that is scheduled for first light in 2012 and designed to study solar magnetic fields in particular. The latter though will use adaptive optics to overcome turbulence, and its large size is needed to achieve high-angular (0.03″—20 km on the Sun) and time-resolution (a few seconds) observations of the Sun.

Optically solar telescopes are fairly conventional (Section 1.1) except that very long focal lengths can be used to give large image scales since there is plenty of available light. It is generally undesirable to fold the light paths since the extra reflections will introduce additional scattering and distortion into the image. Thus, the instruments are often very cumbersome, and many are fixed in position and use a coelostat to acquire the solar radiation.

Conventional large optical telescopes can be used for solar infrared observations if they are fitted with a suitable mask. The mask covers the whole of the aperture of the telescope and protects the instrument from the visible solar radiation. It also contains one or more apertures filled with infrared transmission filters. The resulting combination provides high-resolution infrared images without running the normal risk of thermal damage when a large telescope is pointed toward the Sun.

Smaller instruments may conveniently be mounted onto a solar spar. This is an equatorial mounting that is driven so that it tracks the Sun, and which has a general-purpose mounting bracket in the place normally reserved for a telescope. Special equipment may then be attached as desired. Often several separate instruments will be mounted on the spar together and may be in use simultaneously.

Numerous spacecraft, including many manned missions, have carried instruments for solar observing. Amongst the more recent spacecraft whose primary mission was aimed at the Sun, we may pick out as examples the *SMM, Ulysses, Yokhoh,* and *Solar and Heliospheric Observatory (SOHO)*.

SMM operated from 1980 to 1989 and observed the Sun from the ultraviolet through to γ-rays, as well as monitoring the total solar luminosity. *Ulysses* (1992–to date) has an orbit that takes it nearly over the solar poles enabling high-solar latitude regions to be observed clearly for the first time. As well as monitoring solar x-rays, *Ulysses* also studies the solar wind. *Ulysses* started its third pass over the Sun's northern regions in November 2007. *Yokhoh* (1991–2002) was designed to detect high-energy radiation from flares and the corona, while *SOHO* (1995–to date) monitors solar oscillations (see Section 5.3.7) and the solar wind. *SOHO*'s 12th anniversary since its launch was celebrated in December 2007. The two *Solar Terrestrial Relations Observatory* (*STEREO*) spacecraft launched in 2006 with a planned two-year life span are in orbits just inside and just outside that of the earth around the Sun. They therefore gain and lose on the position of the earth in its orbit by 22° $year^{-1}$ (60 million km $year^{-1}$). Their spatial separation of tens of millions of kilometers enables them to obtain 3D images of the Sun and in particular of coronal mass ejections. *Hinode* (meaning "sunrise" in Japanese) was also launched in 2006 carrying a 0.5 m optical telescope, the largest solar telescope in space. Additionally, *Hinode* has an x-ray telescope and an extreme ultraviolet (EUV) spectroscope for studying the solar corona.

However, spacecraft are not the only option. The *Sunrise* (not to be confused with Hinode—above) telescope has recently been flown on a test flight to a height of 37 km by balloon. It is a 1 m telescope designed to observe the Sun in the ultraviolet. Its first science mission is expected in 2009 and it will be flown from the Arctic during the northern summer so that continuous solar observations are possible.

5.3.3 Spectrohelioscope

This is a monochromator that is adapted to provide an image of the whole or a substantial part of the Sun. It operates as a normal spectroscope, but with a second slit at the focal plane of the spectrum that is aligned with the position of the desired wavelength in the spectrum (Figure 5.24). The spectroscope's entrance slit is then oscillated so that it scans across the solar image. As the entrance slit moves, so also will the position of the spectrum. Hence, the second slit must move in sympathy with the first if it is to continue to isolate the required wavelength. Some mechanical arrangements for ensuring that the slits remain in their correct mutual alignments are shown in Figure 5.25. In the first of these, the slit's movements are in anti-phase with each other, and this is probably the

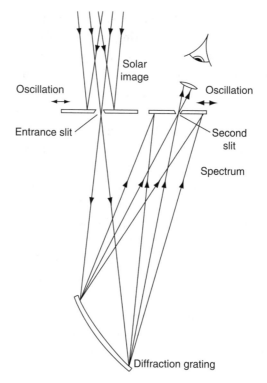

FIGURE 5.24 Principle of the spectrohelioscope.

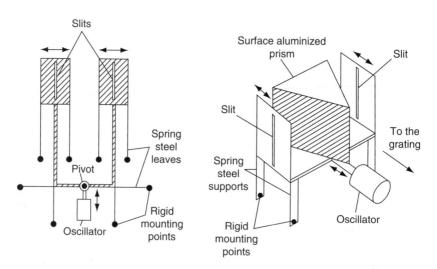

FIGURE 5.25 Mechanical arrangements for co-moving the two slits of a spectrohelioscope.

commonest arrangement. In the second system, the slits are in phase and the mechanical arrangements are far simpler. As an alternative to moving the slits, rotating prisms can be used to scan the beam over the grating and the image plane. If the frequency of the oscillation is higher than 15–20 Hz, then the monochromatic image may be viewed directly by eye. Alternatively, an image may be taken and a spectroheliogram produced.

Usually, the wavelength that is isolated in a spectrohelioscope is chosen to be that of a strong absorption line, so that particular chromospheric features such as prominences, plages, mottling, etc. are highlighted. In the visual region, birefringent filters (see next section) can be used in place of spectrohelioscopes, but for ultraviolet imaging and scanning, satellite-based spectrohelioscopes are used exclusively.

5.3.4 Narrowband Filters

The spectrohelioscope produces an image of the Sun over a very narrow range of wavelengths. However, a similar result can be obtained by the use of a very narrowband filter in the optical path of a normal telescope. Since the bandwidth of the filter must lie in the region of 0.01–0.5 nm if the desired solar features are to be visible, normal dye filters or interference filters (Section 4.1.4.1) are not suitable. Instead, a filter based upon a birefringent material (Section 5.2) has been developed by Bernard Lyot. It has various names: quartz monochromator, Lyot monochromator, or birefringent filter being amongst the commonest.

Its operational principle is based upon a slab of quartz or other birefringent material that has been cut parallel to its optic axis. As we saw in Section 5.2, the extraordinary ray will then travel more slowly than the ordinary ray and so the two rays will emerge from the slab with a phase difference (Figure 5.26). The rays then pass through a sheet of Polaroid film whose axis is midway between the directions of polarization of the ordinary and extraordinary rays. Only the components of each ray that lie along the Polaroid's axis are transmitted by it. Thus, the rays emerging from the Polaroid film have parallel polarization directions and a constant phase difference and so they can mutually interfere. If the original electric vectors of the radiation, E_o and E_e, along the directions of the ordinary and extraordinary rays' polarizations, respectively, are given at some point by

$$E_o = E_e = a \cos(2\pi\nu t) \tag{5.52}$$

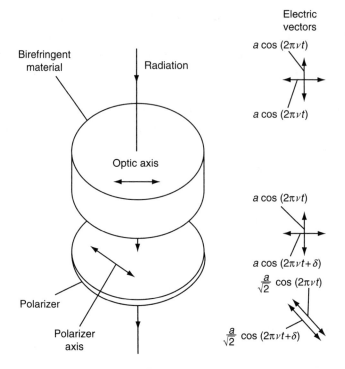

FIGURE 5.26 Basic unit of a birefringent monochromator.

then after passage through the quartz, we will have at some point

$$E_o = a \cos (2\pi\nu t) \qquad (5.53)$$

and

$$E_e = a \cos (2\pi\nu t + \delta) \qquad (5.54)$$

where δ is the phase difference between the two rays. After passage through the Polaroid, we then have

$$E_{45,o} = \frac{a}{\sqrt{2}} \cos (2\pi\nu t) \qquad (5.55)$$

$$E_{45,e} = \frac{a}{\sqrt{2}} \cos (2\pi\nu t + \delta) \qquad (5.56)$$

where $E_{45,x}$ is the component of the electric vector along the Polaroid's axis for the xth component. Thus, the total electric vector of the emerging radiation, E_{45}, will be given by

$$E_{45} = \frac{a}{\sqrt{2}} [\cos(2\pi\nu t) + \cos(2\pi\nu t + \delta)] \tag{5.57}$$

and so the emergent intensity of the radiation, I_{45}, is

$$I_{45} = 2a^2 \quad \text{for } \delta = 2n\pi \tag{5.58}$$

$$I_{45} = 0 \quad \text{for } \delta = (2n+1)\pi \tag{5.59}$$

where n is an integer. Now

$$\delta = \frac{2\pi c \Delta t}{\lambda} \tag{5.60}$$

where Δt is the time delay introduced between the ordinary and extraordinary rays by the material. If v_o and v_e are the velocities of the ordinary and extraordinary rays in the material, then

$$\Delta t = \frac{T}{v_o} - \frac{T}{v_e} \tag{5.61}$$

$$= T\left(\frac{v_e - v_o}{v_e v_o}\right) \tag{5.62}$$

$$= \frac{TJ}{c} \tag{5.63}$$

where
J is the birefringence of the material (Section 5.2)
T is the thickness of the material

Thus,

$$\delta = \frac{\left[2\pi c\left(\frac{TJ}{c}\right)\right]}{\lambda} \tag{5.64}$$

$$= \frac{2\pi TJ}{\lambda} \tag{5.65}$$

The emergent ray therefore reaches a maximum intensity at wavelengths, λ_{max}, given by

$$\lambda_{max} = \frac{TJ}{n} \tag{5.66}$$

and is zero at wavelengths, λ_{min}, given by

$$\lambda_{min} = \frac{2TJ}{(2n+1)} \tag{5.67}$$

Now if we require the eventual filter to have a whole bandwidth of $\Delta\lambda$ centerd on λ_c, then the parameters of the basic unit must be such that one of the maxima in Figure 5.27 coincides with λ_c, and the width of one of the fringes is $\Delta\lambda$; that is,

$$\lambda_c = \frac{TJ}{n_c} \tag{5.68}$$

$$\Delta\lambda = TJ\left(\frac{1}{n_c - 1/2} - \frac{1}{n_c + 1/2}\right) \tag{5.69}$$

Since selection of the material fixes J, and n_c is deviously related to T, the only truly free parameter of the system is T, and thus for a given filter, we have

$$T = \frac{\lambda_c}{2J}\left[\frac{\lambda_c}{\Delta\lambda} + \left(\frac{\lambda_c^2}{\Delta\lambda^2} + 1\right)^{1/2}\right] \tag{5.70}$$

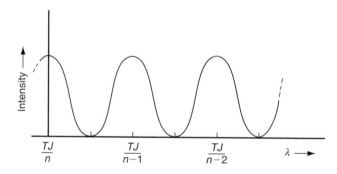

FIGURE 5.27 Spectrum of the emerging radiation from a basic unit of a birefringent monochromator.

and

$$n_c = \frac{TJ}{\lambda_c} \tag{5.71}$$

Normally, however,

$$\lambda_c \gg \Delta\lambda \tag{5.72}$$

and quartz is the birefringent material, for which in the visible region

$$J = +0.0092 \tag{5.73}$$

so that

$$T \approx \frac{109\lambda_c^2}{\Delta\lambda} \tag{5.74}$$

Thus, for a quartz filter to isolate the Hα line at 656.2 nm with a bandwidth of 0.1 nm, the thickness of the quartz plate should be about 470 mm.

Now from just one such basic unit, the emergent spectrum will contain many closely spaced fringes as shown in Figure 5.26. But if we now combine it with a second basic unit oriented at 45° to the first, and whose central frequency is still λ_c, but whose bandwidth is $2\Delta\lambda$, then the final output will be suppressed at alternate maxima (Figure 5.28). From Equation 5.74, we see that the thickness required for the second unit, for it to behave in this way, is just half the thickness of the first unit. Further basic units may be added, with thicknesses 1/4, 1/8, 1/16, etc. that of the first, whose transmissions continue to be centerd upon λ_c, but whose bandwidths are 4, 8, 16, etc. times that of the original unit. These continue to suppress additional unwanted maxima. With six such units, the final spectrum has only a few narrow maxima that are separated from λ by multiples of $32\Delta\lambda$ (Figure 5.29). At this stage, the last remaining unwanted maxima are sufficiently well separated from λ_c for them to be eliminated by conventional dye or interference filters, so that just the desired transmission curve remains. One further refinement to the filter is required and that is to ensure that the initial intensities of the ordinary and extraordinary rays are equal, since this was assumed in obtaining Equation 5.52. This is easily accomplished, however, by placing an

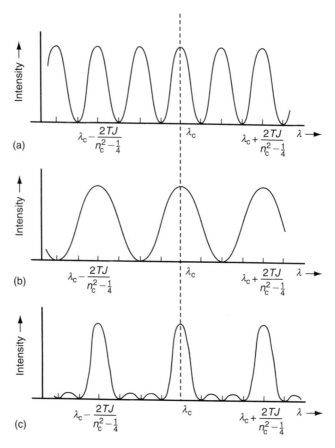

FIGURE 5.28 Transmission curves of birefringent filter basic units. (a) Basic unit of thickness T; (b) basic unit of thickness, $T/2$; (c) combination of the units in (a) and (b).

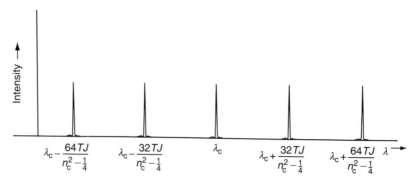

FIGURE 5.29 Transmission curve of a birefringent filter comprising six basic units with a maximum thickness T.

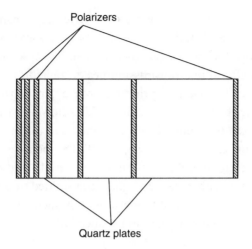

Polarizers

Quartz plates

FIGURE 5.30 Six-element quartz birefringent filter.

additional sheet of Polaroid before the first unit that has its transmission axis at 45° to the optic axis of the first unit. The complete unit is shown in Figure 5.30. Neglecting absorption and other losses, the peak transmitted intensity is half the incident intensity because of the use of the first Polaroid. The whole filter though uses such a depth of quartz that its actual transmission is a few percent. The properties of quartz are temperature dependent and so the whole unit must be enclosed and its temperature controlled to a fraction of a degree when it is in use. However, the temperature dependence allows the filter's central wavelength to be tuned slightly by changing the operating temperature. Much larger changes in wavelength may be encompassed by splitting each retarder into two opposed wedges. Their combined thicknesses can then be varied by displacing one set of wedges with respect to the other in a direction perpendicular to the axis of the filter. The wavelength may also be decreased by slightly tilting the filter to the direction of the incoming radiation, so increasing the effective thicknesses of the quartz plates. For the same reason, the beam of radiation must be collimated for its passage through the filter, or the bandwidth will be increased.

A closely related filter is due to Solč.* It uses only two polarizers placed before and after the retarders. All the retarders are of the same thickness, and they have their optic axes alternately oriented with respect to the axis of the first polarizer in clockwise and anticlockwise directions at some

---

* Pronounced "sholts."

specified angle. The output consists of a series of transmission fringes whose spacing in wavelength terms increases with wavelength. Two such units of differing thicknesses can then be combined so that only at the desired wavelength do the transmission peaks coincide.

Another device for observing the Sun over a very narrow wavelength range is the magneto-optical filter (MOF). This, however, can only operate at the wavelengths of strong absorption lines due to gases. It comprises two polarizers on either side of a gas cell. The polarizers are oriented orthogonally to each other if they are linear or are left- and right-handed if circular in nature. In either case, no light is transmitted through the system. A magnetic field is then applied to the gas cell, inducing the Zeeman effect. The Zeeman components of the lines produced by the gas are then linearly and/or circularly polarized (Section 5.4.1.1) and so permit the partial transmission of light at their wavelengths through the whole system. The gases currently used in MOFs are vapors of sodium or potassium.

Relatively inexpensive Hα filters can be made as solid Fabry–Perot etalons (Section 4.1.4.1). A thin fused-silica spacer between two optically flat dielectric mirrors is used together with a blocking filter. Peak transmissions of several tens of percent and bandwidths of better than a tenth of a nanometer can be achieved in this way.

5.3.5 Coronagraph

This instrument enables observations of the corona to be made at times other than during solar eclipses. It does this by producing an artificial eclipse. The principle is very simple; an occulting disk at the prime focus of a telescope obscures the photospheric image while allowing that of the corona to pass by. The practice, however, is considerably more complex, since scattered and/or diffracted light, etc. in the instrument and the atmosphere can still be several orders of magnitude brighter than the corona. Extreme precautions have therefore to be taken in the design and operation of the instrument to minimize this extraneous light.

The most critical of these precautions lies in the structure of the objective. A single simple lens objective is used to minimize the number of surfaces involved and it is formed from a glass blank that is as free from bubbles, striae, and other imperfections as possible. The surfaces of the lens are polished with extreme care to eliminate all scratches and other surface markings. In use, they are kept dust-free by being tightly sealed when not in operation, and by the addition of a very long tube lined with grease, in front of the objective to act as a dust trap.

The occulting disk is a polished metal cone or an inclined mirror so that the photospheric radiation may be safely reflected away to a separate light and heat trap. Diffraction from the edges of the objective is eliminated by imaging the objective with a Fabry lens after the occulting disk, and by using a stop that is slightly smaller than the image of the objective to remove the edge effects. Alternatively, the objective can be apodized (Section 2.5); its transparency decreases from its center to its edge in a Gaussian fashion. This leads to full suppression of the diffraction halo although with some loss in resolution. A second occulting disk before the final objective removes the effects of multiple reflections within the first objective. The final image of the corona is produced by this second objective, and this is placed after the diffraction stop. The full system is shown in Figure 5.31.

The atmospheric scattering can only be reduced by a suitable choice of observing site. Thus, the early coronagraphs are to be found at high-altitude observatories, while more recently they have been mounted on spacecraft to eliminate the problem of the earth's atmosphere entirely.

The use of a simple lens results in a chromatic prime focus image. Therefore, a filter must normally be added to the system. This is desirable in any case since the coronal spectrum is largely composed of emission lines superimposed upon a diluted solar photospheric spectrum. Selection of a narrowband filter that is centerd upon a strong coronal emission line therefore considerably improves the contrast of the final image. White light or wideband imaging of the corona is only possible using earth-based instruments on rare occasions, and it can only be attempted under

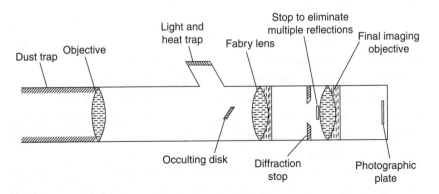

FIGURE 5.31 Schematic optical arrangement of a coronagraph.

absolutely optimum observing conditions. Spacecraft-borne instruments may be so used more routinely.

Improvements to the basic coronagraph may be justified for balloon or spacecraft-borne instruments, since the sky background is then less than the scattered light in a more normal device and they have taken two different forms. A reflecting objective can be used. This is formed from an uncoated off-axis parabola. Most of the light passes through the objective and is absorbed. Bubbles and striations in the glass are of much less importance since they cause scattering primarily in the forward direction. The mirror is uncoated since metallic films are sufficiently irregular to cause a considerable amount of scattering. In other respects, the coronagraph then follows the layout of Figure 5.31. The second approach to the improvement of coronagraphs is quite different and consists simply of producing the artificial eclipse outside the instrument rather than at its prime focus. An occulting disk is placed well in front of the first objective of an otherwise fairly conventional coronagraph. The disk must be large enough to ensure that the first objective lies entirely within its umbral shadow. The inner parts of the corona will therefore be badly affected by vignetting. However, this is of little importance since these are the brightest portions of the corona, and it may even be advantageous since it will reduce the dynamic range that must be covered by the detector. A simple disk produces an image with a bright central spot because of diffraction, but this can be eliminated by using a disk whose edge is formed into a zigzag pattern of sharp teeth, or by the use of multiple occulting disks. By such means, the instrumentally scattered light can be reduced to 10^{-4} of that of a basic coronagraph.

The final image in a coronagraph may be imaged directly, but more commonly is fed to a spectroscope, photometer, or other ancillary instrument. From the earth, the corona may normally only be detected out to about one solar radius, but satellite-based coronagraphs have been used successfully for detections out to six solar radii or more.

Devices similar to the coronagraph are also sometimes carried on spacecraft so that they may observe the atmospheres of planets, while shielding the detector from the radiation from the planet's surface. The contrast is generally far smaller than in the solar case so that such planetary coronagraphs can be far simpler in design.

Terrestrial and spacecraft-borne stellar coronagraphs have also recently been constructed. These are designed to allow the detection of faint objects close to much brighter objects, such as stellar companions and accretion disks, by eliminating the bright stellar image in a similar manner to the removal of the solar image in a solar coronagraph. With the addition of suitable apodization stops (Section 4.1.4.2) and adaptive optics (Section 1.1), it may even be possible to detect directly extrasolar planets. The coronagraphic imager with adaptive optics (CIAO) for the 8 m Subaru telescope, for example, was a near-infrared coronagraph using a 1024×1024 indium antimonide array detector. It reduced the intensity by a factor of 30–40 within a few tens of a second of arc of the central object. It was replaced in 2007 by HiCIAO that benefits from the use of adaptive optics and an improved design to increase the contrast compared with CIAO by a factor of 10–100. The nulling interferometer (Section 2.5) similarly reduces the intensity of the central object.

The proposed *New Worlds Observer* spacecraft has been mentioned in Section 1.1.19. This would use an external shield like more conventional coronagraphs, but at a distance of some 30,000 km from the telescope. The shield would be petal-shaped to eliminate diffraction effects and could be used with an existing or future spacecraft, such as the James Webb Space Telescope (JWST), to image extrasolar planets. A completely different approach to producing a stellar coronagraph has recently been suggested and this is to use an optical vortex. The optical vortex looks like a 360° turn of the steps of a spiral staircase. It is a helical mask with steps of progressively increasing thickness constructed from a transparent material. When the phase delay of the radiation passing through the thinnest step is two wavelengths less than that through the thickest step, destructive interference occurs such that the radiation passing through the center of the mask is eliminated. The operating wavelength of the optical vortex depends upon the step heights, and so it is essentially a monochromatic device, nonetheless it may provide the possibility for future stellar coronagraphs that are much more compact than existing devices.

5.3.6 Pyrheliometer

This is an instrument intended to measure the total flux of solar radiation at all wavelengths. In practice, current devices measure the energy from the microwave to the soft x-ray region. Modern pyrheliometers are active cavity radiometers. The radiation is absorbed within a conical cavity. The cavity is held in contact with a heat sink and maintained at a temperature about 1 K higher than the heat sink by a small heater. The difference between the power used by the heater when the cavity is exposed to the Sun and that when a shutter closes it off provides the measure of the solar energy.

5.3.7 Solar Oscillations

Whole-body vibrations of the Sun reveal much about its inner structure. They are small-scale effects and their study requires very precise measurements of the velocities of parts of the solar surface. The resonance scattering spectrometer originated by the solar group at the University of Birmingham is capable of detecting motions of a few tens of millimeters per second. It operates by passing the 770 nm potassium line from the Sun through a container of heated potassium vapor. A magnetic field is directed through the vapor and the solar radiation is circularly polarized. Depending upon the direction of the circular polarization either the light from the long-wave wing (I_L) or the short wave wing (I_S) of the line is resonantly scattered. By switching the direction of the circular polarization rapidly, the two intensities may be measured nearly simultaneously. The line of sight velocity is then given by

$$v = k \, \frac{I_S - I_L}{I_S + I_L} \tag{5.75}$$

where k is a constant with a value around 3 km s^{-1}.

The Birmingham solar oscillations network (BISON) has six automated observatories around the world so that almost continuous observations of the Sun can be maintained. It measures the average radial velocity over the whole solar surface using resonance scattering spectrometers. The GONG project (Global Oscillation Network Group) likewise has six observatories, and measures radial velocities on angular scales down to 8″ using Michelson interferometers to monitor the position of the nickel 676.8 nm spectrum line. The *SOHO* spacecraft carries three helioseismology instruments: global oscillations at low frequencies (GOLF), Michelson Doppler interferometer (MDI), and variability of solar irradiance and gravity oscillations (VIRGO). GOLF is a resonance scattering device based upon sodium D line that detects motions to better than 1 mm s^{-1} over the whole solar surface. MDI is a Michelson Doppler imager that measures magnetic fields as well as velocities, while VIRGO measures the solar constant and detects oscillations via variations in the Sun's brightness.

5.3.8 Other Solar Observing Methods

Slitless spectroscopes (Section 4.2.1) are of considerable importance in observing regions of the Sun whose spectra consist primarily of emission lines. They are simply spectroscopes in which the entrance aperture is large enough to accept a significant proportion of the solar image. The resulting spectrum is therefore a series of monochromatic images of this part of the Sun in the light of each of the emission lines. They have the advantage of greatly improved speed over a slit spectroscope combined with the obvious ability to obtain spectra simultaneously from different parts of the Sun. Their most well-known application is during solar eclipses, when the flash spectrum of the chromosphere may be obtained in the few seconds immediately after the start of totality or just before its end. More recently, they have found widespread use as spacecraft-borne instrumentation for observing solar flares in the ultraviolet part of the spectrum.

One specialized satellite-borne instrument based upon slitless spectroscopes is the Ly α camera. The Lyman α line is a strong emission line. If the image of the whole disk of the Sun from an ultraviolet telescope enters a slitless spectroscope, then the resulting Ly α image may be isolated from the rest of the spectrum by a simple diaphragm, with very little contamination from other wavelengths. A second identical slitless spectroscope whose entrance aperture is this diaphragm, and whose dispersion is perpendicular to that of the first spectroscope will then provide a stigmatic spectroheliogram at a wavelength of 121.6 nm.

A solar 'instrument' that has recently become available is the virtual solar observatory (see also Section 5.5). This is a software system linking various solar data archives and which provides tools for searching and analyzing the data. It may be accessed by anyone who is interested at http://vso.nascom.nasa.gov/cgi-bin/search.

In the radio region, solar observations tend to be undertaken by fairly conventional equipment although there are now a few dedicated solar radio telescopes. One exception to this, however, is the use of multiplexed receivers to provide a quasi-instantaneous radio spectrum of the Sun. This is of particular value for the study of solar radio bursts, since their emitted bandwidths may be quite narrow, and their wavelengths drift rapidly with time. These radio spectroscopes and the acousto-optical radio spectroscope were discussed more fully in Section 1.2. The data from them is usually presented as a frequency/time plot, from whence the characteristic behavior patterns of different types of solar bursts may easily be recognized.

This account of highly specialized instrumentation and techniques for solar observing could be extended almost indefinitely, and might encompass all the equipment designed for parallax observations, solar radius determinations, oblateness determinations, eclipse work, tree rings, and C^{14} determination, etc. However, at least in the author's opinion, these are becoming too specialized for inclusion in a general book like this, and so the reader is referred to more specialized texts and the scientific journals for further information upon them.

Exercise 5.5

Calculate the maximum and minimum thicknesses of the elements required for an Hα birefringent filter based upon calcite, if its whole bandwidth is to be 0.05 nm, and it is to be used in conjunction with an interference filter whose whole bandwidth is 3 nm. The birefringence of calcite is −0.172.

5.4 MAGNETOMETRY

5.4.1 Background

The measurement of astronomical magnetic fields is accomplished in two quite separate ways. The first is direct measurement by means of apparatus carried by spacecraft, while the second is indirect and is based upon the Zeeman effect of a magnetic field upon spectrum lines (or more strictly upon the inverse Zeeman effect, since it is usually applied to absorption lines). A third approach suggested recently is to observe the x-rays arising from the solar wind interactions with the earth's magnetosheath and thereby infer the shape of the magnetic fields involved (see MagEX, Section 1.3.4.1), but this has yet to be tried in practice.

5.4.1.1 Zeeman Effect

The Zeeman effect describes the change in the structure of the emission lines in a spectrum when the emitting object is in a magnetic field. The simplest change arises for singlet lines—that is lines arising from transitions between levels with a multiplicity of one, or a total spin quantum number, M_s of zero for each level. For these lines, the effect is called the normal Zeeman effect. The line splits into two or three components depending on whether the line of sight is along, or perpendicular to the magnetic field lines. An appreciation of the basis of the normal Zeeman effect may be obtained from a classical approach to the problem. If we imagine an electron in an orbit around an atom, then its motion may be resolved into three simple harmonic motions along the three coordinate axes. Each of these in turn we may imagine to be the sum of two equal but opposite circular motions (Figure 5.32). If we now imagine a magnetic field applied along the z-axis, then these various motions may be modified

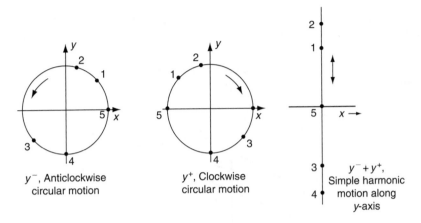

FIGURE 5.32 Resolution of simple harmonic motion along the y-axis into two equal but opposite circular motions.

by it. First, the simple harmonic motion along the z-axis will be unchanged since it lies along the direction of the magnetic field. The simple harmonic motions along the x and y axes, however, are cutting across the magnetic field and so will be altered. We may best see how their motion changes by considering the effect upon their circular components. When the magnetic field is applied, the radii of the circular motions remain unchanged, but their frequencies alter. If ν is the original frequency of the circular motion, and H is the magnetic field strength, then the new frequencies of the two resolved components ν^+ and ν^- are

$$\nu^+ = \nu + \Delta\nu \qquad (5.76)$$

$$\nu^- = \nu - \Delta\nu \qquad (5.77)$$

where

$$\Delta\nu = \frac{eH}{4\pi m_e c} \qquad (5.78)$$

$$= 1.40 \times 10^{10}\ H\ \text{Hz T}^{-1} \qquad (5.79)$$

Thus, we may combine the two higher frequency components arising from the x and y simple harmonic motions to give a single elliptical motion in the xy plane at a frequency of $\nu + \Delta\nu$. Similarly, we may combine the lower frequency components to give another elliptical

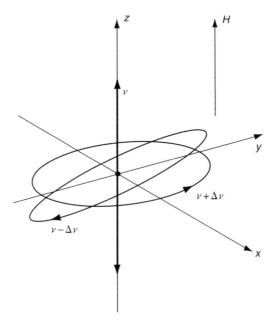

FIGURE 5.33 Components of electron orbital motion in the presence of a magnetic field.

motion in the xy plane at a frequency of $\nu - \Delta\nu$. Thus, the electron's motion may be resolved, when it is in the presence of a magnetic field along the z-axis, into two elliptical motions in the xy plane, plus simple harmonic motion along the z-axis (Figure 5.33), the frequencies being $\nu + \Delta\nu$, $\nu - \Delta\nu$, and ν, respectively. Now if we imagine looking at such a system, then only those components that have some motion across the line of sight will be able to emit light toward the observer, since light propagates in a direction perpendicular to its electric vector. Hence looking along the z-axis (i.e., along the magnetic field lines), only emission from the two elliptical components of the electron's motion will be visible. Since the final spectrum contains contributions from many atoms, these will average out to two circularly polarized emissions, shifted by $\Delta\nu$ from the normal frequency (Figure 5.34), one with clockwise polarization, and the other with anticlockwise polarization. Lines of sight perpendicular to the magnetic field direction, that is within the xy plane, will in general view the two elliptical motions as two collinear simple harmonic motions, while the z-axis motion will remain as simple harmonic motion orthogonal to the first two motions. Again the spectrum is the average of many

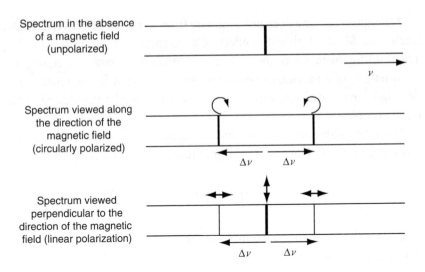

FIGURE 5.34 Normal Zeeman effect.

atoms, and so will therefore comprise of three linearly polarized lines. The first of these is at the normal frequency of the line and is polarized parallel to the field direction. It arises from the z-axis motion. The other two lines are polarized at right angles to the first and are shifted in frequency by $\Delta\nu$ (Figure 5.34) from the normal position of the line. When observing the source along the line of the magnetic field, the two spectrum lines have equal intensities, while when observing perpendicular to the magnetic field, the central line has twice the intensity of either of the other components. Thus, imagine the magnetic field progressively reducing; then as the components remix an unpolarized line results, as one would expect. This pattern of behavior for a spectrum line originating in a magnetic field is termed the normal Zeeman effect.

In astronomy, absorption lines, rather than emission lines, are the main area of interest, and the inverse Zeeman effect describes their behavior. This, however, is precisely the same as the Zeeman effect except that emission processes are replaced by their inverse absorption processes. The above analysis may therefore be equally well applied to describe the behavior of absorption lines. The one major difference from the emission line case is that the observed radiation remaining at the wavelength of one of the lines is preferentially polarized in the opposite sense to that of the Zeeman component, since the Zeeman component is being subtracted from unpolarized radiation.

If the spectrum line does not originate from a transition between singlet levels (i.e., $M_s \neq 0$), then the effect of a magnetic field is more complex. The resulting behavior is known as the anomalous Zeeman effect, but this is something of a misnomer since it is anomalous only in the sense that it does not have a classical explanation. Quantum mechanics describes the effect completely.

The orientation of an atom in a magnetic field is quantized in the following manner. The angular momentum of the atom is given by

$$[J(J+1)]\frac{h}{2\pi} \tag{5.80}$$

where J is the inner quantum number, and its space quantization is such that the projection of the angular momentum onto the magnetic field direction must be an integer multiple of $h/2\pi$ when J is an integer, or a half-integer multiple of $h/2\pi$ when J is a half-integer. Thus, there are always $(2J+1)$ possible quantized states (Figure 5.35). Each state may be

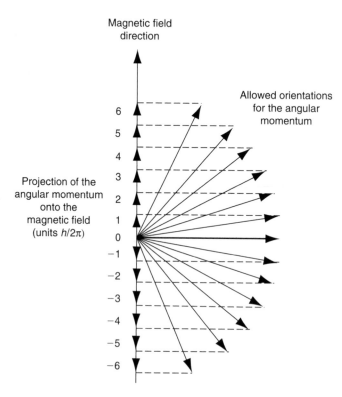

FIGURE 5.35 Space quantization for an atom with $J=6$.

described by a total magnetic quantum number, M, which for a given level can take all integer values from $-J$ to $+J$ when J is an integer, or all half-integer values over the same range when J is a half-integer. In the absence of a magnetic field, electrons in these states all have the same energy (i.e., the states are degenerate), and the set of states forms a level. Once a magnetic field is present, however, electrons in different states have different energies, with the energy change from the normal energy of the level, ΔE, being given by

$$\Delta E = \frac{eh}{4\pi m_e c} MgH \qquad (5.81)$$

where g is the Landé factor, which is given by

$$g = 1 + \frac{J(J+1) + M_s(M_s+1) + L(L+1)}{2J(J+1)} \qquad (5.82)$$

where L is the total azimuthal quantum number. Thus, the change in the frequency, $\Delta\nu$, for a transition to or from the state is

$$\Delta\nu = \frac{e}{4\pi m_e c} MgH \qquad (5.83)$$

$$= 1.40 \times 10^{10} \, MgH \text{ Hz T}^{-1} \qquad (5.84)$$

Now for transitions between such states we have the selection rule that M can change by 0 or ± 1 only. Thus, we may understand the normal Zeeman effect in the quantum mechanical case simply from the allowed transitions (Figure 5.36) plus the fact that the splitting of each level is by the same amount since for the singlet levels

$$M_s = 0 \qquad (5.85)$$

so that

$$J = L \qquad (5.86)$$

and so

$$g = 1 \qquad (5.87)$$

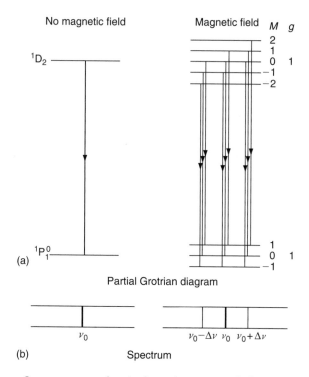

FIGURE 5.36 Quantum mechanical explanation of the normal Zeeman effect.

Each of the normal Zeeman components for the example shown in Figure 5.36 is therefore triply degenerate, and only three lines result from the nine possible transitions. When M_s is not zero, g will in general be different for the two levels and the degeneracy will cease. All transitions will then produce separate lines, and hence we get the anomalous Zeeman effect (Figure 5.37). Only one of many possible different patterns is shown in the figure; the details of any individual pattern will depend upon the individual properties of the levels involved.

As the magnetic field strength increases, the pattern changes almost back to that of the normal Zeeman effect. This change is known as the Paschen–Back effect. It arises as the magnetic field becomes strong enough to decouple L and M_s from each other. They then no longer couple together to form J, which then couples with the magnetic field, as just described, but couple separately and independently to the magnetic field. The pattern of spectrum lines is then that of the normal Zeeman effect, but with each component of the pattern formed from a narrow doublet,

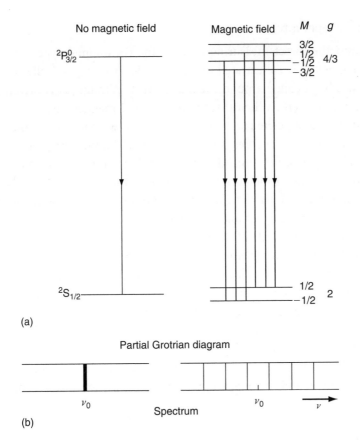

FIGURE 5.37 Quantum mechanical explanation of the anomalous Zeeman effect.

triplet, etc., accordingly, as the original transition was between doublet, triplet, etc. levels. Field strengths of around 0.5 T or more are usually necessary for the complete development of the Paschen–Back effect.

At very strong magnetic field strengths ($>10^3$ T), the quadratic Zeeman effect will predominate. This displaces the spectrum lines to higher frequencies by an amount $\Delta\nu$, which is given by

$$\Delta\nu = \frac{\varepsilon_0 h^3}{8\pi^2 m_e^3 c^2 e^2 \mu_0} n^4(1 + M^2)H^2 \tag{5.88}$$

$$= 1.489 \times 10^4 n^4(1 + M^2)H^2 \text{ Hz} \tag{5.89}$$

where n is the principal quantum number.

5.4.2 Magnetometers

Amongst the direct measuring devices, the commonest type is the flux-gate magnetometer illustrated in Figure 5.38. Two magnetically soft cores have windings coiled around them as shown. Winding A is driven by an alternating current. When there is no external magnetic field, winding B has equal and opposite currents induced in its two coils by the alternating magnetic fields of the cores, and so there is no output. The presence of an external magnetic field introduces an imbalance into the currents, giving rise to a net alternating current output. This may then easily be detected and calibrated in terms of the external field strength. Spacecraft usually carry three such magnetometers oriented orthogonally to each other so that all the components of the external magnetic field may be measured.

Another type of magnetometer that is often used on spacecraft is based upon atoms in a gas oscillating at their Larmor frequency. It is often used for calibrating flux-gate magnetometers in orbit. The vector helium magnetometer operates by detecting the effect of a magnetic field upon the population of a metastable state of helium. A cell filled with helium is illuminated by 1.08 μm radiation that pumps the electrons into the metastable state. The efficiency of the optical pump is affected by the magnetic field, and the population of that state is monitored by observing

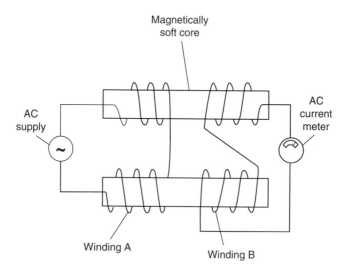

FIGURE 5.38 Flux-gate magnetometer.

the absorption of the radiation. A variable artificially generated magnetic field is swept through the cell until the external magnetic field is nullified. The strength and direction of the artificial field are then equal and opposite to the external field. Any magnetometer on board a spacecraft usually has to be deployed at the end of a long boom after launch to distance it from the magnetic effects of the spacecraft itself.

An alternative technique altogether is to look at the electron flux. Since electron paths are modified by the magnetic field, their distribution can be interpreted in terms of the local magnetic field strength. This enables the magnetic field to be detected with moderate accuracy over a large volume, in comparison with the direct methods that give very high accuracies, but only in the immediate vicinity of the spacecraft.

Most of the successful indirect work in magnetometry based upon the (inverse) Zeeman effect has been undertaken by Harold and Horace Babcock or uses instruments based upon their designs. For stars, the magnetic field strength along the line of sight may be determined via the longitudinal Zeeman effect. The procedure is made very much more difficult than in the laboratory by the widths of the spectrum lines. For a magnetic field of 1 T, the change in the frequency of the line is 1.4×10^{10} Hz (Equation 5.79), which for lines near 500 nm corresponds to a separation for the components in wavelength terms of only 0.02 nm. This is smaller than the normal line width for most stars. Thus, even strong magnetic fields do not cause the spectrum lines to split into separate components, but merely to become somewhat broader than normal. The technique is saved by the opposite senses of circular polarization of the components that enables them to be separated from each other. Babcock's differential analyzer therefore consists of a quarter-wave plate (Section 5.2.2.3) followed by a doubly refracting calcite crystal. This is placed immediately before the entrance aperture of a high-dispersion spectroscope (Figure 5.39). The quarter-wave plate converts the two circularly polarized components into two mutually perpendicular linearly polarized components. The calcite is suitably oriented so that one of these components is the ordinary ray and the other the extraordinary ray. The components are therefore separated by the calcite into two beams. These are both accepted by the spectroscope slit and so two spectra are formed in close juxtaposition, with the lines of each slightly shifted with respect to the other because of the Zeeman splitting. This shift may then be measured and translated back to give the longitudinal magnetic field intensity.

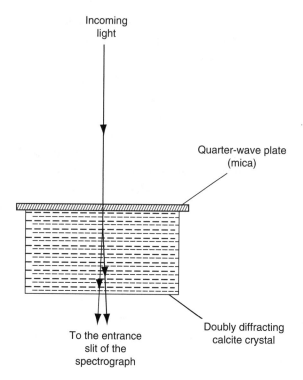

FIGURE 5.39 Babcock's differential analyzer.

A lower limit on the magnetic field intensity of about 0.02 T is detectable by this method, while the strongest stellar fields found have strengths of a few teslas. By convention, the magnetic field is positive when it is directed toward the observer.

The relatively new technique of Zeeman–Doppler imaging potentially enables maps of the distribution of magnetic fields over the surfaces of stars to be constructed, though often with considerable ambiguities left in the results. The basis of the technique is best envisaged by imagining a single magnetic region on the surface of a star. The Zeeman effect will induce polarization into the lines emitted or absorbed from this region, as already discussed. However, when the rotation of the star brings the region first into view, it will be approaching us. The features from the region will therefore be blue-shifted (i.e., Doppler shifted) compared with their normal wavelengths. As the rotation continues to move the region across the disk of the star, the spectral features will change their wave-lengths, going through their normal positions to being red-sifted just before they disappear around the opposite limb of the star. If there are several magnetic regions, then each will usually have a different Doppler shift at any given moment and the changing Doppler shifts can be interpreted in terms of the relative positions of the magnetic regions on the star's surface (i.e., a map). The reduction of the data from this type of observations is a complex procedure and maximum entropy method (MEM) is usually needed to constrain the range of possible answers. Spectropolarimeters such as SemelPol* on the Anglo-Australian telescope (AAT), ESPaDons (echelle spectropolarimetric device for the observation of stars) on the Canada–France–Hawaii telescope (CFHT), and Narval on the Pic du Midi's 2 m Lyot telescope have been used in this manner recently, but the technique is still in its infancy.

On the Sun much greater sensitivity is possible, and fields as weak as 10^{-5} T can be studied. George Ellery Hale first detected magnetic fields in sunspots in 1908, but most of the modern methods and apparatus were invented by the Babcocks. Their method relies upon the slight shift in the line position between the two circularly polarized components, which causes an exaggerated difference in their relative intensities in the wings of the lines (Figure 5.40). The solar light is passed through a differential

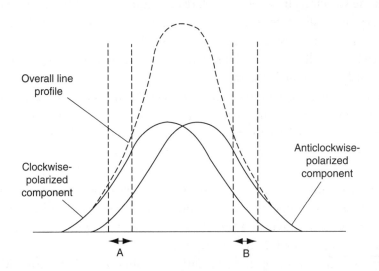

FIGURE 5.40 Longitudinal Zeeman components.

* Not an anagram—named after its designer, Meir Semel.

analyzer as before. This time, however, the quarter-wave plate is composed of ammonium dihydrogen phosphate, and this requires an electric potential of some 9 kV across it, to make it sufficiently birefringent to work. Furthermore, only one of the beams emerging from the differential analyzer is fed to the spectroscope. Originally, a pair of photomultipliers that accepted the spectral regions labeled A and B in Figure 5.40 detected the line, now CCDs or other array detectors are used. Thus, the wing intensities of, say, the clockwise-polarized component are detected when a positive voltage is applied to the quarter-wave plate, while those of the anticlockwise component are detected when a negative voltage is applied. The voltage is switched rapidly between the two states, and the outputs are detected in phase with the switching. Since all noise except that in phase with the switching is automatically eliminated, and the latter may be reduced by integration and/or the use of a phase-sensitive detector (Section 5.2.3.2), the technique is very sensitive. The entrance aperture to the apparatus can be scanned across the solar disk, so that a magnetogram can be built up with a typical resolution of a few seconds of arc and of a few microteslas. Specialized CCD detectors can be used that have every alternate row of pixels covered over. The accumulating charges in the pixels are switched between an exposed row and a covered row in phase with the switching between the polarized components. Thus, at the end of the exposure, alternate rows of pixels contain the intensities of the clockwise and anticlockwise components. Similar instruments have been devised for stellar observations, but so far are not as sensitive as the previously mentioned system.

Several other types of magnetometer have been devised for solar work. For example, at the higher field strengths, Leighton's method provides an interesting pictorial representation of the magnetic patterns. Two spectroheliograms are obtained in opposite wings of a suitable absorption line, and through a differential analyzer that is arranged so that it only allows the passage into the system of the stronger circularly polarized component in each case. A composite image is then made of both spectroheliograms, with one as a negative, and the other as a positive. Areas with magnetic fields less than about 2 mT are seen as gray, while areas with stronger magnetic fields show up light or dark according to their polarity. More modern devices use video recording, but the basic technique remains unaltered. Vector-imaging magnetographs image the Sun over a very narrow wavelength range, and then rapidly step that image through a magnetically sensitive spectrum line. This enables both the direction and strength of the magnetic field to be determined over the selected region. Usually, this region is just a small part of the whole solar disk because of the enormous amounts of data that such instruments can generate.

In the radio region, the Zeeman effect affects lines such as the 1.42 GHz emission from hydrogen (usually better known as the 21 cm line). Their Zeeman splitting is found in a similar manner to Babcock's solar method, by observing in the steepest part of the wings. Interstellar magnetic fields of about 10^{-9} T are detectable in this way.

The very strongest fields, up to 10^8 T, are expected to exist in white dwarfs and neutron stars. However, there is usually little spectral structure in such objects from which the fields might be detected. The quadratic Zeeman effect, however, may then be sufficient for circular polarization of the continuum to be distinguished at points where it is varying rapidly in intensity. Field strengths of upward of 1000 T are found in this way.

A few devices have been constructed to detect electric fields via the Stark effect upon the Paschen hydrogen lines. They are basically similar to magnetometers and for solar studies can reach sensitivities of 500 V m^{-1}.

5.4.3   Data Reduction and Analysis

Once a flux-gate magnetometer has been calibrated, there is little involved in the analysis of its results except to convert them to field strength and direction. However, it may be necessary to correct the readings for influence from the spacecraft, and for the effects of the solar wind and the solar or interplanetary magnetic fields. Measurements with accuracies as high as 10^{-11} T are possible when these corrections are reliably known.

With the first of Babcock's methods, in which two spectra are produced side by side, the spectra are simply measured as if for radial velocity. The longitudinal magnetic field strength can then be obtained via Equation 5.84. The photoelectric method's data are rather more complex to analyze. No general method can be given since it will depend in detail upon the experimental apparatus and method and upon the properties of the individual line (or continuum) being observed.

With strong solar magnetic fields, such as those found in sunspots, information may be obtainable in addition to the longitudinal field strength. The line in such a case may be split into two or three components, and when these are viewed through a circular analyzer (i.e., an analyzer that passes only one circularly polarized component), the relative intensities of the three lines of the normal Zeeman effect are given by Seare's equations

$$I_v = \tfrac{1}{4}(1 \pm \cos\theta)^2 I \tag{5.90}$$

$$I_c = \tfrac{1}{2}(\sin^2\theta)I \tag{5.91}$$

$$I_r = \tfrac{1}{4}(1\mu\cos\theta)^2 I \tag{5.92}$$

where

I_v, I_c, and I_r are the intensities of the high-frequency, central, and low-frequency components of the triplet

I is the total intensity of all the components

θ is the angle of the magnetic field's axis to the line of sight

Thus, the direction of the magnetic field as well as its total strength may be found.

5.5 COMPUTERS AND THE INTERNET

5.5.1 Introduction

Computers and their applications are now essential to just about every aspect of astrophysics, and indeed to most of the rest of science and life in general. Many of the uses of computers, such as controlling telescopes, processing data, etc., have already been mentioned within other sections of this book, but two promising applications deserve a section to themselves. These are the availability of digital sky surveys, and the related development of virtual observatories.

A major concern of astronomy, dating back to Hipparchus and earlier, has always been the cataloging or listing of objects in the sky together with their positions and properties. The quantum changes that have occurred in the last few years are that many of these catalogs are now on open access through the World Wide Web, and personal computers affordable by individuals are now powerful enough to enable them to process the vast amounts of data present in the catalogs. The possibility of conducting real research is therefore open to anyone with the interest and some spare time, no longer just to professional scientists. Of course there are still problems, the main one being data links. Some of the surveys contain a terabyte (10^{12} bytes) of data, and using a 1 MHz broadband connection, this would take 3–4 months to download. Data archives such as the set of identifications, measurements, and bibliography for astronomical data (SIMBAD) and the multimission archive at the Space Telescope Science Institute (MAST) (see next section) contain 10 TB at the time of writing and are being added to at a rate of 5 or more TB year^{-1}. The cumulative information content within astronomy is now hundreds of terabytes, and one proposal alone (the large synoptic survey telescope) could be generating 5 TB of data per day within a few years. Fortunately, for many purposes, it is not necessary to download a whole survey to work with it, a small subset can be sufficient.

5.5.2 Digital Sky Surveys

We have already seen (Section 5.1) that the *Hipparcos*, *Tycho*, and other astrometric catalogs are available via the Web. Other large surveys presented on the web include the following:

- *HST GSC2* containing 500 million objects.

- 2MASS containing 300 million entries.

- SDSS containing 100 million objects.

- DPOSS containing 2000 million objects.

- Second issue of the USNO's *TAC 2.0* based upon photographic plates obtained with the twin astrograph and containing over 700,000 stellar positions to between ±50 and ±120 mas accuracy.

- USNO's *A2.0* catalog containing 526 million entries.

- USNO's *B1.0* catalog containing data on a billion stars with 200 mas positional accuracy.

While the still-being-completed *UCAC* aims to measure the positions of some 40 million stars in the 10^m to 14^m range to ±20 mas accuracy and is expected to be distributed in 2008.

However, not all digital sky surveys are the prerogative of the professional astronomer. The Amateur all-Sky Survey (TASS, http://www.tass-survey.org/) uses small cameras and CCD detectors to record some 200 Mb of data per night while searching for comets and variable stars.

A list of surveys and catalogs available online, that currently contains 6500 entries, is given by VizieR at the Centre de Données Astronomiques de Strasbourg (CDS) (see below).

Anyone with access to the Internet and a computer can join in the work for some surveys. The data produced by search for extraterrestrial intelligence (SETI) and the laser interferometer gravitational-wave observatory (LIGO) and GEO600 gravitational wave detectors are partly processed by volunteers whose computers tackle the processing when they would otherwise be idle. SETI@home and Einstein@home, as the two projects are known, both have close to 200,000 participants at the time of writing. If you are interested in joining either project then an Internet search for "SETI@home" or "Einstein@home" will take you straight to home pages. There is also a distributed computing project aimed at simulating large numbers of cosmological models called Cosmology@Home and several biologically based projects. At the time of writing, there are around 40 such projects, though most are not astronomy oriented. A partial list can be found at http://boinc.berkely.edu/.

Many of these surveys have been gathered together to be available through single sites such as CDS* that operates SIMBAD (http://simbad. u-strasbg.fr/Simbad), NSSDC (National Space Science Data Center, http:// nssdc.gsfc.nasa.gov), and MAST (http://archive.stsci.edu). Anyone who is interested can easily look up the details on an individual object within these catalogs via the Web. However, if more than a few tens objects are needed, then this approach becomes cumbersome and time consuming, larger projects therefore need more powerful search and processing software. Some of the sites provide this on an individual basis, but with little consistency in what is available and how to use it. Recently therefore, virtual observatories have started to come onstream providing much greater software and other support, though for some of them their use is limited to accredited astronomers.

5.5.3 Virtual Observatories

Virtual observatories are interfaces whereby huge data sets such as the digital sky surveys, and others, such as collections of spectra from the *International Ultraviolet Explorer* (*IUE*) and the *HST* spacecraft, results from other spacecraft such as *ROSAT*, *ISO*, *Chandra*, etc., interferometric data from radio and optical telescopes, etc. can be handled and mined for new information. The virtual observatory is a set of data archives, software tools, hardware, and staff that enables data in archives to be found and

---

* CDS also archives bibliographical information, enabling searches for publications relating to objects of interest to be conducted.

processed in many different ways. Amongst the type of functions that a virtual observatory can provide, there are

- Standardizing the formats of differing data sets
- Finding different observations of a given object, or a set of objects, and sorting through them
- Comparing and combining the information about objects available in various catalogs and other databases
- Comparing and combining observations taken at different times
- Combining archive data with new data obtained from telescopes
- Correlating data from different sources
- Classifying objects
- Performing statistical analyses
- Measuring images
- Image processing

While the types of scientific study that are possible include

- Multi-wavelength studies
- Large statistical studies
- Searches for changing, moving, rare, unusual, or new objects and types of objects
- Combining data in new ways leading to unexpected discoveries

Virtual observatories started becoming available a few years ago and are now proliferating rapidly. AstroVirtel started operating in 2000. It was a virtual observatory operated by the ST-ECF (Space Telescope European Coordinating Facility) on behalf of the ESO and the European Space Agency (ESA). It contained in excess of 10 TB of data from the *HST* and the ESO ground-based telescopes and may still be accessed at http://ecf.hq.eso.org/astrovirtel/. The Astrophysical Virtual Observatory (AVO, http://cdsweb.u-strasbg.fr/avo.htx) is the successor to AstroVirtel. The United Kingdom's AstroGrid project (http://www.astrogrid.org/) and the United States's National Virtual Observatory (NVO, http://www.us-vo.org/) are other currently available virtual observatories (and form a part of the AVO), and additional projects are under development in Australia, Germany, and Japan. The virtual solar observatory is a software system linking various solar data archives that provides tools for searching and analyzing the data. It may be accessed by anyone who is interested at http://vso.nascom.nasa.gov/cgi-bin/search. Much of the information on virtual observatories is collated by the International Virtual Observatory Alliance and that may be accessed at http://www.ivoa.net/.

Finally, the most widely used virtual observatory of all is Google Sky. This forms a part of Google Earth that may be accessed at http://earth.google.com/. Google Sky provides a complete map of the sky containing some 100 million stars and 200 million galaxies. The more prominent nebulae and galaxies can be selected for higher resolution images and for notes about the natures of the objects. Best of all, it is available to anyone and it is free!

Appendix A: Julian Date

The Julian date is the number of days elapsed since noon on November 24, 4714 BC (on the Gregorian calendar) or since noon on January 1, 4713 BC (on the Julian calendar). The modified Julian date is a variation of the Julian date that starts at midnight on November 17, 1858. The modified Julian date is thus the Julian date minus 2,400,000.5 days.

Date (January 1 at Noon Gregorian Reckoning)	Julian Day Number	Date (January 1 at Noon Gregorian Number Reckoning)	Julian Day Number
2050	2,469,808.0	1600	2,305,448.0
2025	2,460,677.0	1200	2,159,351.0
2000	2,451,545.0	800	2,013,254.0
1975	2,442,414.0	400	1,867,157.0
1950	2,433,283.0	0	1,721,060.0
1925	2,424,152.0	400 BC	1,574,963.0
1900	2,415,021.0	800 BC	1,428,866.0
1875	2,405,890.0	1200 BC	1,282,769.0
1850	2,396,759.0	1600 BC	1,136,672.0
1825	2,387,628.0	2000 BC	990,575.0
1800	2,378,497.0	2400 BC	844,478.0
1775	2,369,366.0	2800 BC	698,381.0
1750	2,360,234.0	3200 BC	552,284.0
1725	2,351,103.0	3600 BC	406,187.0
1700	2,341,972.0	4000 BC	260,090.0
1675	2,332,841.0	4400 BC	113,993.0
1650	2,323,710.0	4714 BC (November 24)	0.0
1625	2,314,579.0		

For days subsequent to January 1, the following number of days should be added to the Julian day number:

Date (Noon)	Number of Days (Non-Leap Years)	Number of Days Leap Years
February 1	31	31
March 1	59	60
April 1	90	91
May 1	120	121
June 1	151	152
July 1	181	182
August 1	212	213
September 1	243	244
October 1	273	274
November 1	304	305
December 1	334	335
January 1	365	366

Appendix B: Answers to the Exercises

1.1 69.21″ and 21.4″ (do not forget the Purkinje effect)

1.2 1.834 and −3.387 m

1.3 0.31 mm

1.4 94 mm, 850× (for the exit pupil to be smaller than the eye's pupil)

1.5 About 75° (see Equation 1.72)

1.6 84 m along the length of its arms

1.9 0.002 m^3

1.11 250 s^{-1}

1.13 1.6×10^{-23}

1.14 About 44 $^{37}_{18}$Ar atoms

1.16 Calculated data are tabulated below:

Planet	Mass (M_p/M_{Sun})	Period (Days)	L_G (W)
Mercury	1.6×10^{-7}	88	62
Venus	2.4×10^{-6}	225	600
Earth	3.0×10^{-6}	365	180
Mars	3.2×10^{-7}	687	0.25
Jupiter	9.6×10^{-4}	4330	5300
Saturn	2.9×10^{-4}	10760	20
Uranus	4.3×10^{-5}	30700	0.015
Neptune	5.3×10^{-5}	60200	0.002
Pluto	2.5×10^{-6}	90700	10^{-6}

2.1 (a) 35 mm (b) 52 mm

2.2 3.1×10^{16} m (or 1 pc)

2.3 5×10^{17} m (or 16 pc)

2.4 28 mW, 72 km

3.1 $+25.9$, -20.6

3.2 3050 pc

3.3 $(U - B) = -0.84$, $(B - V) = -0.16$

   $Q = -0.72$

   $(B - V)_0 = -0.24$, $E_{B-V} = 0.08$

   $E_{U-B} = 0.06$, $(U - B)_0 = -0.90$

   Spectral type $=$ B3 V

   Temperature $= 20{,}500$ K

   Distance (average) 230 pc

   $U_0 = 2.82$, $B_0 = 3.72$

   $V_0 = 3.96$, $M_V = -2.8$

4.1 -80.75 km s^{-1}

4.2 One prism, 0.79 nm mm^{-1}

4.4 $f' = 25$, $W_\lambda = 10^{-12}$ m, $\lambda = 625$ nm

   $R = 6.3 \times 10^5$, $L = 0.42$ m, $D = 0.38$ m

   $d\theta/d\lambda = 1.66 \times 106$ rad m^{-1}

   $f_1 = 13.3$ m, $f_2 = 12.1$ m

   $D_1 = 0.533$ m, $D_2 = 0.533$ m

   $S = 22$ μm

   Problems to be solved if possible in the next iteration through the exercise. Grating is too large: a normal upper limit on L is 0.1 m. Slit width is too small. The overall size of the instrument will lead to major thermal control problems.

4.5 $+5.3$ (taking $q = 0.002$ and $\alpha = 5 \times 10^{-6}$)

4.6 64.96°

5.2 0.26 mm

5.5 50.6, 0.79 mm

Appendix C: Acronyms

Early editions of this book tried to avoid the use of acronyms, believing that except to those people working in the particular field, they obscure understanding without actually saving much time or effort. Unfortunately, few other astronomers seem to follow this belief, instead appearing to delight in inventing ever more tortuous acronyms to name their instruments, techniques, etc. Second- and third-level (i.e., an acronym that contains an acronym that contains yet another acronym) acronyms are now being encountered. The use of acronyms in astronomy is nowadays so prevalent, and the distortions so often introduced to the actual names, that they were reluctantly included in the fourth edition and have now been extended to the fifth edition. The acronyms are defined when first encountered or within their main section, but are also defined below, so that there meaning can be retrieved after the original definition can no longer be found within the text.

A list of thousands of mostly astronomical acronyms can be found at the time of writing at

http://www.maa.mhn.de/FAQ/acronyms.html,
http://cfa-www.harvard.edu/~gpetitpas/Links/Astroacro.html,
http://faqs.org/faqs/space/acronyms/,

and many other Web sites—alternatively a search for "astronomy acronyms" of the acronym itself will usually be successful. There are also some 200,000 general acronyms listed at http://www.acronymfinder.com/ and numerous lists of acronyms used within special projects such as the *HST*, AAO, ESO, etc. that may easily be found by a Web search.

List of acronyms to be encountered within this book

2dF	2° field (AAT)
2MASS	2-Micron All Sky Survey
6dF	6° field (UKST)
AAT	Anglo-Australian telescope
AC	Astrographic Catalog
ACIS	advanced CCD imaging spectrometer
ACoRνE	acoustic cosmic ray neutrino experiment
ADEPT	advanced dark energy physics telescope
AGASA	Akeno giant air shower array
AGK	Astronomische Gesellschaft Katalog
AIPS	astronomical image processing system
AMANDA	Antarctic muon and neutrino detector array
ANITA	Antarctic impulsive transient antenna
ANTARES	astronomy with a neutrino telescope and abyssal environmental research
AOS	acousto-optical radio spectrometer
APD	avalanche photodiode
APEX	Atacama pathfinder experiment
APM	automatic plate measuring machine
ATST	advanced technology solar telescope
AVO	Astrophysical Virtual Observatory
AzTEC	astronomical thermal emission camera
BAT	burst alert telescope
BATCHES	basic echelle spectrograph
bHROS	bench-mounted high-resolution optical spectrometer
BIB	blocked impurity band device
BISON	Birmingham solar oscillations network
BWFN	beam width at first nulls
BWHP	beam width at half-power points
CANGAROO	collaboration of Australia and Nippon for a gamma ray observatory in the outback
CCAT	Cornell Caltech Atacama telescope
CCD	charge-coupled device
CDMS	cryogenic dark matter search
CDS	Centre de Données Astronomiques de Strasbourg
CELT	California extremely large telescope—now known as the TMT
CFHT	Canada–France–Hawaii telescope
CHARA	Center for High Angular Resolution Astronomy
CIAO	coronagraphic imager with adaptive optics
CID	charge injection device

List of acronyms to be encountered within this book   (continued)

CMB	cosmic microwave background radiation
CMT	Carlsberg meridian telescope
COAST	Cambridge optical aperture synthesis telescope
COBE	Cosmic Background Explorer satellite
CORONAS-F	Russian acronym for complex orbital observatory for a study of the active Sun
COS	cosmic origins spectrograph
COSMOS	coordinates, sizes, magnitudes, orientations, and shapes
CPCCD	column parallel CCD
CRIRES	cryogenic infrared echelle spectrograph
CW	continuous wave (radar)
CZT	cadmium–zinc–telurium detectors
DAMA	dark matter
DASI	degree angular scale interferometer
DEIMOS	deep imaging multi-object specrograph
DENIS	Deep Near Infrared Survey
DIN	Deutsches Institut Für Normung
DPOSS	Digital Palomar Observatory Sky Survey
EBCCD	electron bombarded charge-coupled device
EBS	electron bounce silicon (TV camera)
E-ELT	European extremely large telescope
EFOSC2	ESO's faint object spectrograph and camera
EGRET	Energetic Gamma Ray Experiment Telescope
EISCAT	European incoherent scatter
EMCCD	electron mutliplying charge-coupled device
ESA	European Space Agency
ESO	European Southern Observatory
ESPaDOn	echelle spectropolarimetric device for the observation of stars
EUV	extreme ultraviolet (also XUV)
FAME	Full-Sky Astrometric Mapping Explorer
FCRAO	Five College Radio Astronomy Observatory
FGS	fine guidance sensors
FIR	far infrared
FIRST	Far Infrared and Submillimetre Telescope—now called Herschel
FITS	flexible image transport system
FK	Fundamental Katalog
FLAIR	fiber-linked array-image reformatter
FLAMES	fiber large array multi-object spectrograph
FOC	faint object camera
FUSE	Far Ultraviolet Spectroscopic Explorer

(*continued*)

List of acronyms to be encountered within this book   (continued)

GI2T	Grand Interféromètre à 2 Télescopes
GIF	graphic interchange format
GLAST	Gamma-Ray Large Area Space Telescope
GMOS	Gemini multi-object spectroscopes
GMT	Giant Magellan telescope
GNO	Gallium Neutrino Observatory
GODs	giant optical devices
GOLF	global oscillations at low frequencies
GONG	Global Oscillation Network Group
GSC	(Hubble) Guide Star Catalog
GSMT	giant segmented mirror telescope
GRB	γ-ray burst
GTC	Gran Telescopio Canarias
GZK	Greisen Zatsepin Kuzmin cut-off
HARP	heterodyne array receiver program
HARPS	high-accuracy radial-velocity planet searcher
HAWK-I	high-acuity, wide-field K-band imaging
HEB	hot electron bolometer
HEFT	high-energy focusing telescope
HEMT	high electron mobility transistors
HESS	high-energy stereoscopic system
HFET	heterostructure field-effect transistor
HiCIAO	high-contrast instrument for the Subaru next generation adaptive optics
Hipparcos	High-Precision Parallax Collecting Satellite
HST	Hubble Space Telescope
IBC	impurity band conduction device
ICARUS	imaging cosmic and rare underground signals
ICCD	intensified charge-coupled device
ICRS	International Celestial Reference System
IF	intermediate frequency
IMB	Irvine–Michigan–Brookhaven
InFOCμS	international focusing optics collaboration for μ–Crab sensitivity
INTEGRAL	International Gamma Ray Astrophysics Laboratory
IPCS	image photon counting system
IRAF	Image Reduction and Analysis Facility
IRAM	Institut de Radio Astronomie Millimétrique
IRAS	InfraRed Astronomy Satellite
ISAAC	infrared spectrometer and array camera
ISI	infrared spatial interferometer
ISO	(1) Infrared Space Observatory (2) International Organization for Standardization

List of acronyms to be encountered within this book   (continued)

ISS	International Space Station
IUE	International Ultraviolet Explorer
JCG	Johnson–Cousins–Glass photometric system
JCMT	James Clerk Maxwell telescope
JEDI	joint efficient dark-energy investigation
JPEG	Joint Photographic Experts Group
JWST	James Webb Space Telescope
Kamiokande	Kamioka neutrino detector
KAO	Kuiper Airborne Observatory
KAOS	kilo-aperture optical spectrograph
KAP	potassium (K) acid phthalate
KELT	kilodegree extremely little telescope
LABOCA	large APEX bolometer camera
LAMA	large aperture mirror array
LAMOST	large sky area multi-object fiber spectroscopic telescope
LAT	large area telescope
LBT	large binocular telescope
LCGT	large-scale cryogenic gravity wave telescope
LENS	low-energy neutrino spectroscope
LHC	large hadron collider
LIGO	Laser Interferometer Gravitational-Wave Observatory
LIRIS	long-slit intermediate resolution infrared spectrograph
LISA	laser interferometer space antenna
LOFAR	low-frequency array
LUCIFER	LBT NIR utility with camera and integral-field unit for extragalactic research
LZT	large zenith telescope
MAGIC	major atmospheric gamma imaging Čerenkov telescope
MAMA	(1) multi-anode micro-channel array (2) machine automatique à mesurer pour l'astronomie
MACAO	multi-application curvature adaptive optics
MACHO	massive astrophysical compact halo objects
MagEX	magnetosheath explorer in x-rays
MAP	multichannel astrometric photometer
mas	milliarcsecond
MAST	multi-mission archive for the Space Telescope
MCAO	multi-conjugate adaptive optics
MCP	micro-channel plate
MDI	Michelson Doppler interferometer
MEM	maximum entropy method

(*continued*)

List of acronyms to be encountered within this book   (continued)

MERLIN	multielement radio-linked interferometer
MINOS	main injector neutrino oscillation search
MIR	mid infrared
MIT	Massachusetts Institute of Technology
MMIC	monolithic microwave integrated circuits
MOF	magneto-optical filter
MOS	metal oxide–silicon transistor
MOSFIRE	multi-object spectrometer for infrared exploration
MTF	modulation transfer function
MUSE	Multi-Unit Spectroscopic Explorer
NA	numerical aperture
NASA	National Aeronautics and Space Administration
NESTOR	Neutrino Extended Submarine Telescope with Oceanographic Research
NIR	near infrared
NNLS	nonnegative least squares
NOAO	National Optical Astronomy Observatories
NODO	NASA Orbital Debris Observatory
NPOI	navy prototype optical interferometer
NRAO	National Radio Astronomy Observatory
NSSDC	National Space Science Data Center
NTD	neutron transmutation doping
NuSTAR	Nuclear Spectroscopic Telescope Array
NVO	National Virtual Observatory
OASIS	optically adaptive system for imaging spectroscopy
OTCCD	orthogonal transfer CCD
OWL	overwhelmingly large telescope
PAIRITEL	Peters automated infrared imaging telescope
PAMELA	payload for anti-matter exploration and light nuclei astrophysics
PMM	precision measuring machine
PROMPT	panchromatic robotic optical monitoring and polarimetry telescope
PRVS	precision radial velocity spectrometer
PSF	point spread function
PSST	planet search survey telescope
PZT	photographic zenith tube
Quatratran	quasiparticle trapping transistor
QUEST	Quasar Equatorial Survey Team
QWIPS	quantum well infrared photodetectors
RAND	radio array neutrino detector
RAVE	radial velocity experiment
RL	Richardson Lucy algorithm

List of acronyms to be encountered within this book   (continued)

ROSAT	Röntgen Satellite
RXTE	Rossi X-ray Timing Explorer
SAGE	Soviet-American gallium experiment
SALT	South African large telescope
SAR	synthetic aperture radar
SCUBA	submillimeter common user bolometer array
SDSS	Sloan Digital Sky Survey
SEC	secondary electron conduction (TV camera)
SETI	search for extraterrestrial intelligence
SIM	Space Interferometry Mission
SIMBAD	set of identifications, measurements and bibliography for astronomical data
SINFONI	spectrograph for integral-field observations in the near infrared
SIRTF	Space Infrared Telescope Facility—now the Spitzer spacecraft
SIS	superconductor–insulator–superconductor device
SKA	square kilometer array
SLR	single lens reflex (camera)
SMM	Solar Maximum Mission
SNAP	Supernova/Acceleration Probe
SNIFS	supernova integral-field spectrograph
SNO	Sudbury Neutrino Observatory
SNU	solar neutrino unit
SOFIA	Stratospheric Observatory for Infrared Astronomy
SOHO	Solar and Heliospheric Observatory
SPIFFI	spectrometer for infrared faint field imaging
SPIRAL	segmented pupil image reformatting array lens
SPIRE	spectral and photometric imaging receiver
SSD	silicon strip detector
STARE	stellar astrophysics and research on exoplanets
ST-ECF	Space Telescope European Coordinating Facility
STIS	space telescope imaging spectrograph
STJ	superconducting tunnel junction
SUSI	Sydney University stellar interferometer
SuperWASP	super wide-angle search for planets
TAC	Twin Astrographic Catalog
TASS	The Amateur all-Sky Survey
TAURUS	Taylor Atherton uariable-resolution radial-uelocity system
TDI	time-delayed integration
TES	transition edge sensor

(continued)

List of acronyms to be encountered within this book   (continued)

THEMIS	télescope héliographique pour l'etude du magnétisme et des instabilités solaires
TIFF	tagged image file format
TMT	thirty meter telescope—previously known as the California extremely large telescope (CELT)
TPF	terrestrial planet finder
TRACE	Transition Region And Corona Explorer
TrES	Trans-Atlantic Exoplanet Survey
UCAC	USNO CCD Astrograph Catalog
UIST	UKIRT imager spectrometer
UKIRT	United Kingdom infrared telescope
UKST	United Kingdom Schmidt telescope
ULE	ultra-low-expansion
USNO	U.S. Naval Observatory
UVES	Ultraviolet/visual echelle spectroscope
VIRGO	variability of solar irradiance and gravity oscillations
VERITAS	very energetic radiation imaging telescope array system
VIMOS	visible multi-object spectrograph
VISIR	VLT imager and spectrometer for mid-infrared
VISTA	visible and infrared survey telescope for astronomy
VLA	very large array
VLBA	very long baseline array
VLOT	very large optical telescope
VLT	very large telescope
VLTI	very large telescope interferometer
VPHG	volume-phase holographic grating
VTT	vacuum tower telescope
WARC	World Administrative Radio Conference
WET	whole earth telescope
WFCAM	wide field camera
WFC	wide field camera
WFPC2	wide field planetary camera 2
WHT	William Herschel telescope
WIMP	weakly interacting massive particle
WMAP	Wilkinson Microwave Anisotropy Probe
WRC	World Radio Conference
WSRT	Westerbork synthesis radio telescope
WUPPE	Wisconsin ultraviolet photo-polarimeter experiment
XEUS	x-ray evolving universe spectrometer
XUV	extreme ultraviolet (also EUV)
ZEPLIN	zoned proportional scintillation in liquid noble gases

Appendix D: Bibliography

Some selected journals, books, and articles that provide further reading for various aspects of this book, or from which further information may be sought, are listed below.

RESEARCH JOURNALS

Astronomical Journal
Astronomy and Astrophysics
Astrophysical Journal
Icarus
Monthly Notices of the Royal Astronomical Society
Nature
Publications of the Astronomical Society of the Pacific
Science
Solar Physics

POPULAR JOURNALS

Astronomy
Astronomy Now
Ciel et Espace
New Scientist
Scientific American
Sky and Telescope

EPHEMERIDES

Astronomical Almanac (Published each year), HMSO/U.S. Government Printing Office

Handbook of the British Astronomical Association (Published each year), British Astronomical Association

Yearbook of Astronomy (Published each year), Macmillan

STAR AND OTHER CATALOGS, ATLASES, AND SKY GUIDES

Arnold, H., Doherty, P., and Moore, P., *Photographic Atlas of the Stars*, 1999 (IoP Publishing).

Bakich, M.E., *The Cambridge Guide to the Constellations*, 1995 (Cambridge University Press).

Coe, S.R., *Deep Sky Observing*, 2000 (Springer).

Dunlop, S., Tirion, W., and Ruki, A., *Collins' Atlas of the Night Sky*, 2005 (Harper-Collins).

Hirshfield, A. and Sinnott, R.W., *Sky Catalogue 2000*, Volumes 1 and 2, 1992 (Cambridge University Press).

Inglis, M., *Field Guide to the Deep Sky Objects*, 2001 (Springer).

Inglis, M., *Astronomy of the Milky Way: Observer's Guide to the Southern Sky*, Part 2, 2003 (Springer).

Inglis, M., *Astronomy of the Milky Way: Observer's Guide to the Northern Sky*, Part 1, 2004 (Springer).

James, N. and North, G., *Observing Comets*, 2002 (Springer).

Jones, K.G., *Messier's Nebulae and Star Clusters*, 1991 (Cambridge University Press).

Karkoschka, E., *Observer's Sky Atlas*, 1999 (Springer).

Kitchin, C.R., *Galaxies in Turmoil*, 2007 (Springer).

Kitchin, C.R., *Photo-Guide to the Constellations*, 1997 (Springer).

Kitchin, C.R. and Forrest, R., *Seeing Stars*, 1997 (Springer).

Leavitt, H.S., *Annals of the Harvard College Observatory*, 71(3), 49–52.

Moore, P., *Observer's Year*, 1998 (Springer).

Moore, P., *Astronomy Encyclopaedia*, 2002 (Philip's).

O'Meara, S.J., *The Caldwell Objects*, 2002 (Cambridge University Press).

Privett, G. and Parsons, P., *Deep Sky Observer's Year*, 2001 (Springer).

Ratledge, D., *Observing the Caldwell Objects*, 2000 (Springer).

Ridpath, I. (Ed.), *Norton's 2000*, 19th Edition, 1998 (Longman).

Tirion, W., *Sky Atlas 2000*, 2000 (Sky Publishing Corporation).

REFERENCE BOOKS

Allen, C.W., *Allen's Astrophysical Quantities*, 2001 (Springer).

Kitchin, C.R., *Illustrated Dictionary of Practical Astronomy*, 2002 (Springer).

Lang, K.R., *Astrophysical Formulae: Space, Time, Matter in Cosmology*, 1998 (Springer).

Lang, K.R., *Astrophysical Formulae: Radiation, Gas Processes and High-energy Astrophysics*, 1998 (Springer).

Lang, K.R., *Cambridge Encyclopaedia of the Sun*, 2001 (Cambridge University Press).

Murdin, P. (Ed.), *Encyclopaedia of Astronomy and Astrophysics*, 2001 (Nature and IoP Publishing).

Ridpath, I., *Oxford Dictionary of Astronomy*, 1997 (Oxford University Press).

Shirley, J.H. and Fairbridge, R.W., *Encyclopaedia of Planetary Sciences*, 2000 (Kluwer Academic Publishers).

Woodruff, J. (Ed.), *Philip's Astronomy Dictionary*, 2005 (Philip's).

INTRODUCTORY ASTRONOMY BOOKS

Clark, S., *The Sun Kings*, 2007 (Princeton University Press).

Engelbrektson, S., *Astronomy through Space and Time*, 1994 (WCB).

Freedman, R.A. and Kaufmann, W.J., III, *Universe*, 2001 (WH Freeman).

Halliday, K., *Introductory Astronomy*, 1999 (Wiley).

Inglis, M., *Astrophysics is Easy!: An Introduction for the Amateur Astronomer*, 2005 (Springer).

Moche, D.L., *Astronomy: A Self-Teaching Guide*, 1993 (Wiley).

Montenbruck, O. and Pfleger, T., *Astronomy on the Personal Computer*, 2000 (Springer).

Moore, P., *Eyes on the Universe*, 1997 (Springer).

Moore, P. and Watson, J., *Astronomy with a Budget Telescope*, 2002 (Springer).

Nicolson, I., *Unfolding Our Universe*, 1999 (Cambridge University Press).

Nicolson, I., *Dark Side of the Universe*, 2007 (Canopus Publishing).

Tonkin, S.F., *AstroFAQs*, 2000 (Springer).

Zeilik, M., *Astronomy: The Evolving Universe*, 1994 (Wiley).

Zeilik, M., Gregory, S.A., and Smith, E.V.P., *Introductory Astronomy and Astrophysics*, 1992 (Saunders).

INTRODUCTORY PRACTICAL ASTRONOMY BOOKS

Arditti, D., *Setting up a Small Observatory: From Concept to Construction*, 2007 (Springer).

Argyle, R., *Observing and Measuring Visual Double Stars*, 2004 (Springer).

Bone, N., *Observing Meteors Comets and Supernovae*, 1999 (Springer).

Bone, N., *Aurora: Observing and Recording Nature's Spectacular Light Show*, 2007 (Springer).

Buick, T., *How to Photograph the Moon and Planets with Your Digital Camera*, 2006 (Springer).

Byrne, C.J., *The Far Side of the Moon: A Photographic Guide*, 2007 (Springer).

Charles, J., *Practical Astrophotography*, 2004 (Springer).

Coe, S., *Deep Sky Observing: The Astronomical Tourist*, 2002 (Springer).

Cooke, A., *Visual Astronomy under Dark Skies: A New Approach to Observing Deep Space*, 2005 (Springer).

Covington, M.A., *How to Use a Computerized Telescope*, 2002 (Cambridge University Press).

Dragesco, J. and McKim, R. (Trans.), *High Resolution Astrophotography*, 1995 (Cambridge University Press).

Dunlop, S. and Tirion, W., *Practical Astronomy*, 2006 (Philip's).

Gainer, M., *Real Astronomy with Small Telescopes: Step-by-Step Activities for Discovery*, 2006 (Springer).

Good, G.A., *Observing Variable Stars*, 2002 (Springer).

Kitchin, C.R., *Solar Observing Techniques*, 2001 (Springer).

Kitchin, C.R., *Telescope & Techniques*, 2nd Edition, 2003 (Springer).

James, N. and North, G., *Observing Comets*, 2002 (Springer).

Lena, P., *A Practical Guide to CCD Astronomy*, 1997 (Cambridge University Press).

Manly, P.L., *The 20-cm Schmidt-Cassegrain Telescope*, 1994 (Cambridge University Press).

Mizon, R., *Light Pollution: Responses and Remedies*, 2001 (Springer).

Mobberley, M., *Astronomical Equipment for Amateurs*, 1999 (Springer).

Mobberley, M., *Lunar and Planetary Webcam User's Guide*, 2006 (Springer).

Mobberley, M., *Total Solar Eclipses and How to Observe Them*, 2007 (Springer).

Mobberley, M., *Supernovae and How to Observe Them*, 2007 (Springer).

Mollise, R., *Choosing and Using a Schmidt-Cassegrain Telescope*, 2001 (Springer).

Monks, N., *Astronomy with a Home Computer*, 2004 (Springer).

Mullaney, J., *A Buyer's and User's Guide to Astronomical Telescopes and Binoculars*, 2007 (Springer).

Mullaney, J., *The Herschel Objects and How to Observe Them*, 2007 (Springer).

North, G., *Observing Variable Stars, Novae and Supernovae*, 2004 (Cambridge University Press).

Pepin, M.B., *The Care of Astronomical Telescopes and Accessories: A Manual for the Astronomical Observer and Amateur Telescope Maker*, 2004 (Springer).

Ratledge, D., *Art and Science of CCD Astronomy*, 1997 (Springer).

Ratledge, D., *Software and Data for Practical Astronomers*, 1999 (Springer).

Ratledge, D., *Digital Astrophotography: The State of the Art*, 2005 (Springer).

Robinson, K., *Spectroscopy, the Key to the Stars: Reading the Lines in Stellar Spectra*, 2007 (Springer).

Roth, G.D., *Compendium of Practical Astronomy*, 1993 (Springer).

Steinicke, W. and Jakiel, R., *Galaxies and How to Observe Them*, 2007 (Springer).

Stuart, A., *CCD Astrophotography: High Quality Imaging from the Suburbs*, 2006 (Springer).

Tonkin, S.F., *Amateur Telescope Making*, 1999 (Springer).

Tonkin, S.F., *Astronomy with Small Telescopes*, 2001 (Springer).

Tonkin, S.F., *Practical Amateur Spectroscopy*, 2002 (Springer).

Tonkin, S.F., *Binocular Astronomy*, 2006 (Springer).

Wlasuk, P., *Observing the Moon*, 2000 (Springer).

Zirker, J.B., *An Acre of Glass*, 2005 (John Hopkins University Press).

ADVANCED PRACTICAL ASTRONOMY BOOKS (OR BOOKS WITH A SIGNIFICANT SECTION ON ADVANCED PRACTICAL ASTRONOMY)

Birney, D.S., *Observational Astronomy*, 1991 (Cambridge University Press).

Bode, M.F., *Robotic Observatories*, 1995 (Wiley).

Chapman, A., *Dividing the Circle: Development and Critical Measurement of Celestial Angles (1500–1850)*, 1990 (Ellis Horwood).

Comte, G. and Marcelin, M. (Ed.), *Tridimensional Optical Spectroscopic Methods in Astrophysics*, 1995 (Astronomical Society of the Pacific Conference Series Vol. 71).

Cornwell, T.J. and Perley, R.A. (Ed.), *Radio Interferometry: Theory, Techniques and Applications*, 1991 (Astronomical Society of the Pacific Conference Series Vol. 19).

Davies, J.K., *Astronomy from Space: The Design and Operation of Orbiting Observatories*, 1997 (Praxis).

Emerson, D.T. and Payne, J.M. (Ed.), *Multi-Feed Systems for Radio Telescopes*, 1995 (Astronomical Society of the Pacific Conference Series Vol. 75).

Glass, I., *Handbook of Infrared Astronomy*, 1999 (Cambridge University Press).

Goddard, D.E. and Milne, D.K. (Ed.), *Parkes: Thirty Years of Radio Astronomy*, 1994 (CSIRO).

Hearnshaw, J.B., *Measurement of Starlight*, 1996 (Cambridge University Press).

Henry, G.W. and Eaton, J.A. (Ed.), *Robotic Telescopes: Current Capabilities, Present Developments and Future Prospect for Automated Astronomy*, 1995 (Astronomical Society of the Pacific Conference Series Vol. 79).

Howell, S.B., *Handbook of CCD Astronomy*, 2000 (Cambridge University Press).

Kitchin, C.R., *Optical Astronomical Spectroscopy*, 1995 (Institute of Physics Publishing).

Mattok, C. (Ed.), *Targets for Space-Based Interferometry*, 1993 (European Space Agency).

Mizon, R., *Light Pollution: Strategies and Solutions*, 2001 (Springer).

Rieke, G.H., *Detection of Light: From the Ultraviolet to the Submillimeter*, 2002 (Cambridge University Press).

Stix, M., *The Sun*, 2nd Edition, 2002 (Springer).

Zensus, J.A. et al. (Ed.), *Very Long Baseline Interferometry and the VLBA*, 1995 (Astronomical Society of the Pacific Conference Series Vol. 82).

DATA REDUCTION AND ANALYSIS

Bruck, M.T., *Exercises in Practical Astronomy Using Photographs*, 1990 (Institute of Physics).

di Gesù, V. et al. (Ed.), *Data Analysis in Astronomy IV*, 1992 (Plenum).

Emerson, D., *Interpreting Astronomical Spectra*, 1996 (Wiley).

Grandy, W.T. Jr. and Schick, L.H., *Maximum Entropy and Bayesian Methods*, 1991 (Kluwer).

Jaschek, C. and Murtagh, F. (Ed.), *Errors, Bias and Uncertainties in Astronomy,* 1990 (Cambridge University Press).

Privett, G., *Creating and Enhancing Digital Astro Images: A Guide for Practical Astronomers,* 2007 (Springer).

Taff, L.G., *Computational Spherical Astronomy,* 1991 (Krieger).

Wall, J. and Jenkins, C., *Practical Statistics for Astronomers,* 2003 (Cambridge University Press).

Warner, B., *A Practical Guide to Lightcurve Photometry and Analysis,* 2006 (Springer).

Index